T0341236

An Invitation to Knot Theory

Virtual and Classical

An Invitation to Knot Theory

Virtual and Classical

Heather A. Dye

McKendree University
Lebanon, Illinois, USA

CRC Press is an imprint of the
Taylor & Francis Group, an **informa** business
A CHAPMAN & HALL BOOK

CRC Press
Taylor & Francis Group
6000 Broken Sound Parkway NW, Suite 300
Boca Raton, FL 33487-2742

Printed on acid-free paper
Version Date: 20160113

International Standard Book Number-13: 978-1-4987-0164-8 (Hardback)

Visit the Taylor & Francis Web site at
http://www.taylorandfrancis.com

and the CRC Press Web site at
http://www.crcpress.com

To my husband Donovan and my children, Rowan and Gareth,
and my family and friends

Contents

List of Figures

List of Tables

Preface

This book is an introduction to virtual and classical knot theory. The book is designed to introduce key ideas and provide the background for undergraduate research on knot theory. Suggested readings from research papers are included in each chapter. The goal is for students to experience mathematical research. The proofs are written as simply as possible using combinatorial approaches, equivalence classes and linear algebra so that they are accessible to junior-senior level students.

I enjoy explaining virtual knot theory to undergraduate students. In the combinatorial proofs, I find examples of many of the concepts that we to try to introduce students to in a course on proof writing. Undergraduates have also experienced a high level of success in projects relating to virtual knot theory. For example, in 2003 Anatoly Pregal won 3rd place in the 2003 Intel Science Talent Search for a project entitled "Computation of quandle cocycle invariants." (Quandles are one of the topics in this book.) Virtual knot theory is a field where undergraduates can experience research in mathematics in a way that results in peer reviewed, publishable research. Another wonderful aspect of studying virtual knots is that it is current research. Students could contact many of the researchers mentioned in this text and participate in the mathematical community.

To write this book, I've drawn on resources that include research articles on virtual knot theory and classic texts on knot theory. Virtual links are the central topic in this book; classical links are a subset of virtual links. The virtual link invariants defined in this text are invariants of both classical and virtual knots. Proofs of invariance are written for virtual links and as such apply to classical knots. For this reason, the proofs are sometimes different from the proofs in classical texts. Well-known applications and results for classical knots and links are also covered in the text. For example, the fundamental group of knot is defined in this text, but our main attention is on the knot quandle (of which the fundamental group of a knot is a homomorphic image).

Classical knot theory originated in the 1880s with Lord Kelvin's proposal that atoms were knotted vortices in the ether. The subject grew from there with P. G. Tait's construction knot tables. The Reidemeister moves were introduced in the 1920s by Kurt Reidemeister's in his article "Knotentheorie." The Alexander polynomial invariant was introduced in the 1928 article, "Topological invariants of knots and links," published in the *Transactions of the American Mathematical Society.* The Alexander polynomial was the main invariant of knots until the introduction of the Jones polynomial in 1985. Much work was done in the area of knot theory involving hyperbolic geometry and topology, but a main focus of this text is the normalized f-polynomial which is related by a change of variables to the Jones polynomial. The Jones polynomial was introduced in a 1985 article, "A Polynomial Invariant for Knots via von Neumann Algebras," published in the *Bulletin of the American Mathematical Society* by V. Jones. This polynomial was quickly followed by the HOMFLY-PT polynomial and a skein relation introduced by Louis Kauffman in 1985. The development of this work encompassed a hundred year span.

In comparison, the development of virtual knot theory spans only twenty years. Virtual knot theory was first defined by Louis Kauffman as an extension of classical knot theory in 1996. Kauffman's paper "Virtual knot theory" was first published in the *European Journal of Combinatorics* in 1999. Kauffman's paper immediately introduced extensions of the fundamental group and the f-polynomial to virtual knots and links. Topological interpretations of virtual knots as embeddings of links in thickened surfaces modulo isotopy and stablization have followed. In 2003, Greg Kuperberg proved that each virtual link has a unique minimal genus surface in which the virtual link embeds. Additionally, there are other topological interpretations. Virtual knots have also led to the development of different invariants related to the biquandle and quandle structures, as well as finite type invariants of virtual knots. There are deep connections between virtual knots and algebraic and topological structures. However, the combinatorial structure of virtual knots opens the field to the undergraduates. The fun of virtual knots is working on these multiple levels; invariants and results can be developed independently of these connections (although the knowledge of the connections certainly inspires the results).

This book is written for students who have completed the first two years of a mathematics major. I assume that students have completed linear algebra and a three semester calculus sequence. The book is intended as a gentle introduction to the field of virtual knot theory and mathematical research. For this reason, references are formatted somewhat differently from a research monograph — a list of undergraduate-friendly references is given at the end of each chapter. The complete list of references for each chapter is given as an appendix in the back of the book. I've also explicitly written out references and information about original papers in the body of the text.

The text is divided into four sections: an introduction to virtual knots and counted invariants, an introduction to the normalized f-polynomial (Jones polynomial) and other skein invariants, and an introduction to algebraic invariants such as the quandle and biquandle. The last section is a brief introduction to two applications of virtual knots: textiles and quantum computation.

Readers of this text should begin with the introduction to virtual knot theory in Chapter 1. After Chapter 1, readers can continue with the remainder of Part I, or move on to either Part II or III depending on the focus of the course. Each section has a particular focus and builds toward a major result or question in the field of virtual knot theory.

Part I introduces link invariants that are calculated by counting subsets of crossings weighted by their crossing sign. Chapter 2 introduces linking invariants of ordered, oriented link diagrams that are ultimately used to define an invariant of oriented knot diagrams. In Chapter 3, we introduce additional equivalences classes of link diagrams and their motivations. The next chapter examines unknotting numbers and related invariants of virtual link diagrams. These invariants are computed by taking the minimum number in a set of non-negative integers that also possibly contain the improper element ∞. Next, we examine how to construct families of diagrams through various methods. We explore whether or not the crossing and unknotting invariants defined earlier in the book differentiate between members of this family. Students should naturally conclude that the invariants defined so far do not differentiate between many of the diagrams constructed in the last section.

In Part II, we explore the bracket polynomial and the Kauffman-Murasugi-Thistlethwaite theorem. Our focus on is on extending this theorem to virtual knot diagrams. The normalized f-polynomial is introduced in Chapter 6 and a theorem giving a weak bound on the span of the f-polynomial. To improve this bound, we introduce background information about 2-dimensional surfaces and the Euler characteristic in Chapter 7. The focus is on being able to compute the genus of an abstract link diagram constructed from a virtual link diagram. In Chapter 8, we use the information about the genus of a virtual link diagram to improve the bound on f-polynomials for checkerboard colorable virtual link diagrams. In Chapter 9, we introduce cut points and checkerboard framings of virtual link diagrams. Each topic is introduced as needed; for example, students familiar with 2-dimensional surfaces can skip the majority of the chapter on surfaces. Chapter 9 concludes with an extension of the KMT Theorem. Part II concludes with Chapter 10; we introduce specializations of the bracket polynomial. This chapter can also be omitted for time.

In Part III, the focus is on algebraic invariants. The first chapter introduces the quandle, an algebraic structure from which we can construct many knot invariants. Using quandles, we construct the fundamental group, the Alexander polynomial, and the determinant of a virtual link. In the first chapter of this section, we see that tricoloring (a type of Fox coloring) distinguishes Kishino's knot from its flip. This partially answers a question about distinguishing a link diagram from its flip. However, we are not able to obtain a general result about a knot diagram and its symmetries. The biquandle and the Alexander biquandle remedy this issue, and we see that these are very effective invariants of non-classical virtual links and yet vanish on classical knots. Finally, we conclude with a section on Gauss diagrams which allow us to easily compute the crossing weight invariants introduced in Part I and the important concept of parity. We use parity to augment the bracket polynomial and immediately detect Kishino's knot. The section concludes with two snapshots of applications: textiles and quantum computation. The chapters on Gauss diagrams (Chapter 14) and applications (Chapter 15) can either be omitted or read as stand-alone chapters.

Acknowledgments

I would like to thank all the researchers who responded to my questions about their most undergraduate-friendly papers. An incomplete list of people who corresponded with me includes: Scott Carter, Micah Chrisman, Allison Henrich, Aaron Kaestner, Slavik Jablan, Seiichi Kamada, Naoko Kamada, Louis Kauffman, Adam Lowrance, Sam Nelson, and Radmila Sazdanovic. I would also like to give special thanks to Micah Chrisman and Gail Reed for their comments.

In addition, I would like to thank my parents, Harold and Elaine Dye, and my sister, Kathryn Gresh.

About the author

Heather A. Dye is an Associate Professor of Mathematics at McKendree University in Lebanon, Illinois. Her favorite courses to teach are linear algebra, probability, graph theory, and knot theory. She regularly works with Honors students on their senior thesis.

Her research focuses on virtual knot theory. She has published articles on virtual knot theory in the *Journal of Knot Theory and its Ramifications*, *Algebraic and Geometric Topology*, and *Topology and its Applications*. She enjoys organizing special sessions on knot theory at sectional meetings of the American Mathematical Society. She is a member of both the American Mathematical Society and the Mathematical Association of America.

Her favorite aspect of virtual knot theory is the opportunity to work with undergraduates and meet fellow mathematicians.

In her free time, she enjoys spending time with her family and quilting.

Symbol List

$\mathbb{Z}$ The set of integers

$\mathcal{D}_{i,j}(L)$ The difference linking number of components C_i and C_j in L

$\mathcal{E}_{i,j}(F)$ The flat linking number of F

$\mathcal{FR}$ The set of free virtual links

$\mathcal{F}$ The set of flat virtual links

$\mathcal{L}_j^i(L)$ The linking number of component C_i over component C_j in L

$\mathcal{P}_d(D)$ The minimum number of cut points in any framing of the diagram D

$\mathsf{us}(K)$ The unknotting sequence number of K

$\sim$ The equivalence relation on link diagrams

$[x]$ The equivalence class of x

$\mathcal{F}_d$ The set of flat (or free) virtual link diagrams

$\mathcal{P}(D, B)$ The number of cuts points in checkerboard framing B of the oriented diagram D

$\mathsf{vu}(K)$ The virtual unknotting number of K

$\mathsf{w}(c)$ The weight of the crossing c

$\chi(F)$ The Euler characteristic of the surface F

$\langle K \rangle$ The bracket polynomial of K

$\langle K \rangle_a$ The arrow bracket polynomial of K

$\leftrightarrow$ A single move relation between two link diagrams

$\mathcal{K}$ The set of classical knots

$\mathcal{P}(K)$ The minimal number of cut points in any framing of any diagram equivalent to K

$\mathcal{V}$ The set of virtual knots and links

$\mathcal{V}_d$ The set of virtual knot and link diagrams

$\mathcal{W}_a(K)$ The set of crossings in K with weight a

$g(S)$ The genus of the surface S

$\det(K)$ The determinant of K

$\operatorname{sgn}(c)$ The sign of crossing c

$\mathsf{c}(K)$ The crossing number of K

$\mathsf{u}(K)$ The unknotting number of K

$\mathsf{vus}(K)$ The virtual unknotting sequence number

$\mathsf{v}(K)$ The virtual crossing number of K

$\neg P$ The negation of the statement P

$Ar_K(A, K_i)$ The normalized arrow polynomial of K

$K_1 \sharp K_2$ The connected sum of K_1 and K_2

$P \Rightarrow Q$ P implies Q

$P_K(A)$ The normalized parity bracket polynomial of K

$\langle K \rangle_p$ The parity bracket polynomial of K

$\mathcal{C}_a(K)$ The a crossing weight number of the knot K

$\mathcal{I}$ A link invariant

$G(K)$ The virtual genus of a virtual K

$a(s)$ The arrow number of the state s

$AB_K(s, t)$ The generalized Alexander polynomial of K

$ABM(K)$ The Alexander biquandle matrix of K

$BQ(K)$ The knot biquandle of K

$Col_X(K)$ The number of colorings of K with X

$F(K)$ The flat f-polynomial of K

$f_K(A)$ The f-polynomial of K

K^F The flip of K

K^I The inverse of K

K^M The mirror of K

K^S The switch of K

$Q(K)$ The knot quandle of K

R_n The dihedral quandle of order n

s_β The all B state

s_α The all A state

$w(K)$ The writhe of K

I

Knots and crossings

CHAPTER 1

Virtual knots and links

In this chapter, we introduce knot and link diagrams. From a physical viewpoint, we construct a knot by taking a segment of rope, knotting the rope, and fusing the endpoints. A more mathematical description of a knot is that a knot is a closed loop in three-dimensional space. The study of knots dates to the 1880s, when Lord Kelvin proposed that atoms were knotted vortices in the ether. The idea of a virtual knot was introduced in 1996 by Louis Kauffman, who published "Virtual Knot Theory" in the *Journal of Knot Theory and its Ramifications* in 1999. This article introduced a more combinatorial view where knots and links are defined by crossing information. We begin this chapter with a discussion of curves in the plane and gradually build toward a description of oriented virtual links.

1.1 CURVES IN THE PLANE

Imagine that you've been given a collection of "nodes" and "wires" and a list of instructions on how to connect the nodes and wires to a circuit board. You are supposed to attach the nodes to a board (the plane) and then attach the wires to the nodes as indicated. The wires need to lie flat on the plane and can only meet at nodes. Four wires can be connected to each node. Each node is labeled by a ordered list: (a, b, c, d). This ordered list indicates how the wire segments are connected to the node in a clockwise order. If two nodes share a label then they are connected by a wire. Your first set of instructions is:

$$(a, b, c, d),\ (d, c, e, f),\ (f, e, b, d). \tag{1.1}$$

After some effort, you construct the diagram shown in Figure 1.1. You attach your nodes and wire to the board following the diagram.

Your second set of instructions is:

$$(a, b, c, d) \text{ and } (c, d, b, a). \tag{1.2}$$

However, you can't complete this diagram without introducing an additional crossing (see

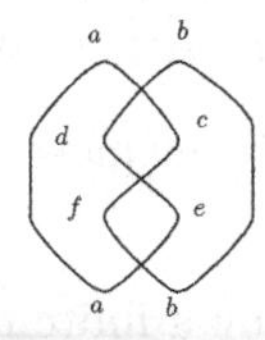

FIGURE 1.1: Circuit diagram 1

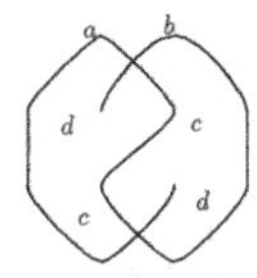

(a) Incomplete circuit diagram

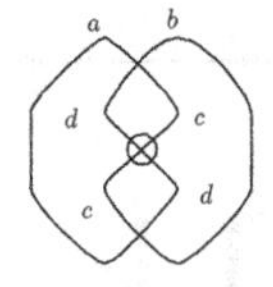

(b) Completed circuit diagram 1

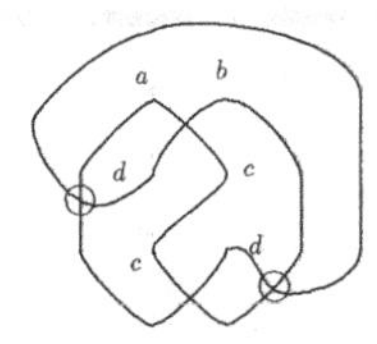

(c) Completed circuit diagram 2

FIGURE 1.2: Circuit diagrams

Figure 1.2a). You determine two possible diagrams, but both require extra crossings (which you indicate by circling the crossings). These diagrams are shown in Figures 1.2b and 1.2c. These two circuit diagrams look different, but represent the same set of instructions.

But now you have questions—when do I have to add these circled crossings? How will I know when two diagrams represent the same set of instructions? What happens if the instructions include the node (a, b, b, c)? What happens if I re-arrange the position of my nodes? Can I always avoid circled crossings?

The diagrams constructed in Figure 1.2 are flat virtual knot diagrams. Virtual knot and link diagrams are drawn in the plane and represent the possible lists of instructions described above. This includes diagrams that are projected images of knots and links in three dimensional space as well as "virtual" knots and links that do not exist in three dimensional space, even though we can describe them using lists of nodes. In this chapter, we study how to draw virtual link diagrams in a way that avoids ambiguity and define when two diagrams represent the same set of "instructions".

To do this, we first recall terms that describe curves in the plane. (The diagrams that we constructed in the example above are planar curves with decorated double points.) A **closed curve** in the plane forms a loop with no endpoints. A **simple curve** has no points of self-intersections. A **simple, closed curve** has no endpoints and no points of self-intersection. Informally, a simple closed curve is a circle that has been stretched out of shape. (A mathematician would say that the circle has been deformed.) Examples of these curves are shown in Figure 1.3. A simple closed curve in the plane represents the simplest possible knot—the **unknot**—an unknotted loop.

The planar curves in a virtual link diagram have the following properties:

1. The curve is smooth. There are no cusps and every point on a closed planar curve has a tangent line. At a double point, we see two different tangent lines. (This eliminates the possibility of a double point where the curve does not actually cross itself.)

2. Points of intersection are formed by at most two segments of the curve. These points are called double points.

3. Each curve can be approximated by a finite number of line segments.

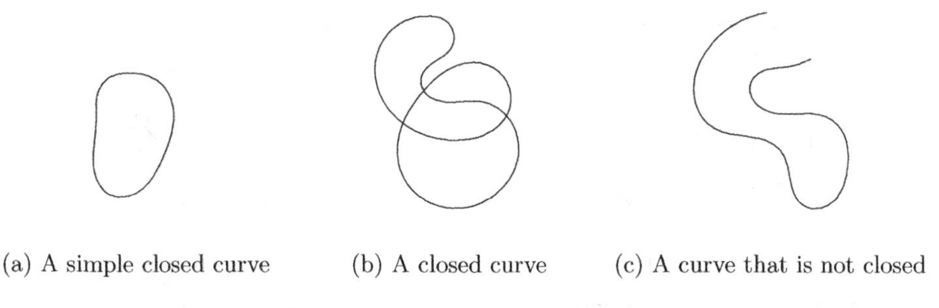

(a) A simple closed curve (b) A closed curve (c) A curve that is not closed

FIGURE 1.3: Curves in the plane

A collection of n curves with these properties is an **underlying diagram**, M. Two diagrams that do not have these properties are shown in Figure 1.4. In Figure 1.4a, we see a diagram with a triple point. Our restrictions don't allow a diagram with triple (or higher) points. In Figure 1.4b, the arrow indicates a double point with a shared tangent line. The arcs of the curve intersect at this point and then separate again. This type of double point introduces an ambiguity into our diagrams. We can not distinguish from the diagram whether the arcs cross or simply meet at a point and share a tangent. We eliminate the ambiguity by banning this type of double point.

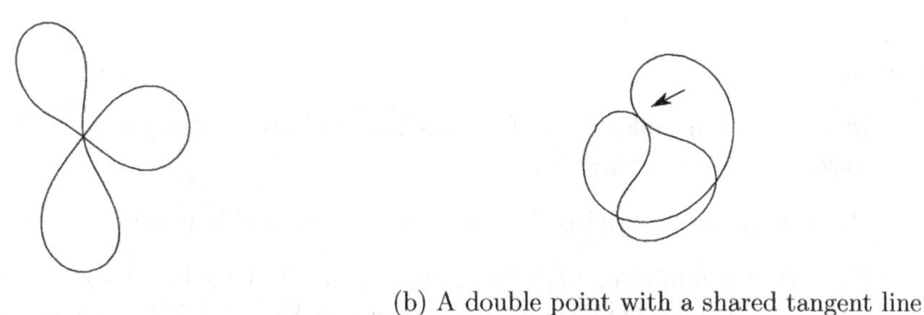

(a) Curve with triple point

(b) A double point with a shared tangent line

FIGURE 1.4: Curves that are not underlying diagrams

A **component** of an underlying diagram M is an individual closed curve. An **edge** of an underlying diagram is a curve segment bounded by double points in the underlying diagram.

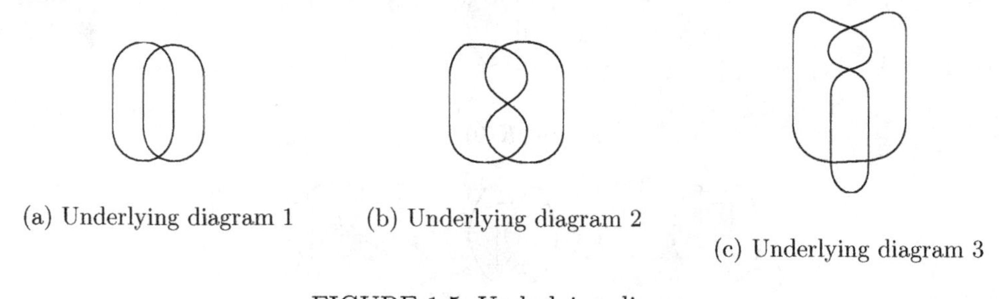

(a) Underlying diagram 1 (b) Underlying diagram 2 (c) Underlying diagram 3

FIGURE 1.5: Underlying diagrams

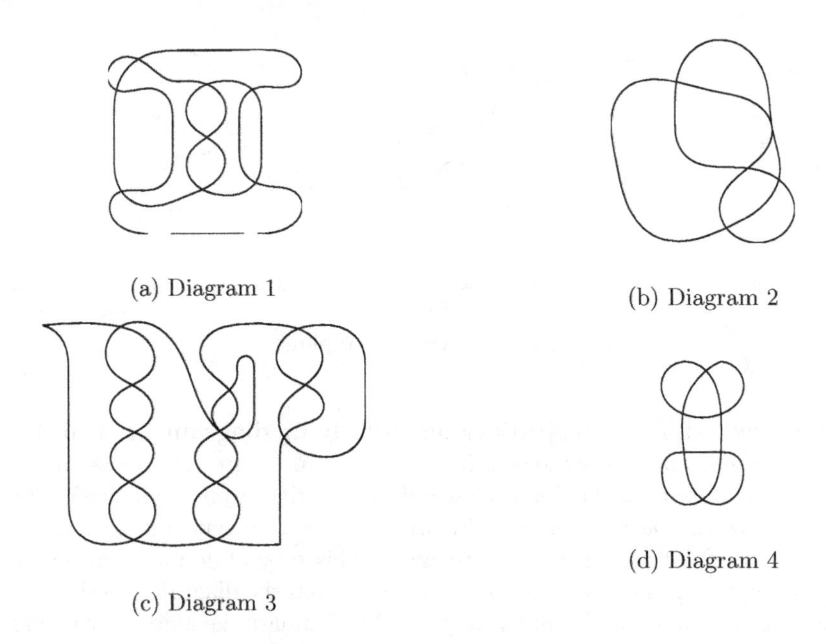

(a) Diagram 1

(b) Diagram 2

(c) Diagram 3

(d) Diagram 4

FIGURE 1.6: Which are underlying diagrams?

Exercises

1. Identify which collections of curves are underlying diagrams in Figure 1.6. Which satisfy the three properties?

2. Prove that an underlying diagram with $n > 0$ double points has $2n$ edges.

3. Construct a family of underlying diagrams $K(n)$ by inserting a vertical string of n crossings in the schematic diagram shown in Figure 1.7. Give a conjecture about the relationship between the number of crossings and the number of components. Prove this conjecture.

4. Construct a family of underlying diagrams $K(n, m)$ by inserting a vertical string

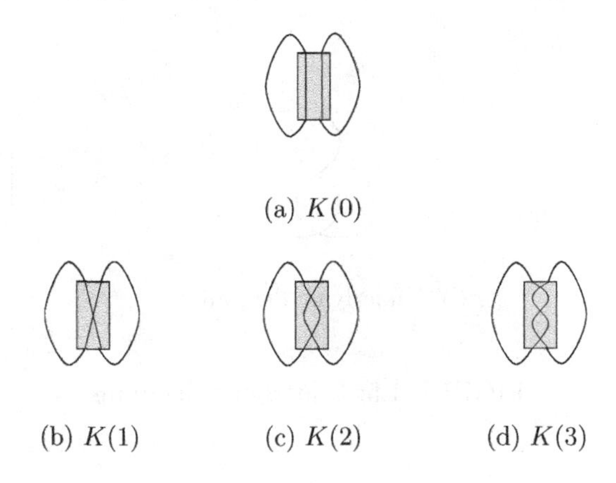

(a) $K(0)$

(b) $K(1)$ (c) $K(2)$ (d) $K(3)$

FIGURE 1.7: Constructing the family $K(n)$

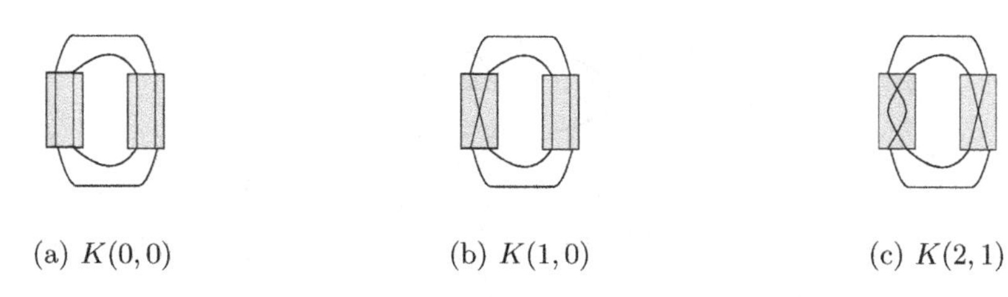

(a) $K(0,0)$ (b) $K(1,0)$ (c) $K(2,1)$

FIGURE 1.8: Constructing the family $K(n,m)$

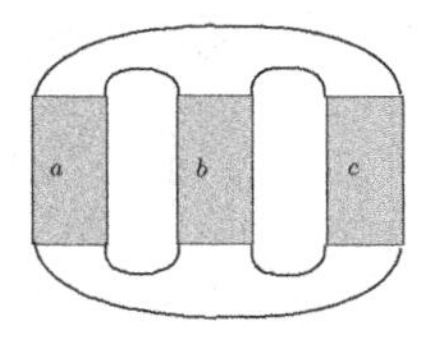

FIGURE 1.9: $K(n,m,p)$

of crossings in each of the gray boxes shown in the diagram in Figure 1.8. Insert n crossings in the left hand box and m crossings in the right hand box. Give a conjecture about relationship between the value of n and m and the number of components. Prove this conjecture.

5. Construct a family of underlying diagrams $K(n,m,p)$ by inserting a vertical string of crossings in each gray box. What is the relationship between the values n, m, and p and the number of components (see Figure 1.9)?

1.2 VIRTUAL LINKS

We construct virtual link and knot diagrams by decorating the double points of an underlying diagram with crossing information. We refer to a decorated double point as a **crossing**. A **virtual link diagram** is a collection of n closed curves (an underlying diagram) where the crossings are marked as a classical crossing or a virtual crossing. A **classical crossing** is a double point marked with under/overpassing information. A **virtual crossing** is a double point marked by a circle. Both crossing types are shown in Figure 1.10.

(a) Classical crossing (b) Virtual crossing

FIGURE 1.10: Crossing types

A **virtual knot diagram** is a 1-component virtual link diagram. Examples of virtual knot diagrams are shown in Figure 1.11. Virtual link diagrams with multiple components are shown in Figure 1.12.

An **arc** of a virtual link diagram begins and ends at the undermarkings at classical crossings. A knot with n classical crossings has n arcs. An arc may contain one or more edges.

We defined the set of virtual knot and link diagrams. Next, we need to define when

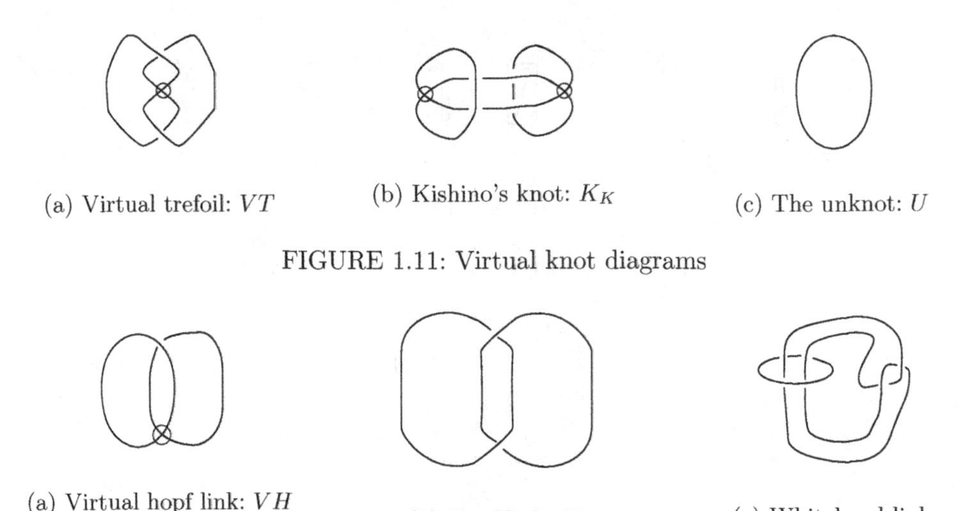

(a) Virtual trefoil: VT (b) Kishino's knot: K_K (c) The unknot: U

FIGURE 1.11: Virtual knot diagrams

(a) Virtual hopf link: VH (b) Hopf link: H (c) Whitehead link

FIGURE 1.12: Virtual link diagrams

two diagrams are equivalent (equality between two diagrams). When do we consider two different diagrams to represent the same object? It is clear that if a diagram's size is reduced (or increased) then the diagrams should be considered equivalent. But what else should be included? How can we define **equivalence** of diagrams? The equivalence relation between link diagrams is based on a set of diagrammatic moves.

Review: Equivalence classes and relations

We review the definition of equivalence relation and equivalence classes. An **equivalence relation** $\sim$ on a set X is a subset of $X \times X$. If (x, y) is an element of the subset then we write $x \sim y$. For all x, y, and z in the set X, an equivalence relation has three properties

$x \sim x$ (reflexive property),

if $x \sim y$ then $y \sim x$ (symmetric property),

if $x \sim y$ and $y \sim z$ then $x \sim z$ (transitive property).

Equivalence relations occur frequently in mathematics—the most intuitive equivalence relation is the notion of even and odd integers. Remember that an integer is even if the integer is a multiple of two. (If the integer is not a multiple of two then the integer is odd.) The set of integers is denoted as $\mathbb{Z}$. We define an equivalence relation, $\equiv_2$, on $\mathbb{Z}$ using the definition of even. For $x, y \in \mathbb{Z}$, we say that $x \equiv_2 y$ if and only if $x - y$ is a multiple of two. Under this equivalence relation, any two even integers are equivalent. We consider an example; $2 \equiv_2 4$ since $2 - 4 = -2$ but $2 \not\equiv_2 3$ since $2 - 3 = 5$. Notice that any two odd integers are considered equivalent. As an example, $3 - 5 = -2$ so that $3 \equiv_2 5$.

Theorem 1.1. *The relation $\equiv_2$ is an equivalence relation.*

Proof. Let x, y, and z be elements of $\mathbb{Z}$. The relation $\equiv_2$ is reflexive since $x - x = 0$ for all $x \in \mathbb{Z}$. Next, the relation $\equiv_2$ is symmetric since $x - y = -(y - x)$. If $x \equiv_2 y$ then $y \equiv_2 x$. Finally, we prove that the relation is transitive. If $x \equiv_2 y$ and $y \equiv_2 z$ then $x - y = 2k$ and $y - z = 2j$ for some $j, k \in \mathbb{Z}$. Then $x - z = 2(k + j)$ and by definition, $x \equiv_2 z$. □

Given an equivalence relation $\sim$ on a set X, for $x \in X$, the **equivalence class** of x, denoted $[x]$, is the set $\{y \in X | y \sim x\}$. The equivalence classes partition the set X. Namely, the equivalence classes form a collection of disjoint subsets and the union of these subsets is X. Returning to our example, we see that there are exactly two equivalence classes determined by $\equiv_2$: the set of even integers and the set of odd integers. These sets partition the set of integers.

We describe these sets using set builder notation:

$$[0] = \{x \in \mathbb{Z} | x - 0 \text{ is even}\}$$
$$[1] = \{x \in \mathbb{Z} | x - 1 \text{ is even}\}.$$

Another equivalence relation is found in geometry. Two triangles, T and T', are congruent if the following measurements: angle, side, angle (ASA) were equal. Congruence determines an equivalence relation on triangles and satisfies all three axioms: reflexive, symmetric, and transitive. The congruence (or equivalence) class of a triangle is specified by stating the measurements (angle, side, angle).

The equivalence relation on virtual link diagrams is based on small changes to the virtual link diagram. For example, "wiggling" or "stretching" the arcs of a diagram does not fundamentally affect the underlying diagram. These types of changes are called **planar isotopy moves** and do not interact with the crossings. Local planar isotopy moves are shown in Figure 1.13. Local modifications involving the classical crossings are called the **Reidemeister moves** and these moves are shown in Figure 1.14. Consider the neighborhood of the diagram to which we apply the Reidemeister moves. If this neighborhood was constructed from string, we could actually perform the Reidemeister moves with the string. The Reidemeister moves are diagrammatic versions of changes that could be made if a diagram was made of actual string. The moves drawn in Figure 1.14 are examples. They represent one possible configuration of crossings. For example, in the Reidemeister III move, we can switch the over/undermarkings in the center crossing and still obtain a valid move. Last, we have the **virtual Reidemeister moves** in Figure 1.15. These moves are visually analogous to the Reidemeister moves except for move IV. Move IV involves both virtual and classical crossings and allows an edge with only virtual crossings to slide past a classical crossing. The virtual Reidemeister moves change the placement of the curves in the plane without altering any of the classical crossings in the diagram. We will refer to the Reidemeister and virtual Reidemeister moves as the **diagrammatic moves**.

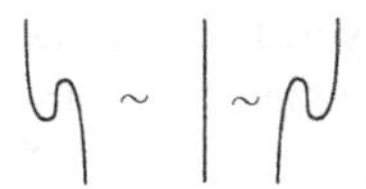

FIGURE 1.13: Planar isotopy

Two virtual link diagrams K and K' are said to be **equivalent** ($K \sim K'$) if one can be changed into the other by a finite sequence of planar isotopy moves, Reidemeister moves, and virtual Reidemeister moves. This means that if $K \sim K'$ then there is a finite sequence of diagrams $K = K_0, K_1, \ldots K_n = K'$ and a diagrammatic move takes K_{i-1} to K_i. Labeled diagrammatic moves relate the diagrams in Figure 1.16. If two diagrams, K and K', are related by a single Reidemeister or virtual Reidemeister move, we use the notation $K \leftrightarrow K'$ for clarity.

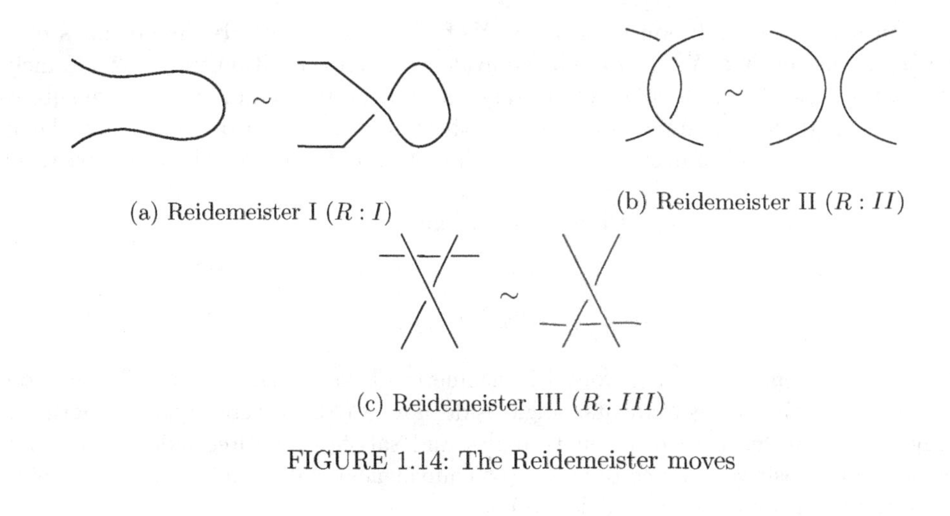

(a) Reidemeister I $(R:I)$

(b) Reidemeister II $(R:II)$

(c) Reidemeister III $(R:III)$

FIGURE 1.14: The Reidemeister moves

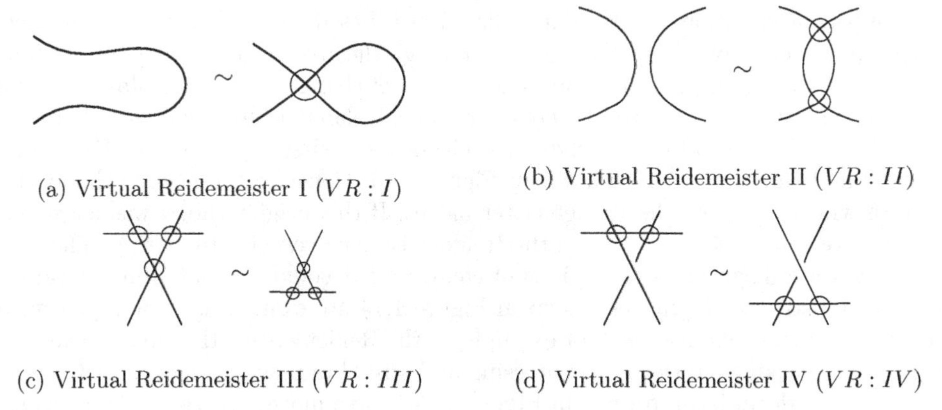

(a) Virtual Reidemeister I $(VR:I)$

(b) Virtual Reidemeister II $(VR:II)$

(c) Virtual Reidemeister III $(VR:III)$

(d) Virtual Reidemeister IV $(VR:IV)$

FIGURE 1.15: The virtual Reidemeister moves

Each pair of adjacent diagrams shown in the sequence in Figure 1.17 is connected by single diagrammatic moves; all the diagrams are equivalent. Notice that the diagrammatic moves can change the number of crossings and dramatically change the appearance of the diagram. An unlabeled sequence of diagrams related by diagrammatic moves is shown in Figure 1.18. In Figure 1.19, we see several diagrams that are equivalent to the unknot. The diagrammatic moves define an equivalence relation on virtual link diagrams. We prove this in Theorem 1.2.

Theorem 1.2. *The diagrammatic moves define an equivalence relation on virtual link diagrams.*

Proof. A virtual link diagram D is related to itself by a sequence of length zero, so the relation is reflexive. If $D_1 \sim D_2$, then the finite sequence of moves transforming D_1 into D_2 can be reversed, so that $D_2 \sim D_1$. Finally, if $D_1 \sim D_2$ and $D_2 \sim D_3$, we can form a finite sequence of moves transforming D_1 into D_3 so that $D_1 \sim D_3$. □

A **virtual link** is an equivalence class of virtual link diagrams. The equivalence class of the unknot is denoted U. If we know that two links L_1 and L_2 are not equivalent then we write $L_1 \nsim L_2$. Note that if two diagrams "look similar", this is not a proof that the

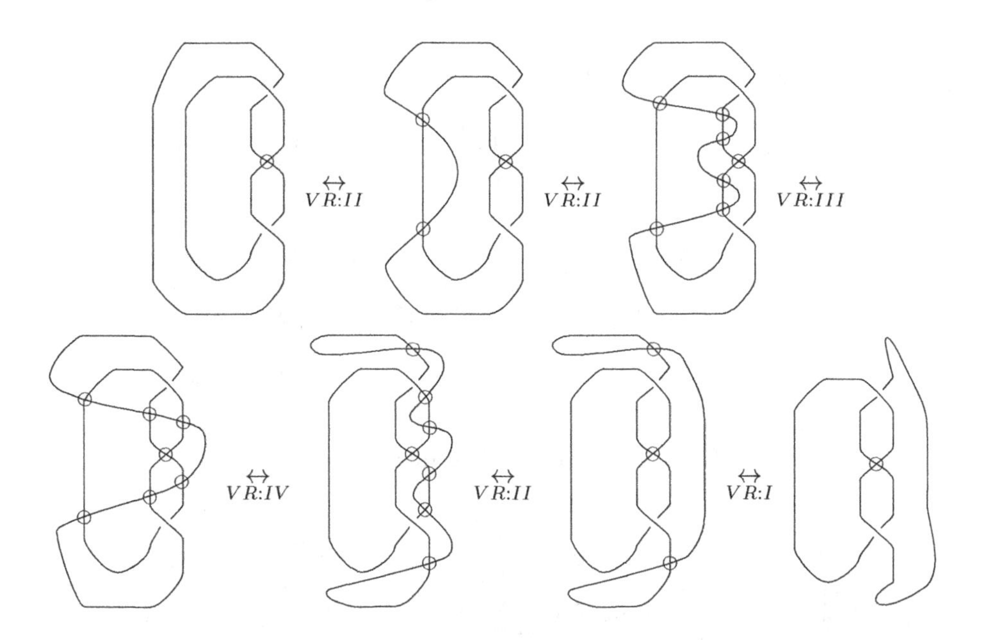

FIGURE 1.16: Equivalent diagram sequence

diagrams are equivalent. To prove that two diagrams are equivalent, we give a finite sequence of diagrammatic moves relating the two diagrams. This task can be very challenging—for example, a diagram and its reflection look very similar but they are not necessarily equivalent. We denote the set of virtual links as $\mathcal{V}$. We use the notation $\mathcal{V}_d$ to specifically refer to the set of virtual link diagrams. A **virtual knot** is a one component virtual link. Knot theorists sometimes abuse notation and use K to indicate both a knot diagram and its equivalence class. If needed, we use K to indicate a virtual link and the notation D to indicate a specific diagram.

The **detour move** is a single diagrammatic move that can replace the entire set of virtual Reidemeister moves. To perform the detour move, we select a section of an arc in the virtual knot diagram. If the section contains only virtual crossings, we may remove the section and redraw the section so that it contains only double points. Any crossings that result when the section of the arc is redrawn are virtual. The detour move is illustrated in Figure 1.20. The selected section of the arc is marked by two solid dots, then the arc is moved and the new crossings are marked as virtual. We view the sequence of moves in Figure 1.16 as a single detour move (see Figure 1.21).

Theorem 1.3. *The virtual Reidemeister moves are detour moves.*

Proof. Each virtual Reidemeister move can be performed by marking off a section of arc, erasing the arc, and drawing a new arc that forms an underlying diagram. □

Remark 1.1. *Knots are used in a variety of ways in topology. For this reason, people may choose to use a restricted set of moves and use different terminology from above. Planar isotopy wiggles and stretches the knot, regular isotopy refers to the Reidemeister II and III moves, and ambient isotopy refers to Reidemeister moves I –III.*

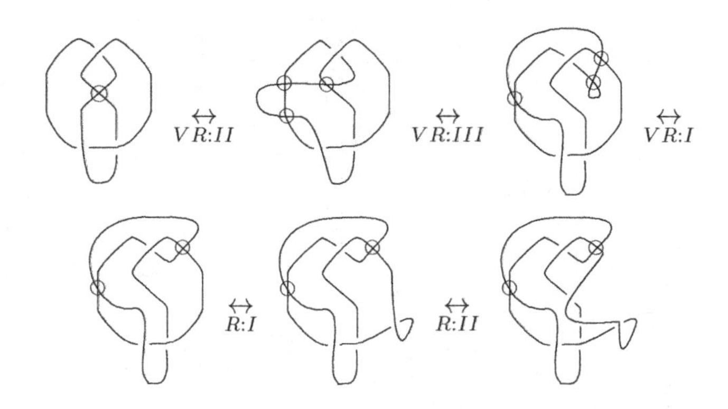

FIGURE 1.17: Equivalent diagram sequence

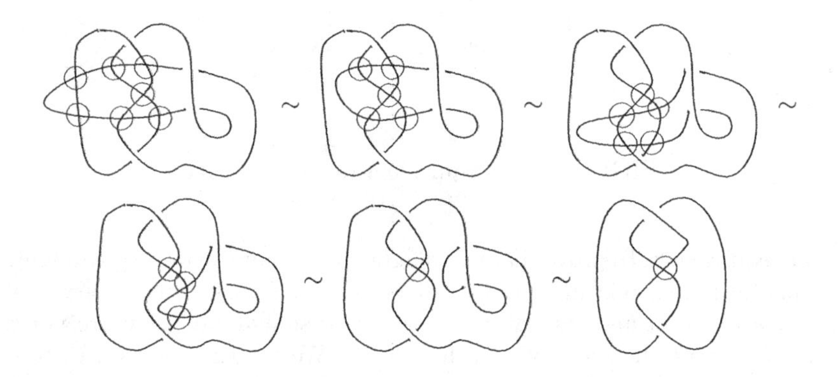

FIGURE 1.18: A sequence of equivalent diagrams

Classical links

A link K is a **classical link** if there is a virtual link diagram D equivalent to K with zero virtual crossings. The set of classical links is a subset of the set of virtual links. A **classical knot** is a one component classical link. A classical link diagram is a diagram with zero virtual crossings. The set of classical knots is denoted as $\mathcal{K}$.

A classical link is a mapping of n copies of the circle into three-dimensional space ($\mathbb{R}^3$)—two classical links are equivalent if they are **ambient isotopic**. Two ambient isotopic links can be nicely deformed into one another without cutting the edges or introducing sharp points. In classical knot theory, classical link diagrams can be obtained from curves in three dimensional space by projecting the image of the loops onto the plane. (Think of projecting a shadow onto the wall!) The double points in the projection are decorated with over- and undermarkings. We can move the loops in three dimensional space and see corresponding changes in the two dimensional projection. The possible changes in the projection result in the planar isotopy moves and the Reidemeister moves. We develop a similar interpretation of virtual links and knots later in the text.

If we eliminate virtual crossings and the virtual Reidemeister moves, a classical link is an equivalence class of link diagrams with no virtual crossings. Any two diagrams in the

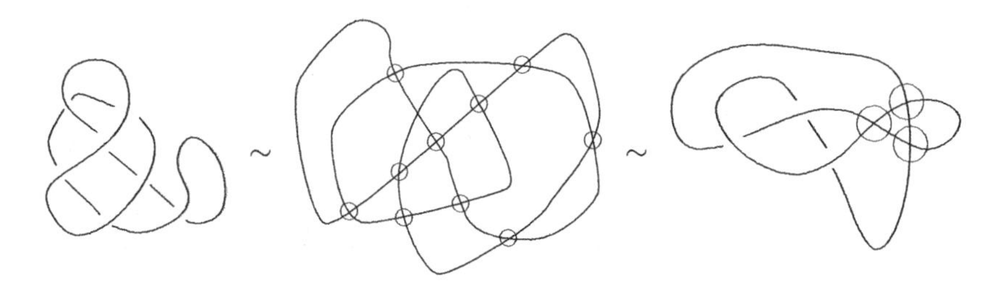

(a) Three diagrams equivalent to the unknot

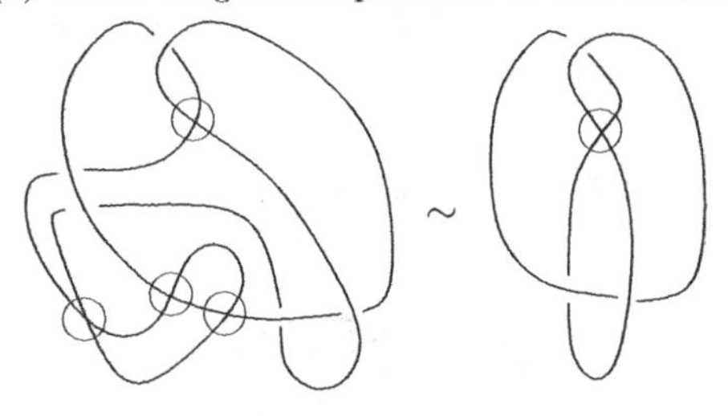

(b) Two equivalent diagrams

FIGURE 1.19: Equivalent diagrams

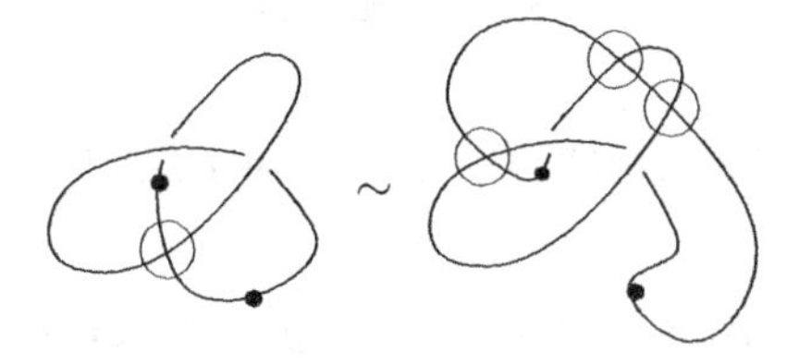

FIGURE 1.20: The detour move

set are related by a finite sequence of planar isotopy and Reidemeister moves. Note that if we allow the virtual Reidemeister moves, we do not add any classical diagrams to the equivalence class.

A classical knot diagram is **alternating** if the knot has a diagram such that overpasses and underpasses at crossings alternate as you trace the curve of the knot.

A classical knot diagram has a **checkerboard coloring**. A classical knot diagram divides the plane into regions bounded by edges and classical crossings. Adjacent regions share an edge of the classical knot diagram. We checkerboard color a classical knot diagram by assigning one of two colors so that two adjacent regions are not assigned the same color. An example of a checkerboard coloring is given in Figure 1.22. In a virtual knot diagram, the regions of the plane are not bounded by edges and classical crossings. The boundary of a region possibly includes both classical and virtual crossings.

Exercises

1. The diagram underlying the trefoil is shown in Figure 1.5b. How many different ways can you choose classical crossing marks for the double points?

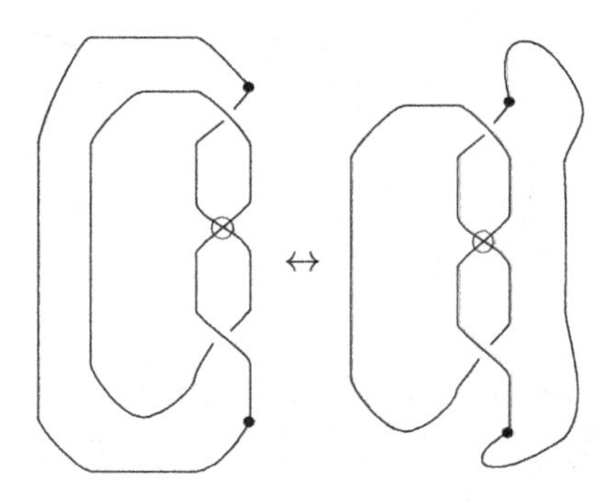

FIGURE 1.21: Detour move example

FIGURE 1.22: Checkerboard coloring of the trefoil

2. For all $x, y \in \mathbb{Z}$, we define $x \equiv_3 y$ if $x - y = 3k$ for some k in $\mathbb{Z}$. Prove that $\equiv_3$ is an equivalence relation on $\mathbb{Z}$ and satisfies the reflexive, symmetric, and transitive properties.

3. Prove that the following relation on $\mathbb{Z}$ is an equivalence relation. Let $x \equiv_5 y$ if and only if $x - y = 5k$ for some k in $\mathbb{Z}$.

4. Use set builder notation to explicitly describe the equivalence classes for the relation $\equiv_5$ where $x \equiv_5 y$ if and only if $x - y = 5k$ for some k in $\mathbb{Z}$.

5. Determine all possible versions of the Reidemeister III move.

6. Label the sequence of moves in Figure 1.18.

7. Find a sequence of moves relating the diagrams in Figure 1.23.

8. Find a sequence of moves relating the diagrams in Figure 1.24.

9. Find a sequence of moves relating the diagrams in Figure 1.19 to the unknot.

10. Prove that a knot diagram with two or fewer classical crossings and zero virtual crossings is equivalent to the unknot.

11. List all virtual link diagrams with two classical crossings.

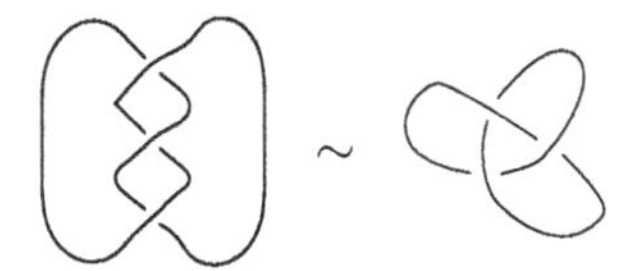

FIGURE 1.23: Find a sequence of diagrammatic moves relating the diagrams

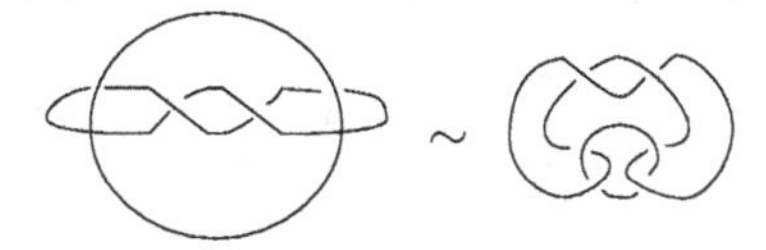

FIGURE 1.24: Find a sequence of diagrammatic moves between the links

1.3 ORIENTED VIRTUAL LINK DIAGRAMS

A link diagram is **oriented** if each component of the diagram has been assigned an orientation or direction. To indicate orientation, we mark each component of the diagram with an arrow indicating the direction of each component of the diagram. Orienting the components increases the number of distinct diagrammatic moves. For example, there are two ways to orient a Reidemeister II move, resulting in two types of oriented Reidemeister II moves. An oriented Reidemeister I move is shown in Figure 1.25.

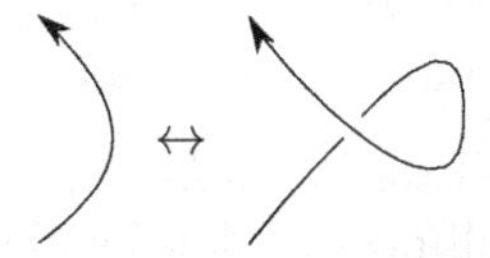

FIGURE 1.25: An oriented Reidemeister I move

The two possible oriented Reidemeister II moves are shown in Figure 1.26. The edges are co-oriented in Figure 1.26a. The edges are contra-oriented (oppositely oriented) in Figure 1.26b. Orientation does not dramatically increase the number of possible diagrams of the Reidemeister I and II moves. However, each Reidemeister III move has three strands. There are eight possible choices of orientation for a Reidemeister III move. Using symmetry, we can reduce the number of orientations from eight cases to three cases as shown in Figure 1.27. A Reidemeister III move with co-oriented strands is shown in Figure 1.28.

The other oriented Reidemeister III moves, the move with circular oriented strands and the move with two co-oriented strands, can be expressed as a sequence of moves involving the co-oriented Reidemeister III and the Reidemeister I and II moves. We apply a sequence of Reidemeister II moves to transform the diagram in Figure 1.27b into a diagram containing the left hand side of a co-oriented Reidemeister III move in Figure 1.29.

The orientation also increases the number of versions of the virtual Reidemeister moves, but each of these moves can be expressed as a single detour move.

A classical knot is said to be **descending** (respectively ascending) if there is a point on the knot such that as you trace the curve in the direction of orientation then the overpassing (respectively underpassing) of a crossing is always encountered first.

(a) Co-oriented Reidemeister II move (b) Contra-oriented Reidemeister II move

FIGURE 1.26: Oriented Reidemeister II moves

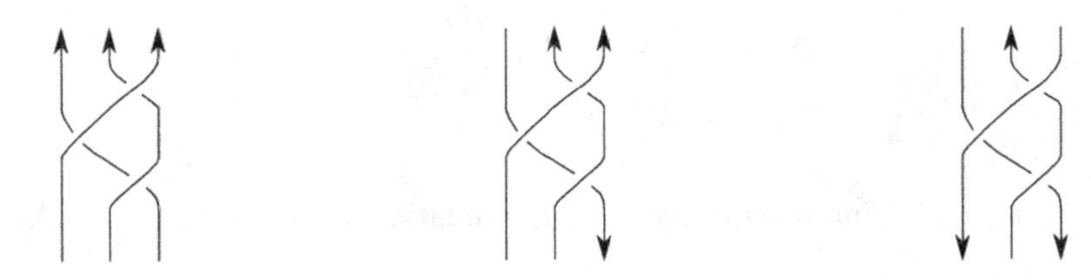

(a) Three co-oriented strands (b) Two co-oriented strands (c) Circular-oriented strands

FIGURE 1.27: Oriented left hand sides of a Reidemeister III move

Exercises

1. Construct all possible Reidemeister I moves based on orientation and over/undermarkings. (Hint: there are two sets of equivalent diagrams.)
2. Construct all possible Reidemeister II moves based on orientation and over/undermarkings. (Hint: there are four sets of equivalent diagrams.)
3. How many different Reidemeister III moves are there are based on orientation and over/undermarkings? Draw all possible left hand sides of a Reidemeister III move.
4. Prove that the Reidemeister III move with circular orientation can be described as a sequence of oriented Reidemeister II moves and the co-oriented Reidemeister III move.
5. Given a classical knot diagram with n classical crossings, what is the maximum number of crossings that need to be switched to construct a descending diagram?
6. Given a classical knot diagram with only classical crossings, argue that a descending diagram is equivalent to the unknot. (The solution to the problem is a physical argument.)

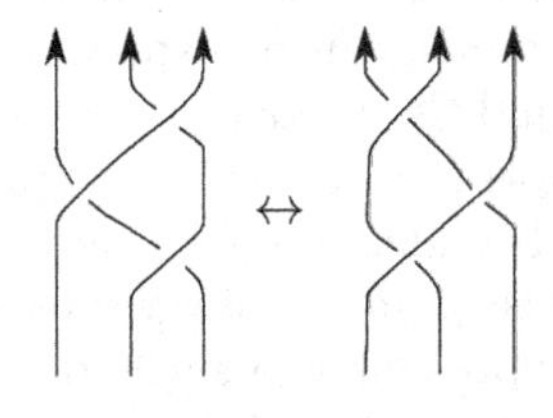

FIGURE 1.28: Co-oriented Reidemeister III move

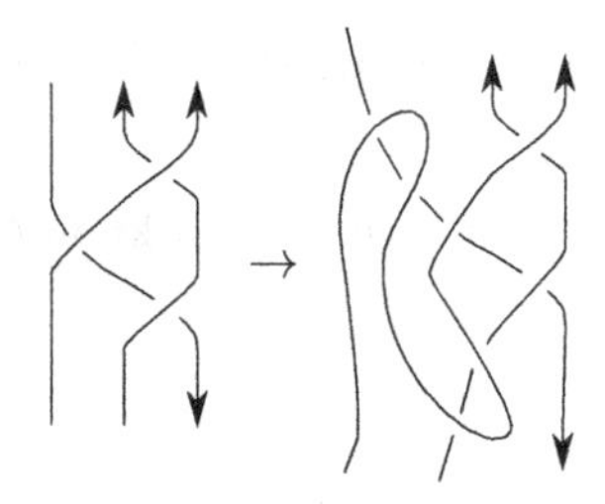

FIGURE 1.29: Transforming a Reidemeister III move

7. Prove that a virtual knot diagram with zero classical crossings is equivalent to the unknot.

8. Prove that an underlying diagram can always be checkerboard colored.

1.4 OPEN PROBLEMS AND PROJECTS

Open problems

What questions drive research into virtual links and knots? Initially, virtual knot theory was studied as an extension of classical knot theory. In this book, we study virtual knot theory as a topic that includes classical knot theory. From this viewpoint, the first goal is to enumerate—give a complete list of the mathematical objects that we are studying. The second goal is to determine a method of classification. This means that if someone was to present you with a random virtual link diagram then you could determine its place in the enumerated list. These questions drive the initial research forward and, in the process, introduce other questions.

Motivation for investigating virtual knots and links can also be based on the possible applications (DNA entanglement, textiles, and quantum physics to name a few) but first we need to develop our understanding and intuition of links and knots. We list some questions.

1. Given two oriented diagrams K_1 and K_2, can we prove that the diagrams are equivalent without finding a sequence of diagrammatic moves? Is there a complete invariant of virtual knots?

2. Is there a way to prove that a virtual link diagram is not equivalent to the unknot? And can this method be easily applied to any link diagram? (There is an invariant that distinguishes virtual knots from the unknot. This invariant is called Khovanov homology and is difficult to compute. It would be nice to have an invariant that is easier to compute.)

3. Are there topological constructions related to virtual links? Can we use virtual links to prove theorems about these topological constructions?

Projects

In this section, we provide a list of suggestions for projects: computational investigations, short papers, or presentations.

1. Construct a family of underlying diagrams using the "gray box" approach from the

exercises. Can you determine and prove the number of components that each diagram contains?

2. Can you define your own equivalence class of knots? What types of crossings? What moves and why?

Further Reading

The following articles investigate the history of knot theory, knot theory in art, and more.

1. Robert Bosch. Simple-closed-curve sculptures of knots and links. *J. Math. Arts*, 4(2):57–71, June 2010: Robert Bosch uses curves and knots - with a mathematically involved design process to create fine art.

2. Peter R. Cromwell. The distribution of knot types in Celtic interlaced ornament. *J. Math. Arts*, 2(2):61–68, June 2008: One source for classical knot diagrams is Celtic artwork. He examines how frequently possible knot types occur in Celtic knot artwork.

3. J. Dennis Lawrence. *A catalog of special plane curves.* Dover, 1972: If you are interested in the curves that form underlying diagrams, this text provides a catalog of parametrized plane curves.

4. Sam Nelson. The combinatorial revolution in knot theory. *Notices Amer. Math. Soc.*, 58(11):1553–1561, 2011: A quick introduction to virtual knot theory and some reasons for studying virtual and classical knots.

5. Daniel Silver. Knot theory's odd origins. *American Scientist*, 94(2):158, 2006 : A historical introduction to knot theory and how physics motivated its study.

CHAPTER 2

Linking invariants

Finding a sequence of Reidemeister moves that changes one virtual link diagram into another virtual link diagram can be challenging. Even if a sequence between two diagrams contains thousands of diagrammatic moves, the diagrams are still equivalent diagrams. Suppose that you believe two diagrams L and L' are equivalent. If you try and fail to produce a sequence of diagrammatic moves relating L and L', does this mean that the diagrams are not equivalent? Sadly, attempting and failing to find the sequence does not prove that the diagrams represent different links. At the same time, we have not eliminated the possibility that the diagrams are equivalent. It is possible that the sequence is just too long (possibly thousands of moves) for us to find.

A **link invariant**, $\mathcal{I}\colon \mathcal{V}_d \to S$, is a map from the the set of virtual link diagrams to a set S. The set S is hopefully a set of objects with a simpler structure—such as the set of integers, $\mathbb{Z}$. (We can certainly determine if two integers are equal!) An invariant $\mathcal{I}$ is well defined on the set of virtual links—which means that for all diagrams L in the equivalence class $[K]$ then $\mathcal{I}(L) = \mathcal{I}(K)$. A link invariant is a map on the set of link diagrams with the property that for all link diagrams L and L', if $L \sim L'$ then $\mathcal{I}(L) = \mathcal{I}(L')$.

Does $\mathcal{I}(L) = \mathcal{I}(L')$ guarantee that $L \sim L'$? Not necessarily. The answer to this question depends on our map, $\mathcal{I}$. It is possible that two different links map to the same element of the set S (just like $f(3) = f(-3)$ when $f(x) = x^2$). More commonly, a link invariant is used to prove that two links are not equivalent. If we calculate that $\mathcal{I}(L) \neq \mathcal{I}(L')$ then we conclude that $L \nsim L'$. This approach doesn't seem to give us much information, but it will prevent us from searching for a non-existent sequence of diagrammatic moves.

In this chapter, we focus on link invariants that are computed by counting sets of crossings. We also learn how to use invariants to make conclusions about a link diagram by focusing on the logic of conditional statements.

2.1 CONDITIONAL STATEMENTS

We defined a link invariant as map with the following property. For all virtual link diagrams L and L', if $L \sim L'$ then $\mathcal{I}(L) = \mathcal{I}(L')$.

The statement has two parts: a quantifier "For all virtual link diagrams L and L'" and a **conditional** or if-then statement. In the if-then statement are two predicate statements. We determine if the predicate statements are true once we have selected values for the variables in the predicate statements from a set specified by the quantifier. For example, if we select specific virtual knot diagrams for L and L' then we can evaluate if the statement $L \sim L'$ is true or false. The quantifier provides a context for L and L'. In practice, the quantifier is sometimes dropped because the possible values for variables in the if-then statement are clear from context.

To discuss the truth value of if-then statements, we use the symbols P and Q to denote the predicate statements. In mathematical notation, we write a conditional statement as $P \Rightarrow Q$ which is read aloud as "*if* P *then* Q" or "P *implies* Q".

The truth table for $P \Rightarrow Q$, shown in Table 2.1, contains all possible combinations of true and false for P and Q. The table also indicates true-false value of the statement $P \Rightarrow Q$ based on the true-false values of P and Q. One common question about truth tables is: Why

P	Q	$P \Rightarrow Q$
T	T	T
T	F	F
F	T	T
F	F	T

TABLE 2.1: Truth table for $P \Rightarrow Q$

are we concerned with the possibility that P is false? If the statement P is false, then the negation of P, denoted $\neg P$, is true. Let's consider a specific example. Suppose that P is the statement "the children behave" and Q is the statement "the children eat ice cream". The statement $P \Rightarrow Q$ is: "if the children behave then the children eat ice cream". Suppose that P is true and the children behave. We have two possibilities to consider: Q is true and Q is false.

1. If Q is a true statement, then "if the children behave then the children eat ice cream" is a true statement.

2. If Q is a false statement, then "if the children behave then the children eat ice cream" is a false statement. (The statement P is true but the statement "the children eat ice cream" has not followed.)

Now suppose that P is false and "the children do not behave" is true. We again have two possibilities: Q is true or Q is false ("the children do not eat ice cream").

1. If Q is a true statement then the statement "if the children behave then the children eat ice cream" is true. The children eat ice cream even though the statement P is not true. The if statement does not occur, so what follows is irrelevant.

2. If Q is a false statement, then the statement "if the children behave then the children eat ice cream" is true. In this case, children did not behave and the children did not eat ice cream. Since the children did not behave then the outcome "the children eat ice cream" is irrelevant.

There are three other statements that are related to the conditional statement:

The contrapositive: $\neg Q \Rightarrow \neg P$

The converse: $Q \Rightarrow P$

The inverse: $\neg P \Rightarrow \neg Q$

We evaluate the contrapositive, converse, and inverse statements in Table 2.2. The statement $P \Rightarrow Q$ and the contrapositive $\neg Q \Rightarrow \neg P$ have the same values in the truth table, as

P	Q	$P \Rightarrow Q$	$Q \Rightarrow P$	$\neg P$	$\neg Q$	$\neg Q \Rightarrow \neg P$	$\neg P \Rightarrow \neg Q$
T	T	T	T	F	F	T	T
T	F	F	T	F	T	F	T
F	T	T	F	T	F	T	F
F	F	T	T	T	T	T	T

TABLE 2.2: The converse, contrapositive, and inverse of $P \Rightarrow Q$

do the pair $Q \Rightarrow P$ and $\neg P \Rightarrow \neg Q$. These pairs of statements are **logically equivalent**. If a statement is true then any logically equivalent statement is also true.

We consider examples of these types of statements. We are interested in the set of integers, so we begin with the quantifier "For all integers x". The quantifier indicates that the variable x is an integer. We let P denote the statement *x is an integer divisible by four* and Q denote the statement *x is an even integer.* We determine the negation of P and Q.

$\neg P$ is the statement: x is not an integer divisible by four.

$\neg Q$ is the statement: x is not an even integer.

The statement $P \Rightarrow Q$ is "if x is an integer divisible by four then x is an even integer".

$P \Rightarrow Q$: If x is an integer divisible by four then x is an even integer.

$\neg Q \Rightarrow \neg P$ (contrapostive): If x is not an even integer then x is an integer not divisible by four.

$Q \Rightarrow P$ (converse): If x is an even integer then x is an integer divisible by four.

$\neg P \Rightarrow \neg Q$ (inverse): If x is an integer not divisible by four then x is not an even integer.

In our example, the conditional and contrapositive statements are true for all integers. We prove these statements using a logical argument or proof. The converse, inverse, and biconditional statements are not true—a single example is sufficient to show that these statements are not true. So make sure that you don't confuse your converse and your contrapositive!

We can also form more complicated statements using **and** and **or**. The statement P **and** Q is denoted $P \wedge Q$. The statement $P \wedge Q$ is true when both P and Q are true. The statement P **or** Q is denoted $P \vee Q$. The "or" statement is inclusive, meaning that if either P or Q is true or both P and Q are true then $P \vee Q$ is true. The biconditional statement $P \Leftrightarrow Q$ is read as P if and only if Q. The biconditional statement is logically equivalent to the statement: $(P \Rightarrow Q) \wedge (Q \Rightarrow P)$.

P	Q	$P \vee Q$	$P \wedge Q$	$P \Leftrightarrow Q$
T	T	T	T	T
T	T	T	F	F
F	T	T	F	F
F	F	F	F	T

TABLE 2.3: $P \vee Q$ and $P \wedge Q$

The most common method of proof that is used to prove that a mapping is a link invariant is direct proof. To construct a direct proof of $P \Rightarrow Q$, you assume that the quantified statement P is true. We then construct a sequence of logical statements that terminate with Q. Since you know both P and Q, it is possible to work the proof from both ends of the statement. Consider a statement that can be proved by direct proof.

Theorem 2.1. *For all integers x, if x is an integer divisible by four then x is an even integer.*

Let x be an integer divisible by four. By the definition of integer and divisible by four, there exists $k \in \mathbb{Z}$ such that $x = 4k$. The conclusion is the statement: x is an even integer. Equivalently, we can write that there exists $n \in \mathbb{Z}$ such that $x = 2n$. We see that the important point is to fill in the steps between the statements $x = 4k$ and $x = 2n$. We write the proof.

Proof. Let x be an integer divisible by four. By definition, there exists $k \in \mathbb{Z}$ such that $x = 4k$. Rewriting, we see that $x = 2(2k)$. Since $2k \in \mathbb{Z}$, let $n = 2k$. Then, there exists $n \in \mathbb{Z}$ such that $x = 2n$. By definition, x is an even integer. □

Another common method of proof is **proof by contradiction**. In proof by contradiction, you begin by assuming that P and $\neg Q$ are true and then construct a sequence of logical statements that end with a statement that is clearly false. (We use this strategy to conclude that pairs of virtual link diagrams are not equivalent.) Here is a sample proof by contradiction.

Theorem 2.2. *For all integers x, if x is not an even integer then x is not divisible by four.*

Proof. If x is not an even integer then x is odd. By definition, there exists $k \in \mathbb{Z}$ such that $x = 2k + 1$. Suppose that x is divisible by four. Then $x = 4n$ for some $n \in \mathbb{Z}$. Combining the statements, we see that $4n = 2k + 1$. After rewriting, we obtain $2(2n - k) = 1$. We conclude that 1 is an even integer. This is a contradiction, since 1 is not an even integer. We conclude that x is not divisible by four. □

We can find infinitely many examples of integers x that satisfy the statement: If x is an integer divisible by four then x is an even integer. For example, $x = 4$ satisfies the conditional statement. But, examples are not a proof since we have not verified that all integers divisible by four are also divisible by two.

Examples can be used to show that a statement with the quantifier "For all" is not true. Consider the converse of our statement: For all integers x, if x is an even integer then x is divisible by four. The integer 14 is an even integer and 14 is not divisible by four. This single example demonstrates that the statement is not true for all integers. This type of example is called a **counterexample**.

We want to use these ideas to prove results about link invariants. By the definition of a link invariant, for all virtual link diagrams L and L', if L and L' are equivalent diagrams then $\mathcal{I}(L) \equiv \mathcal{I}(L')$. By definition of equivalence, there is a sequence of equivalent diagrams $L \sim L_0, L_1, \ldots L_n \sim L'$ where move m_i transforms diagram L_{i-1} into diagram L_i. For all $i, j \in \{0, 1 \ldots n\}$, the statement $\mathcal{I}(L_i) = \mathcal{I}(L_j)$ is true since the diagrams in the sequence are equivalent. Our strategy involves proving that individual diagrammatic moves do not change the value of the mapping. This outlined strategy is a form of direct proof. However, when we apply the invariant, we focus on the contrapositive—calculating the invariant to prove that two link diagrams are not equivalent. In addition to proving that two link diagrams are not equivalent, we can use an invariant to obtain information about the link. We can ask the following questions about a link invariant $\mathcal{I}$.

What values of S are in the image (range) of $\mathcal{I}$?

Can we find a link diagram L such that $\mathcal{I}(L) = x$ for $x \in S$?

Does the invariant provide information about some characteristic of the link?

Exercises

1. Compute the truth table for the converse, contrapostive, and inverse of $(\neg P) \Rightarrow Q$.
2. Compute the truth table for $\neg(P \vee \neg Q)$.
3. Give an example of an integer that for which the following statement is true. For all integers x, if x is not an even integer then x is an integer that is not divisible by six.
4. Give a counterexample to the following statement: For all integers x, if x is an integer that is not divisible by six then x is not an even integer.
5. Use direct proof to prove that: For all integers x, if x is an integer that is divisible by six then x is an even integer.
6. Use proof by contradiction to prove the following statement: For all integers x, if x is an odd integer then x is an integer that is not divisible by six.

2.2 WRITHE AND LINKING NUMBER

The link invariants in this section are calculated by summing the signs of crossings in a subset of the crossings in the diagram. In an oriented diagram, each crossing inherits a positive or negative sign from the orientation and its local configuration. The **sign of a crossing** is defined using the right hand rule from physics as shown in Figure 2.1. For a crossing c in a link diagram L, we denote the sign of c as sgn(c).

(a) $+1$ crossing (b) -1 crossing

FIGURE 2.1: Crossing sign

The **writhe** of a link diagram L, denoted as $w(L)$, is the sum of the sign of all crossings in the diagram. Using a summation symbol,

$$w(K) = \sum_{c \in K} \text{sgn}(c). \tag{2.1}$$

We apply this definition to the oriented knot diagrams in Figure 2.2. To calculate the writhe, label each crossing with its crossing sign and add the signs. It may help to draw a positively signed crossing on a small piece of paper and then compare the sample crossing to each crossing in a diagram to determine the sign. The simplest unknot diagram has zero crossings and the writhe of this diagram is zero. The Reidemeister I move changes the writhe of diagram so that writhe is **not** an invariant of link diagrams.

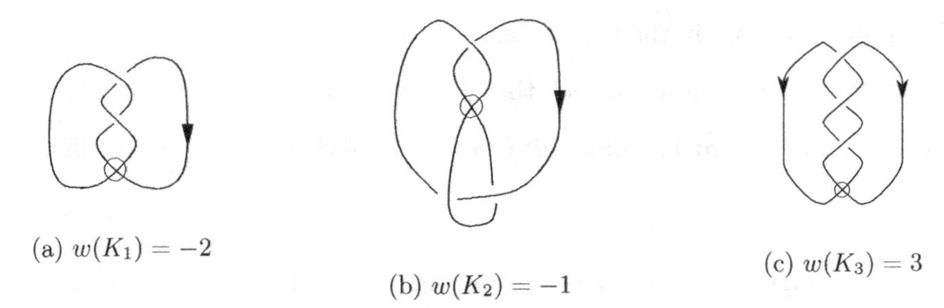

FIGURE 2.2: Examples of writhe

(a) Reidemeister II move with signs

(b) Reidemeister III move with signs

FIGURE 2.3: Writhe and the Reidemeister moves

Theorem 2.3. *Writhe is unchanged by all the diagrammatic moves except Reidemeister I.*

Proof. The planar isotopy moves do not change any crossings and do not change the writhe. Similarly, the virtual Reidemeister moves do not change any classical crossings and do not change the writhe. We need to show that both sides of the Reidemeister II and III moves make the same contribution to the writhe. Let D_L denote the left hand side of a Reidemeister move and let D_R denote the right hand side of a Reidemeister move.

We examine a co-oriented Reidemeister II move in Figure 2.3a. The part of diagram D_L containing the Reidemeister II move has no crossings and makes no contribution to the sum. The diagram D_R contains two oppositely signed crossings with a net contribution of 0. Then $w(D_L) = w(D_R)$. We leave the contra-oriented Reidemeister II move as an exercise.

In the Reidemeister III move, the number of crossings and the sign of the crossing is preserved. The diagram on the left hand side of a Reidemeister III move and the diagram on the right hand side are shown in Figure 2.3b. Now, if c is a crossing on the left hand side of a Reidemeister III move then there is a corresponding crossing c' on the right hand side of the Reidemeister III move such that $\text{sgn(c)} = \text{sgn(c')}$. We compute that both sides of the Reidemeister III move make the same contribution to the writhe. The Reidemeister I move inserts or removes a single crossing, changing the writhe of the diagram by ± 1. □

We use writhe as a starting point for the linking number invariant. Reidemeister I moves involve only 1 component so if we do not count the crossings that only involve 1 component, invariance under all diagrammatic moves will follow. Let L be an oriented, n-component link where the components are assigned an order: $C_1, C_2, \ldots, C_n$. Let $C^i_j(L)$ be the set of crossings where the component C_i overpasses the component C_j. (It is possible for $C^i_j(L)$ to be the empty set.) We define the **linking number** of C_i over C_j, denoted $\mathcal{L}^i_j(L)$, as

$$\mathcal{L}^i_j(L) = \sum_{c \in C^i_j(L)} \text{sgn}(c). \tag{2.2}$$

If $C_j^i(L)$ is the empty set ($C_j^i(L) = \emptyset$) then $\mathcal{L}_j^i(L) = 0$. Notice that a change in the order of the components will change the value of $C_j^i(L)$.

We compute some examples of linking numbers. One way to make this computation easier is to color the components of the link diagram. Color the components of a link diagram L so that C_1 is red and C_2 is blue. This will make it easier to determine the elements of $C_2^1(L)$ and $C_1^2(L)$. To compute $\mathcal{L}_2^1(L)$, we count the set of classical crossings where red and blue meet and the red component overpasses. Similarly, to compute $\mathcal{L}_1^2(L)$, we count the crossings where the blue overpasses the red. In Figure 2.4, $\mathcal{L}_2^1(L) = -1$ and $\mathcal{L}_1^2(L) = -3$.

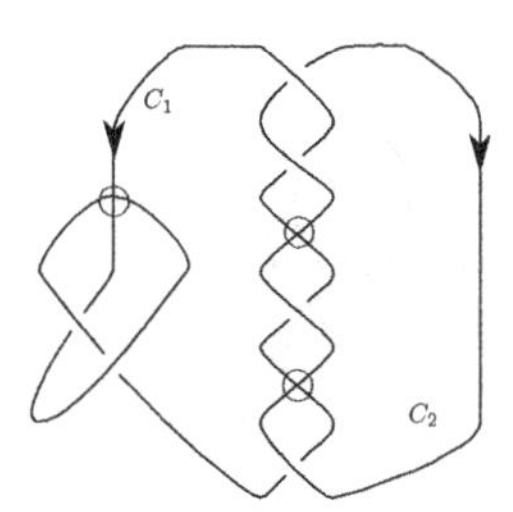

FIGURE 2.4: Linking number example: $\mathcal{L}_2^1(L) = -1$ and $\mathcal{L}_1^2(L) = -3$

We prove that $\mathcal{L}_j^i(L)$ is unchanged by each diagrammatic move and $\mathcal{L}_2^1$ is well defined on the set of ordered, oriented virtual links.

Theorem 2.4. *For all ordered, oriented virtual links, the sum $\mathcal{L}_j^i(L)$ is unchanged by the diagrammatic moves.*

Proof. Let D_L and D_R represent ordered, oriented, virtual link diagrams related by a single diagrammatic move. We assume that D_L represents the left hand side of a single diagrammatic move and that D_R denotes the right hand side of a diagrammatic move. Let C_1 and C_2 (respectively C_1^R and C_2^R) denote components of L (respectively L'). We prove directly that $\mathcal{L}_2^1(D_L) = \mathcal{L}_2^1(D_R)$. The proof that $\mathcal{L}_j^i$ is invariant is analogous.

Under the planar isotopy moves, the sets $C_2^1(D_L)$ and $C_2^1(D_R)$ are in one-to-one correspondence, so $\mathcal{L}_2^1(L) = \mathcal{L}_2^1(L')$. The virtual Reidemeister moves do not change the classical crossings, and $\mathcal{L}_2^1(D_L) = \mathcal{L}_2^1(D_R)$.

The Reidemeister I move introduces or removes a single crossing. However, the Reidemeister I move involves only one component. The crossings introduced by the Reidemeister I move are not counted by $\mathcal{L}_2^1$.

In the Reidemeister II moves, the diagram D_L contains zero crossings. The diagram D_R contains two crossings: c_1 and c_2 with $sgn(c_1) = -sgn(c_2)$. This holds for both the co-oriented and contra-oriented Reidemeister II moves. The crossing c_1 is an element of $C_2^1(D_R)$ if and only if c_2 is also an element of $C_2^1(D_R)$. The sum of the signs of c_1 and c_2 is zero. Then $\mathcal{L}_2^1(D_L) = \mathcal{L}_2^1(D_R)$.

The Reidemeister III move does not change the number of crossings. Let c denote a crossing in D_L (the left hand side of a Reidemeister III move). The diagram D_R (the right hand side of a Reidemeister III move) contains a corresponding crossing c'. Again, c is an element of $C_2^1(D_L)$ if and only if the corresponding crossing c' is also an element of $C_2^1(D_R)$. Since $\text{sgn}(c) = \text{sgn}(c')$ then $\mathcal{L}_2^1(D_L) = \mathcal{L}_2^1(D_R)$. □

$\mathcal{L}_j^i$ is a well defined mapping on the set of virtual links since any two equivalent diagrams have the same image. We obtain the following corollary.

Corollary 2.5. *$\mathcal{L}^i_j$ is an invariant of ordered, oriented virtual links.*

Proof. Let L and L' be equivalent virtual link diagrams. By definition, there is a finite sequence of diagrams relating L and L'. Each pair of consecutive diagrams differ by a single diagrammatic move. The individual moves do not change $\mathcal{L}^i_j$, so that $\mathcal{L}^i_j(L) = \mathcal{L}^i_j(L')$. □

Exercises

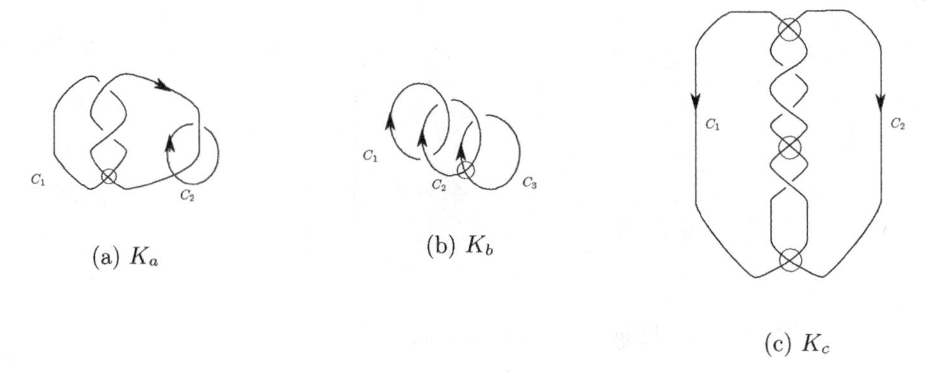

(a) K_a (b) K_b (c) K_c

FIGURE 2.5: Compute the writhe

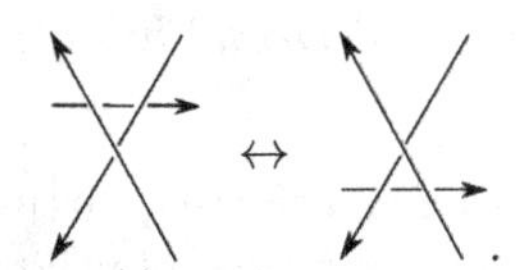

FIGURE 2.6: Identify the corresponding crossings

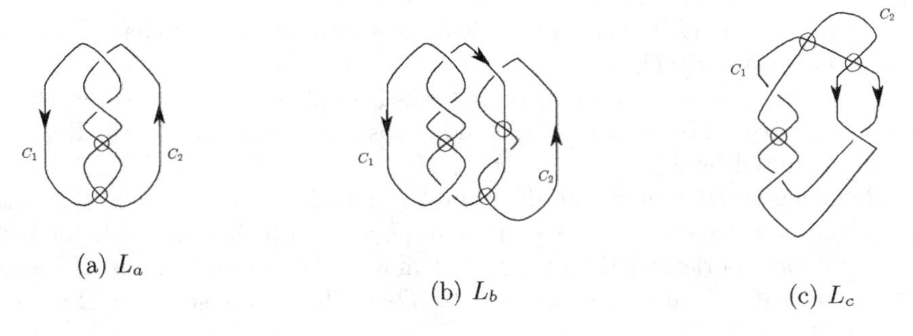

(a) L_a (b) L_b (c) L_c

FIGURE 2.7: Compute L^1_2 and L^2_1

1. Compute the writhe of the knot and link diagrams shown in Figure 2.5.
2. Prove that the contra-oriented Reidemeister II move does not change writhe.
3. Compute the signs of the crossings in the oriented Reidemeister III move and identify the corresponding crossings in Figure 2.6.

FIGURE 2.8: Prove these links are not equivalent

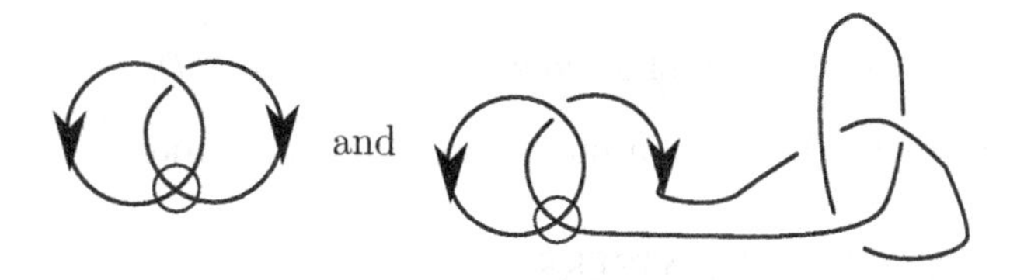

FIGURE 2.9: Non-equivalent links with the same linking numbers

4. Calculate $\mathcal{L}_1^2(L)$ for the links in Figure 2.7.
5. Calculate $\mathcal{L}_2^1(L)$ for the link diagram in Figure 2.7.
6. Prove the links in Figure 2.8 are not equivalent using linking number.
7. Construct a two component, ordered, oriented virtual link diagram K with $\mathcal{L}_2^1(L) = 3$ and $\mathcal{L}_1^2(L) = 1$.
8. Construct a family of two component, oriented link diagrams K_n with $\mathcal{L}_2^1(K_n) = n$ and $\mathcal{L}_1^2(L) = 0$.
9. Prove that $\sum_{i<j} \mathcal{L}_j^i(L)$ is an invariant of ordered, oriented virtual link diagrams.

2.3 DIFFERENCE NUMBER

We consider an invariant related to the linking number: the linking difference number. Let L be a n-component, ordered, oriented link diagram and let $1 \leq i < j \leq n$. We define the **difference number** $\mathcal{D}_{i,j}(L) = \mathcal{L}_j^i(L) - \mathcal{L}_i^j(L)$. If the link L contains only two components, we can write $\mathcal{D}_{1,2}(L)$ simply as $\mathcal{D}(L)$.

Theorem 2.6. *The difference $\mathcal{D}_{i,j}(L)$ is unchanged by the diagrammatic moves and is an invariant of ordered, oriented virtual links.*

Note that $|D_{1,2}(L)|$ is an invariant of two component, oriented virtual link diagrams. With the absolute value, we can relax the condition that the components are ordered. Linking number and the difference number allows us to distinguish some link diagrams. However, we still have no way of distinguishing the diagrams shown in Figure 2.9.

Exercises

1. Compute $\mathcal{D}_{i,j}(L)$ for the links in Figure 2.7.
2. Construct a two component, ordered, oriented link diagram L with $\mathcal{D}_{1,2}(L) = 2$.
3. Construct a two component, ordered, oriented link diagram L with $\mathcal{D}_{1,2}(L) = 2$ and $\mathcal{L}_2^1(L) = 3$.

4. Construct a two component, ordered, oriented link diagram L with $\mathcal{L}_2^1(L) = \mathcal{L}_1^2(L) = 2$ and $\mathcal{D}_{1,2}(L) = 0$.

5. Prove Theorem 2.6.

6. Prove that $\sum_{i<j} \mathcal{D}_{i,j}(L)$ is a link invariant.

7. Prove that changing the sign of a crossing does not alter $\mathcal{D}_{i,j}(L)$.

8. Prove that reversing the orientation of both components does not change $\mathcal{D}_{i,j}(L)$.

2.4 CROSSING WEIGHT NUMBERS

We use the difference number to construct an invariant of knots. First, we associate a difference number to each crossing. Then we use the difference numbers to separate the crossings into sets and sum the signs of the crossings in each set. Let K be an oriented knot diagram. Given a crossing c in K, we obtain an oriented, two component link diagram by removing a small neighborhood of the crossing c and gluing together the endpoints so that orientations agree as shown in Figure 2.10. This is a **vertical smoothing** of a crossing.

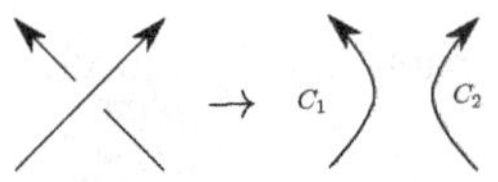

FIGURE 2.10: Vertical smoothing example

Let K_c denote the two component link obtained by vertically smoothing the crossing c in the diagram K. Let C_1 (respectively C_2) denote the left hand (respectively right hand) component. We define the **weight of the crossing** as

$$\begin{aligned} \mathsf{w}(c) &= sgn(c)\mathcal{D}_{1,2}(K_c) \\ &= sgn(c)\left(\mathcal{L}_2^1(K_c) - \mathcal{L}_1^2(K_c)\right). \end{aligned} \tag{2.3}$$

We compute the weight of each crossing in the diagram K. (It may be helpful to label each crossing with its weight.) Next, we use the weight to partition the crossings into sets. We denote the set of crossings with weight a as $\mathcal{W}_a(K)$. Using set builder notation

$$\mathcal{W}_a(K) = \{c \in K | \mathsf{w}(c) = a\}.$$

The a **crossing weight number** of K is denoted as $\mathcal{C}_a(K)$. The value of $\mathcal{C}_a(K)$ is determined by summing the signs of the crossings in $\mathcal{W}_a(K)$

$$\mathcal{C}_a(K) = \sum_{c \in \mathcal{W}_a(K)} sgn(c).$$

Before proving that for $a \neq 0$, $\mathcal{C}_a(K)$ is a knot invariant, we compute some examples. Consider the unknot, U. The simplest diagram of the unknot has no classical crossings and $\mathcal{C}_a(U) = 0$ for all integers a.

The virtual trefoil diagram in Figure 2.11a has two classical crossings and one virtual crossing. The crossings u and d in this knot are positive. Vertically smoothing the crossing

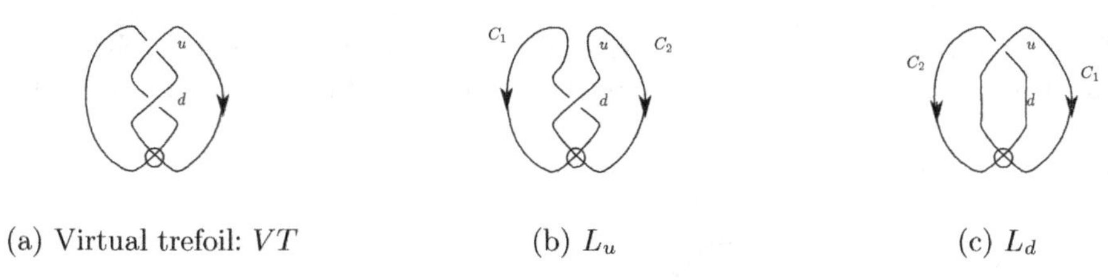

(a) Virtual trefoil: VT (b) L_u (c) L_d

FIGURE 2.11: Virtual trefoil and smoothed diagrams

u results in the two component link diagrams L_u (shown in Figure 2.11b) and smoothing d results in the diagram L_d (shown in Figure 2.11c). We compute $\mathsf{w}(u) = -1$ and $\mathsf{w}(d) = 1$. Now, $\mathcal{W}_1(K) = \{d\}$, $\mathcal{W}_{-1}(K) = \{u\}$ and

$$\mathcal{C}_1(K) = \mathcal{C}_{-1}(K) = 1.$$

All the crossings in the diagram have been accounted for, so $\mathcal{C}_a(K) = 0$ for all $|a| > 1$.

In the next example, we consider the virtual figure eight knot. The virtual figure eight knot, K, has three classical crossings and one virtual crossing (see Figure 2.12a). We cal-

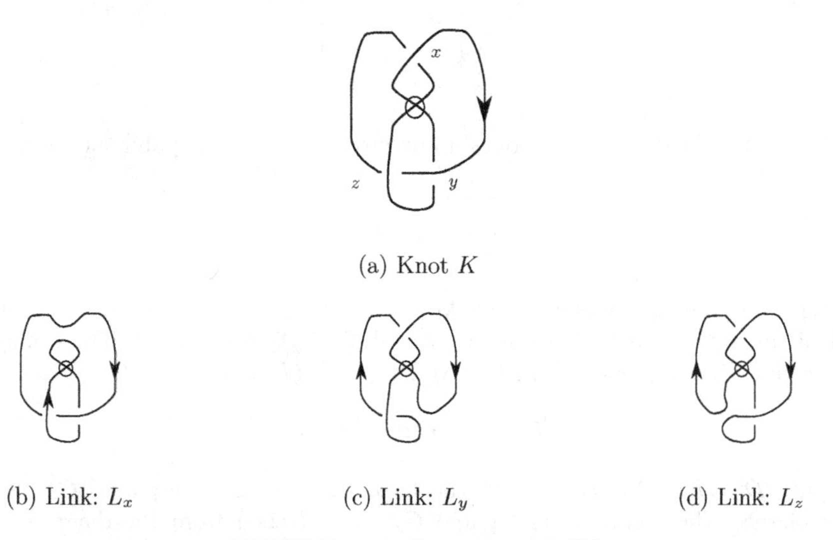

(a) Knot K

(b) Link: L_x (c) Link: L_y (d) Link: L_z

FIGURE 2.12: Computing $\mathcal{C}_a(L)$

culate the weight of each crossing using the difference linking number.

$$\mathsf{w}(x) = 0 \qquad \mathsf{w}(y) = -1 \qquad \mathsf{w}(z) = 1$$

We compute that $\mathcal{C}_1(L) = \mathcal{C}_{-1}(L) = -1$ and $\mathcal{C}_a(L) = 0$ for all $a > 1$.

Based on these calculations, the trefoil and the virtual figure 8 knot are not equivalent: $L \nsim VT$. We use our experience computing $\mathcal{C}_a(K)$ to prove that $\mathcal{C}_a(K)$ is a knot invariant. The proof is a direct proof and centers around showing that the diagrammatic moves do not change the value of $\mathcal{C}_a(K)$.

Theorem 2.7. *Let K be a knot and let $a \neq 0$. Then $\mathcal{C}_a(K)$ is unchanged by the diagrammatic moves and $\mathcal{C}_a(K)$ is a knot invariant.*

Proof. The planar isotopy moves and the virtual Reidemeister moves do not change the set of classical crossings in the diagrams. As a result, we only need to examine the Reidemeister moves.

Consider an oriented Reidemeister I move:

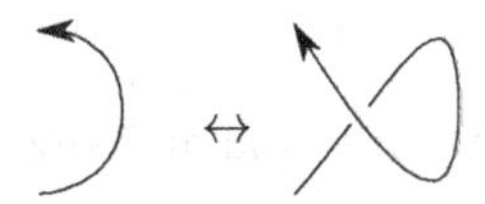

The left hand side does not contain any crossings, so we expand the right hand side crossing.

$$L_c =$$

Here, we see that $\mathsf{w}(c) = 0$. The net effect is that every Reidemeister I move introduces or removes a classical crossing of weight zero. This means that $\mathcal{C}_0(K)$ is not invariant under the Reidemeister I move (the reason for the exclusion $a = 0$).

Consider an oriented Reidemeister II move:

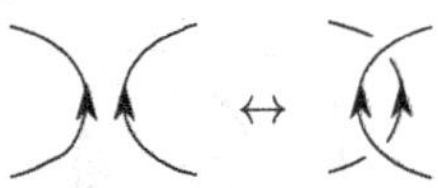

$$\leftrightarrow$$

Again, the left hand side does not contain any crossings, so we expand the right hand side crossings.

$$L_u = \quad \text{and } L_d =$$

In L_u, there are crossings outside of our local picture where C_1 overpasses C_2. We denote the sum of the sign of these crossings as x and $\mathcal{L}_2^1(L_u) = x + (-1)$. There may also be crossings where C_2 overpasses C_1 in L_u. We let $y = L_1^2(L_u)$. Then

$$D_{1,2}(L_u) = x - (y + 1). \tag{2.4}$$

We compute $D_{1,2}(L_d)$. Notice that in the local diagram C_1 overpasses C_2 and outside the local picture, the components C_1 and C_2 are switched from the diagram L_u. Now, $L_2^1(L_d) = y + 1$ and $L_1^2(L_d) = x$. We compute that

$$\mathsf{w}(u) = x - (y + 1) \qquad \mathsf{w}(d) = -[(y + 1) - x]. \tag{2.5}$$

Using the weights in Equation 2.5, we conclude that $\mathsf{w}(u) = \mathsf{w}(d)$ and that u and d are elements of the set $W_a(L)$ for some a. The crossings u and d are oppositely signed and the net contribution to $\mathcal{C}_a(K)$ is zero.

Of the oriented Reidemeister III moves, we need only check the co-oriented Reidemeister III move. In this move, a crossing c on the left hand side corresponds to a crossing c' on the right hand side as shown below.

$$L = \quad c \quad \leftrightarrow \quad c' \quad = R$$

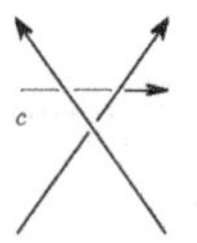

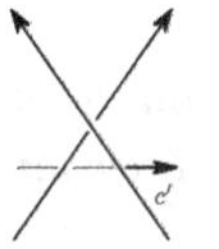

We compute w for the crossing c and its corresponding crossing c'.

$L_c =$ $\qquad\qquad\qquad$ $R_{c'} =$

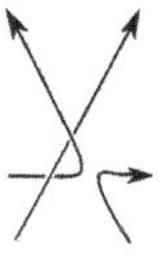

At the site of the smoothing the component on the left hand side is C_1 and the component on the right hand side is C_2. But in this diagram, we do not know whether the diagonal strand lies in C_1 or C_2. To prove that $\mathsf{w}(c) = \mathsf{w}(c')$, we check both possibilities. If the diagonal strand is in C_1, calculation shows that the local contribution to $\mathcal{D}_{1,2}$ from both sides is zero. In both diagrams, $L_2^1(L_c) = L_2^1(R_{c'})$ and $L_1^2(L_c) = L_1^2(R_{c'})$. We conclude that $\mathsf{w}(c) = \mathsf{w}(c')$.

If the diagonal strand is in C_2, we calculate that the local contribution to $\mathcal{D}_{1,2}$ is zero. We are able to conclude that $\mathsf{w}(c) = \mathsf{w}(c') = a$ for some a. After verifying that this is true for the two other crossings in the Reidemeister III move, we conclude that the Reidemeister III move does not change the value of $\mathcal{C}_a$ for all values of a. □

Earlier in this section, we stated that a knot invariant could be used to recover information about the knot or link. Using $\mathcal{C}_a(K)$, we can obtain a lower bound on the number of classical crossings in any diagram of K.

Theorem 2.8. *Let K be an oriented knot diagram and let L be an equivalent oriented knot diagram with m classical crossings. Let*

$$n = \sum_{a \in \mathbb{Z}, a \neq 0} |\mathcal{C}_a(K)| \tag{2.6}$$

then $n \leq m$.

Proof. Let c be a crossing in L. Then

$$m \geq \sum_{c \in L} |\mathrm{sgn(c)}|.$$

Partitioning the crossings into the sets $\mathcal{W}_a(L)$ for $a \neq 0$:

$$\sum_{c \in L} |\mathrm{sgn(c)}| \geq \sum_{a \in \mathbb{Z}, a \neq 0} \left[\sum_{c \in \mathcal{W}_a(L)} |sgn(c)| \right].$$

Next, we consider the absolute value operator:

$$\sum_{a \in \mathbb{Z}, a \neq 0} \left[\sum_{c \in \mathcal{W}_a(L)} |sgn(c)| \right] \geq \sum_{a \in \mathbb{Z}, a \neq 0} \left| \sum_{c \in \mathcal{W}_a(L)} sgn(c) \right| = \sum_{a \in \mathbb{Z}, a \neq 0} |\mathcal{C}_a(L)|.$$

□

Classical linking number

If L is a classical link diagram, then this places some restrictions on the linking numbers and crossing weights. First, we realize that if L is a two component classical link then $D_{1,2}(L) = 0$.

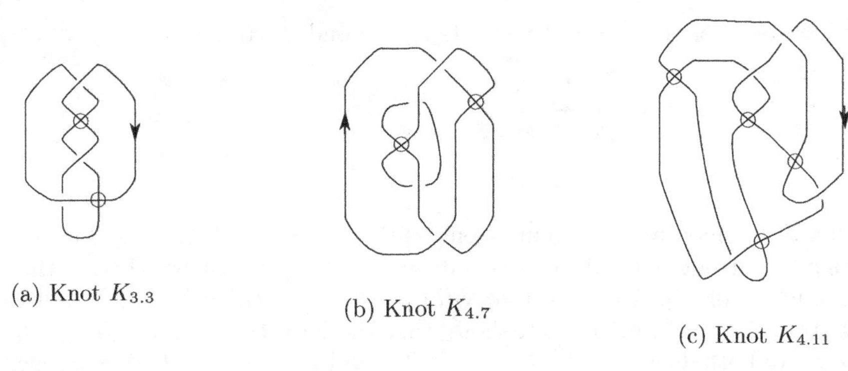

(a) Knot $K_{3.3}$

(b) Knot $K_{4.7}$

(c) Knot $K_{4.11}$

FIGURE 2.13: Compute $C_a(K)$

Theorem 2.9. *For all classical links L, $\mathcal{L}^i_j(L) = \mathcal{L}^j_i(L)$.*

Proof. Let L be a classical link. There is a diagram D with no virtual crossings. We consider components C_1 and C_2 and prove that $\mathcal{L}^1_2(L) = \mathcal{L}^1_2(L)$.

Let c be a crossing in $C^2_1(L)$. If we switch the overpassing strand in the crossing, we form a new classical link diagram D' with a crossing c' in $C^1_2(L)$ that corresponds to c.

If $\text{sgn}(c) = +1$ then $\text{sgn}(c') = -1$. We calculate that

$$\mathcal{L}^1_2(D') = \mathcal{L}^1_2(D) - 1 \text{ and } \mathcal{L}^2_1(D') = \mathcal{L}^2_1(D) - 1. \tag{2.7}$$

If $\text{sgn}(c) = -1$ then $\text{sgn}(c') = 1$. We calculate that

$$\mathcal{L}^1_2(D') = \mathcal{L}^1_2(D) + 1 \text{ and } \mathcal{L}^2_1(D') = \mathcal{L}^2_1(D) + 1. \tag{2.8}$$

Since D is a diagram with no virtual crossings, we can choose a minimum set of crossings $c_1, c_2, \ldots c_n$ in diagram D. We switch these crossing to form a diagram D' where the components C_1 and C_2 are unlinked. We let

$$x = \sum_{i=1}^{n} \text{sgn}(c_i). \tag{2.9}$$

Then,

$$\mathcal{L}^1_2(D') = \mathcal{L}^1_2(D) - x \text{ and } \mathcal{L}^2_1(D') = \mathcal{L}^2_1(D) - x. \tag{2.10}$$

But, $\mathcal{L}^1_2(D') = \mathcal{L}^2_1(D') = 0$ since the components are unlinked. We conclude that $\mathcal{L}^1_2(D) = \mathcal{L}^2_1(D)$. □

Corollary 2.10. *For all virtual link diagrams L, if $\mathcal{L}^i_j(L) \neq \mathcal{L}^j_i(L)$ then L is not classical (every diagram of L has at least one virtual crossing).*

Proof. This is the contrapositive of Theorem 2.9. □

Exercises

1. Calculate all non-zero values of $C_a(K)$ $(a > 0)$ for the knot diagrams in Figure 2.13.
2. Reverse the orientations of the link diagrams in Figure 2.13 and compute $C_a(K)$ for all $a \neq 0$.

3. For the co-oriented Reidemeister III move, prove that the other corresponding pairs of crossings have the same crossing weight.
4. Prove that if K is a classical knot diagram then $\mathcal{C}_a(K) = 0$ for all $a \neq 0$.
5. Construct a knot diagram K where $\mathcal{C}_1(K) = 3$.
6. Construct a knot diagram K where $\mathcal{C}_3(K) = 1$.
7. Construct a family of knot diagrams K_n where $\mathcal{C}_1(K_n) = n$.

2.5 OPEN PROBLEMS AND PROJECTS

Open problems

The following problems involve linking number and difference number and are of interest to active researchers in the field of knot theory.

1. Use sets of crossings with specific weights to define an invariant of virtual knots.
2. Determine a lower bound on the number of virtual and classical crossings in a link L.
3. Describe the set of non-classical knots such that all crossings have weight zero.
4. A crossing is said to be cosmetic if changing the crossing's sign does not change the knot type. Determine when a crossing is cosmetic.

Projects

The projects involve constructing families of virtual link diagrams with specific properties and the computation of linking numbers for families of virtual link diagrams.

1. Construct a family of virtual link diagrams that satisfy specific conditions with regard to linking number and difference linking number. For example, find knots such that $\mathcal{D}_{1,2}(L) = 5$ and $\mathcal{L}_2^1(L) = 2$ and $\mathcal{L}_1^2(L) = -3$.
2. Construct a family of knot diagrams K with specific values for $\mathcal{C}_a(K)$.
3. Given a virtual knot diagram K with a positive crossing c, construct a knot diagram K_- by changing the overpassing arc in crossing c to undercrossing c. Compare the values of $\mathcal{C}_a(K)$ and $\mathcal{C}_a(K_-)$.

Further Reading

Undergraduate-friendly readings that involve crossing weight, linking number, and difference number.

1. H. A. Dye. Smoothed invariants. *J. Knot Theory Ramifications*, 21(13):17,1240003, 2012
2. Lena C. Folwaczny and Louis H. Kauffman. A linking number definition of the affine index polynomial and applications. *J. Knot Theory Ramifications*, 22(12):30,1341004, 2013
3. Allison Henrich. A sequence of degree one Vassiliev invariants for virtual knots. *J. Knot Theory Ramifications*, 19(4):461–487, 2010

4. Louis H. Kauffman. A self-linking invariant of virtual knots. *Fund. Math.*, 184:135–158, 2004

5. Dale Rolfsen. *Knots and links*, volume 7 of *Mathematics Lecture Series.* Publish or Perish, Inc., Houston, TX, 1990 p. 132–136

CHAPTER 3

A multiverse of knots

We defined virtual knots and links as equivalence classes of decorated underlying diagrams. Equivalence is determined by a set of diagrammatic moves, and two equivalent diagrams are related by a finite sequence of moves. With this viewpoint, we are free to consider a multiverse of knot theories. New knot theories are constructed by modifying the set of markings on the underlying diagram, modifying the set of diagrammatic moves, or altering both the diagrams and the diagrammatic moves. In this chapter, we introduce several different definitions of equivalence of diagrams leading to different knot theories, along with the motivation for their construction.

3.1 FLAT AND FREE LINKS

We modify the set of markings on the diagrams to construct flat links. A **flat crossing** is marked by a solid crossing without over- or under-markings. A **flat link diagram** is an underlying diagram with n components and two types of crossing decoration: flat and virtual. The set of flat link diagrams is denoted $\mathcal{F}_d$. Examples of flat virtual link diagrams are shown in Figure 3.1. The set of diagrammatic moves on flat link diagrams is illustrated

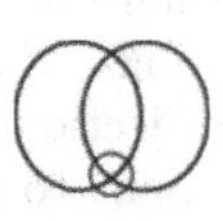

(a) Flat link diagram: F_1

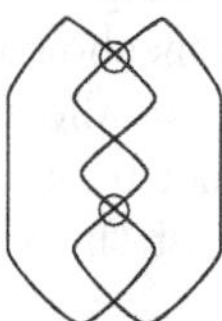

(b) Flat link diagram: F_2

(c) Flat Kishino's knot diagram: F_K

FIGURE 3.1: Flat link diagram examples

in Figure 3.2 and Figure 3.3.

Two flat link diagrams L and L' are **flat equivalent** ($L \sim L'$) if they are related by a finite sequence of **flat diagrammatic moves**. A **flat virtual link** is an equivalence class of flat virtual link diagrams. We denote the set of flat virtual links as $\mathcal{F}$. Two flat equivalent flat link diagrams and a sequence of moves between the diagrams are shown in Figure 3.4. We still use the notation $\sim$ to denote equivalence. The intended definition of equivalence should be clear from the context. If the definition is not clear, we can apply a subscript to the notation such as $\sim_F$ to specify that we mean flat equivalence.

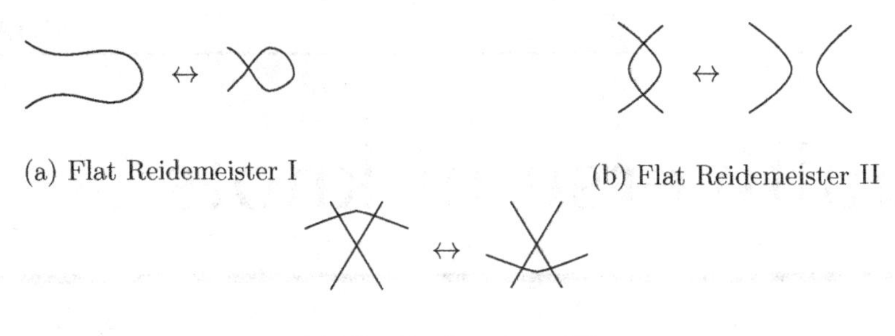

(a) Flat Reidemeister I

(b) Flat Reidemeister II

(c) Flat Reidemeister III

FIGURE 3.2: Flat Reidemeister moves

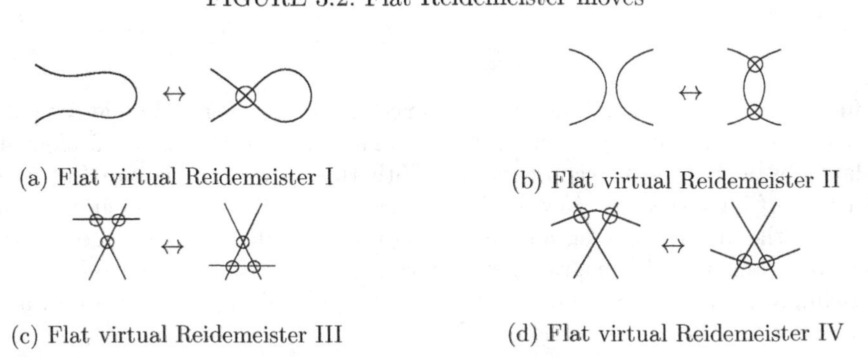

(a) Flat virtual Reidemeister I

(b) Flat virtual Reidemeister II

(c) Flat virtual Reidemeister III

(d) Flat virtual Reidemeister IV

FIGURE 3.3: Flat virtual moves

Review: Maps, images, and quotients

We define a projection map $P\colon \mathcal{V}_d \to \mathcal{F}_d$. We want to show that P defines a link invariant $\mathcal{P}$ that maps $\mathcal{V}$ into the set of free links. The projection map P is defined on the set of virtual link diagrams. We must prove that for all virtual link diagrams L and L', if $L \sim L'$ then $P(L) = P(L')$. Then P defines a map $\mathcal{P}$ on the set of virtual links: $\mathcal{P}\colon \mathcal{V} \to \mathcal{F}$. The map P "flattens" classical crossings as shown in Figure 3.5, removing the over/under-markings. Conceptually, this means that we allow the over/under-markings on crossings to be switched during a sequence of Reidemeister moves. We apply the projection map to the virtual trefoil in Figure 3.6. The image of the virtual trefoil is equivalent to the unknot using flat diagrammatic moves.

Theorem 3.1. *The map $P\colon \mathcal{V}_d \to \mathcal{F}_d$ defines a link invariant: $\mathcal{P}\colon \mathcal{V} \to \mathcal{F}$.*

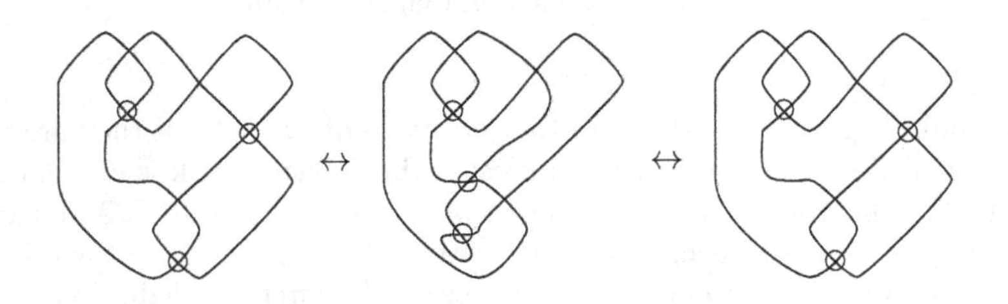

FIGURE 3.4: Sequence of equivalent flat links

FIGURE 3.5: Flattening map $P\colon \mathcal{V}_d \to \mathcal{F}$

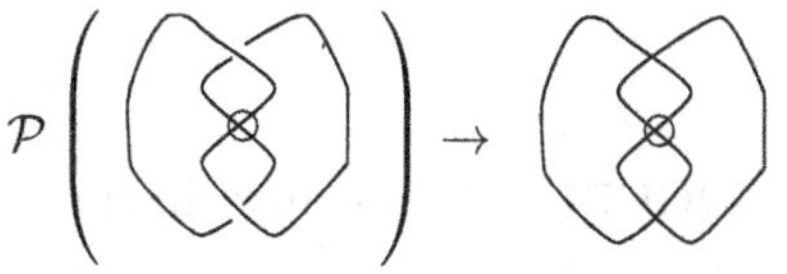

FIGURE 3.6: Applying the projection map $\mathcal{P}$

Proof. Let L and L' be equivalent virtual link diagrams. By definition, there is a finite sequence of virtual link diagrams $L = K_1, K_2, \dots, K_n = L'$ where K_i and K_{i+1} are related by a single diagrammatic move.

If we apply the projection map to the Reidemeister moves and the virtual Reidemeister moves then we obtain the flat Reidemeister moves (shown in Figure 3.2) and the flat virtual Reidemeister moves (shown in Figure 3.3). As a result, there is a sequence of flat virtual diagrams: $P(L) = P(K_1), P(K_2), \dots, P(K_n) = P(L')$ where $P(K_i)$ and $P(K_{i+1})$ are related by a single flat diagrammatic move. We conclude that $P(L) \sim P(L')$ and the mapping $\mathcal{P}$ is well defined on $\mathcal{V}$. □

We investigate the properties of the map $\mathcal{P}$. Given a mapping $f : X \to Y$, we can ask if the map is **surjective** (onto) or **injective** (one to one). A mapping $f : X \to Y$ is surjective if for all elements y in Y, there exists an element x in X such that $f(x) = y$. Let $|\mathcal{V}|$ denote the number of elements in $\mathcal{V}$ and let $|S|$ denote the number of elements in a set S. If a link invariant $I : \mathcal{V} \to S$ is**surjective** then $|\mathcal{V}| \geq |S|$. A mapping $f : X \to Y$ is **injective** if for all elements x_1, x_2 in the set X, if $f(x_1) = f(x_2)$ then $x_1 = x_2$. If a link invariant $\mathcal{I} : \mathcal{V} \to S$ is injective, we could uniquely label the elements of $\mathcal{V}$ with elements of S and $|\mathcal{V}| \leq |S|$. If $\mathcal{I} : \mathcal{V} \to S$ is a surjective and injective invariant then $|\mathcal{V}| = |S|$ and we could label each element of $\mathcal{V}$ with an element of S. In this sense, a surjective and injective link invariant where we can easily count and identify elements of S is our ultimate goal. The mapping $\mathcal{P}$ is surjective, but not injective.

We can also investigate the **image** of the map and **pre-image** of a flat knot or link. The projection of the virtual trefoil (shown in Figure 3.6) is equivalent to the unknot, so both the unknot and the virtual trefoil map to the unknot. In fact, any virtual knot diagram with zero virtual crossings projects to a flat knot diagram that is equivalent to the unknot. Hence, the pre-image of the unknot can be formally described as

$$\mathcal{P}^{-1}(U) = \{L \in \mathcal{V} | \mathcal{P}(L) \sim_F U\}. \tag{3.1}$$

Every classical knot diagram is an element of this set, but there are also non-classical, virtual knot diagrams that map to the unknot under projection.

The map $\mathcal{P}$ is a surjection (an onto map) and $\mathcal{P}(\mathcal{V}) = \mathcal{F}$. We prove that for every flat virtual link F, there exists a virtual link L such that $\mathcal{P}(L) = F$. We can do this by constructing an element of $\mathcal{P}^{-1}(L)$ for an arbitrary flat link diagram F.

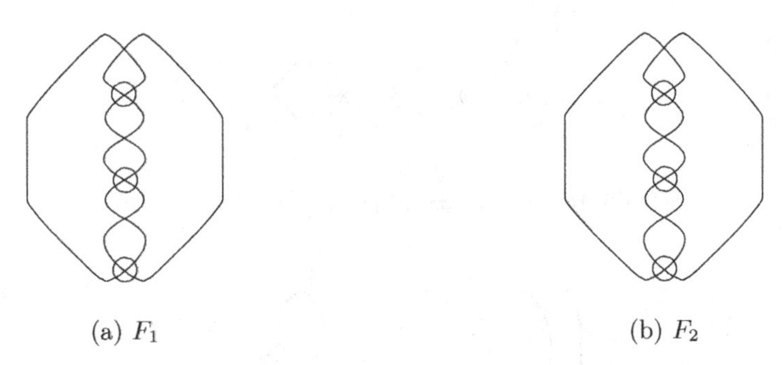

(a) F_1 (b) F_2

FIGURE 3.7: Flat link diagram

Theorem 3.2. *The image of $\mathcal{P}$ is the set $\mathcal{F}$.*

Proof. Let K be an arbitrary flat virtual link diagram. Then we can choose over/under-markings for each flat crossing and construct a virtual link diagram K' so that $\mathcal{P}(K') = K$. □

The flat virtual links are a quotient set of the set virtual links. We could define the flat equivalence class of a link K:

$$[K]_F = \{L \in \mathcal{V} | \mathcal{P}(L) = \mathcal{K}\}. \tag{3.2}$$

Taking this viewpoint, we add a single relation to the diagrammatic moves: switching the crossing from a positive crossing to a negative crossing and vice versa. This additional relationship means that some equivalence classes of virtual links are combined to form equivalence classes of flat links. (Recall that $\mathbb{Z}_2$ is constructed by combining all the even integers into the set [0] and all the odd integers into the set [1].)

We define an invariant of ordered, flat virtual links. Let K be a two component virtual link with components K_1 and K_2. Recall the definition of linking number $\mathcal{L}_2^1(K)$

$$\mathcal{L}_2^1(K) = \sum_{c \in C_2^1(K)} sgn(c).$$

Note that

$$\mathcal{L}_2^1(K) \equiv \sum_{c \in C_2^1(K)} |sgn(c)| \mod 2.$$

That is, $\mathcal{L}_2^1(K)$ is even (respectively odd) if and only if the number of crossings in $C_2^1(K)$ is even (respectively odd). However, if we eliminate the under/over-marking information then we can't determine if a crossing is an element of $C_2^1(K)$ or $C_1^2(K)$. We count all crossings involving both components to define a version of linking number for flat links. Let F denote an ordered, flat link diagram with components $C_1, C_2, \ldots C_n$. There are no over/under-markings and $C_{i,j}(F)$ denotes the set of crossings that involve components C_i and C_j. Let $|C_{i,j}(F)|$ denote the number of crossings in $C_{i,j}(F)$. Then

$$\mathcal{E}_{i,j}(F) = |C_{i,j}(F)| \mod 2. \tag{3.3}$$

The definition does not use the orientation of the components.

We compute $\mathcal{E}_{1,2}$ for the link diagrams shown in Figure 3.7. Assuming that the leftmost component is C_1 in each link, we obtain $\mathcal{E}_{1,2}(F_1) = 1$ and $\mathcal{E}_{1,2}(F_2) = 0$. We prove that $\mathcal{E}_{i,j}$ is a link invariant of ordered, flat links.

(a) $\text{fsgn}(c) = +1$ (b) $\text{fsgn}(c) = -1$

FIGURE 3.8: The flat crossing sign

Theorem 3.3. *For all ordered, flat virtual links K, $\mathcal{E}_{i,j}(K)$ is invariant under the flat diagrammatic moves.*

Proof. By definition,

$$\mathcal{E}_{i,j}(K) \equiv |C_{i,j}(K)| \mod 2. \tag{3.4}$$

We consider the individual Reidemeister moves. The flat Reidemeister I move does not change the set $C_{i,j}(K)$. The flat Reidemeister II move adds (or removes) two crossings to the set $C_{i,j}(K)$, and does not change $\mathcal{E}_{i,j}(K)$. Similarly, the flat Reidemeister III move does not change the number of elements in $C_{i,j}(K)$. The flat virtual Reidemeister moves do not change the set of flat crossings and do not change $\mathcal{E}_{i,j}(K)$. □

Corollary 3.4. *$\mathcal{E}_{i,j}$ is an invariant of ordered, n component flat virtual links.*

Proof. $\mathcal{E}_{i,j}$ is invariant under the diagrammatic moves. If $L \sim_F L'$ then $\mathcal{E}_{i,j}(L) = \mathcal{E}_{i,j}(L')$. □

If L is a two component link, we can relax the condition that the components are ordered. The invariant $\mathcal{E}_{1,2}$ partitions the set of two component flat links into two sets. This partition recalls the partition of the integers into evens or odds. But since the set of flat virtual links is far more complicated than the set of integers, we would (ideally) prefer an invariant that leads to a finer partition of virtual links. This motivates the following definition.

For $i < j$, we define $\mathcal{L}_{i,j}(K)$ on ordered, oriented n component flat virtual links with components $K_1, K_2, \ldots, K_n$. Recall that $C_{i,j}$ denotes the set of crossings involving both components K_i and K_j. For each crossing c in $C_{i,j}$, we define the flat sign (denoted $\text{fsgn}(c)$) as shown in Figure 3.8. Notice that the component K_i is oriented upwards in the diagram.

$$\mathcal{L}_{i,j}(K) = \sum_{c \in C_{i,j}(K)} fsgn(c). \tag{3.5}$$

This definition is dependent on orientation.

We use direct proof to prove that $\mathcal{L}_{i,j}$ is unchanged by the diagrammatic moves. In the proof, we use the phrase "without loss of generality". This means that we focus on a specific case, but the proof of the other cases involves the same steps.

Theorem 3.5. *The value of $\mathcal{L}_{i,j}$ is unchanged by the flat diagrammatic moves and $\mathcal{L}_{i,j}$ is an invariant of ordered, oriented, n component flat virtual link diagrams.*

Proof. Let L and K be equivalent oriented, n component flat virtual link diagrams. We denote the components of L and K as $L_1, L_2, \ldots, L_n$ and $K_1, K_2, \ldots, K_n$, respectively.

Without loss of generality, we assume that L and K are related by a single flat diagrammatic move.

Suppose K is transformed into L by the flat Reidemeister I move. This crossing involves only a single component and makes no contribution to $\mathcal{L}_{i,j}$.

Suppose K is transformed into L by the flat Reidemeister II move. Both crossings in a Reidemeister II move involve the same components, say C_i and C_j, and the net contribution of both sides to $\mathcal{L}_{i,j}$ is zero.

Suppose K is transformed into L by the flat Reidemeister III move. If a crossing c in L (in the Reidemeister III move) involves components L_i and L_j then there is a corresponding crossing c' in K involving components K_i and K_j. Note that $\text{fsgn}(c) = \text{fsgn}(c')$.

Suppose K is transformed into L by the flat virtual Reidemeister IV move. Again, any crossing c in K will have a corresponding crossing in c' in L, and $\text{fsgn}(c) = \text{fsgn}(c')$. □

We define the set of free knots and links using the same approach that we used to define flat virtual links. As diagrams, the set of free link diagrams is equal to the set of flat link diagrams. Since these are the same sets, the notation F_d denotes both the set of flat virtual links and the set of free virtual links, but the equivalence classes are not the same! The **flat virtualization move** is shown in Figure 3.9. Equivalence classes of free links are

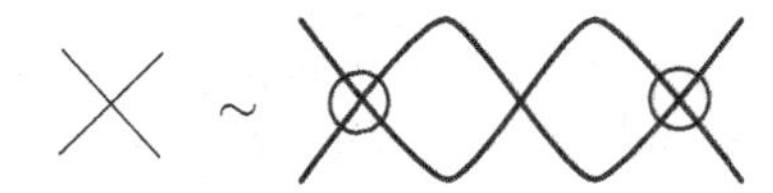

FIGURE 3.9: The flat virtualization move

determined by finite sequences of the flat virtualization move and flat diagrammatic moves. Two free virtual link diagrams are **free equivalent** if there is a finite sequence of flat diagrammatic and flat virtualization moves relating the two diagrams. A **free link** is an equivalence class of flat virtual link diagrams. We denote the set of free virtual links as $\mathcal{FR}$. Free links are a quotient set of the set of flat virtual links. As a result, the set of free links has fewer elements than the set of flat virtual links, but is still an infinite set of equivalence classes of diagrams.

Initially, it may appear that the set of free knots is trivial—that the only free knot is the trivial knot. Some experimentation shows that the set of free knots has more than one element. A non-trivial free knot is shown in Figure 3.10a. However, we will not be able to prove that this free knot is not equivalent to the unknot until much later in the text. A trivial free knot is shown in Figure 3.10b. Additional examples of free links are shown in Figure 3.11. Free links were first introduced by Vassily O. Manturov in the article *On free knots*, which is available at www.arxiv.org.

Exercises

1. Compute $\mathcal{E}_{i,j}$ for the ordered, flat link diagrams in Figure 3.12.
2. Compute $\mathcal{L}_{i,j}$ for the ordered, oriented, flat link diagrams in Figure 3.12.
3. What diagrams does $\mathcal{E}_{1,2}$ distinguish? Describe the possible equivalence classes of two component virtual links determined by $\mathcal{E}_{1,2}$.

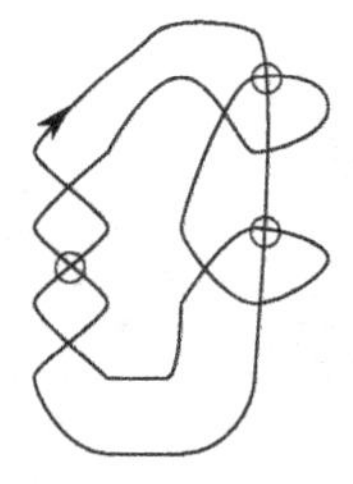

(a) A non-trivial free knot

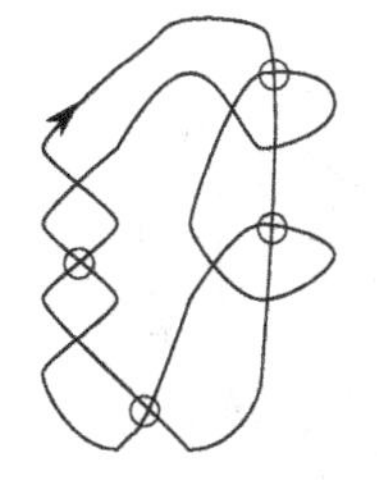

(b) A trivial free knot

FIGURE 3.10: Free knots

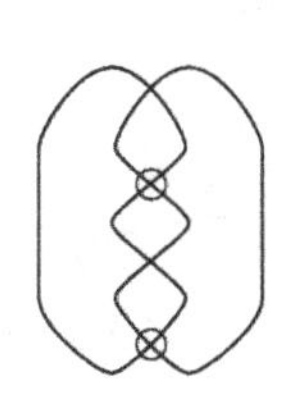

(a) A trivial free link

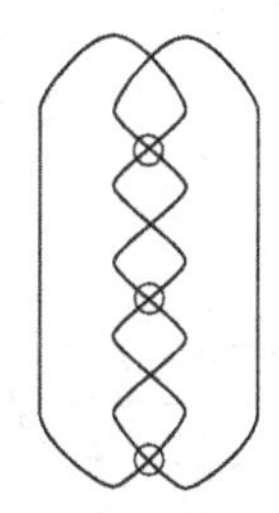

(b) A non-trivial free knot

FIGURE 3.11: Free links

4. What diagrams does $\mathcal{L}_{1,2}$ distinguish? Find two diagrams that are distinguished by $\mathcal{L}_{1,2}$.

5. Construct a family of links K_n such that $\mathcal{L}_{1,2}(K_n) = n$ for all $n > 0$.

6. Prove that $\sum_{i<j} \mathcal{L}_{i,j}$ is a link invariant.

7. Apply $\mathcal{P}$ to the diagrams in Figure 3.13. Are these knots equivalent to the unknot or unlink?

8. Determine which of the knot diagrams shown in Figure 3.14 are elements of $\mathcal{P}^{-1}(U)$. If they are elements of $\mathcal{P}^{-1}(U)$, find a sequence of moves that transforms the projection into the unknot.

9. Prove that the projections of the link diagrams in Figure 3.15 are related by a sequence of flat diagrammatic moves.

10. Given a flat virtual link diagram K with n flat crossings, how many virtual diagrams project directly to K under $\mathcal{P}$?

11. Prove that $\mathcal{F}$ is an infinite set.

12. Find a sequence of moves transforming the knot in Figure 3.10b into the unknot diagram. Flat diagrammatic moves and the virtualization move are allowed.

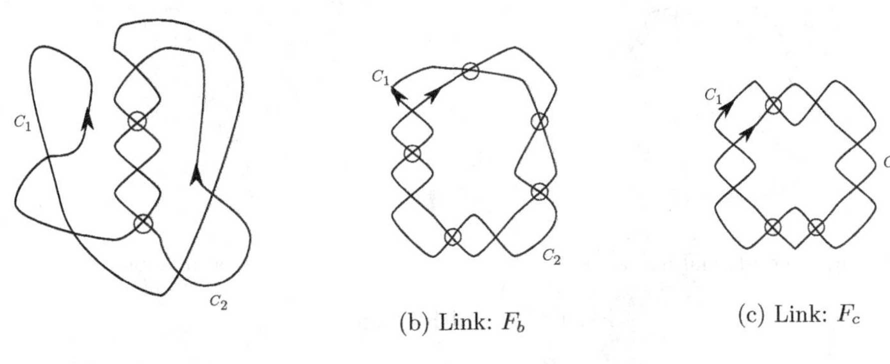

(a) Link: F_a

(b) Link: F_b

(c) Link: F_c

FIGURE 3.12: Compute $\mathcal{E}_{i,j}$ and $\mathcal{L}_{i,j}$

(a) Kishino's knot

(b) Virtual link

FIGURE 3.13: Apply $\mathcal{P}$ to the links

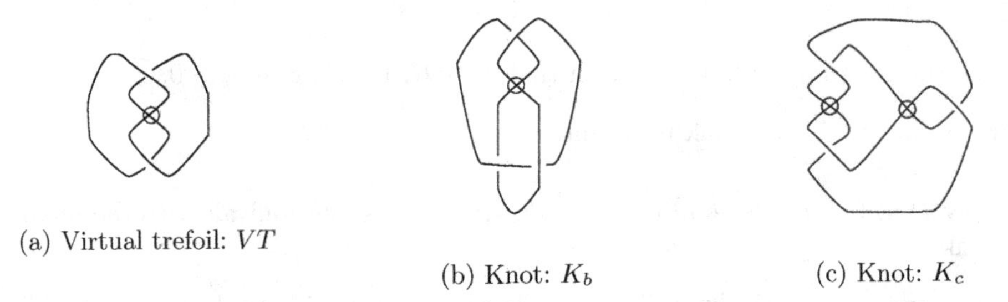

(a) Virtual trefoil: VT

(b) Knot: K_b

(c) Knot: K_c

FIGURE 3.14: Are the knots elements of $\mathcal{P}^{-1}(U)$?

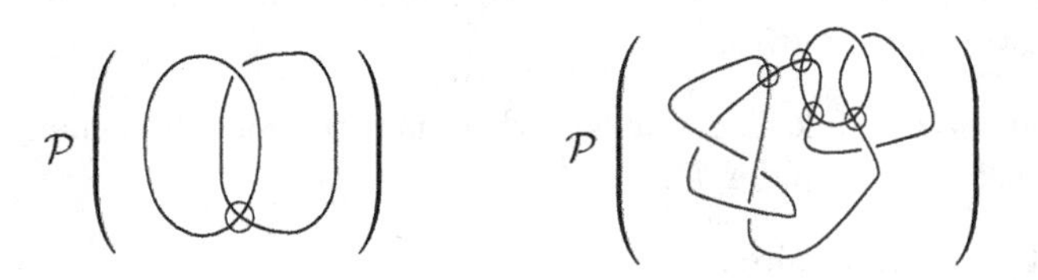

FIGURE 3.15: Show the projections of these knots are flat equivalent

13. Find a sequence of moves transforming the knot in Figure 3.11a into two unlinked copies of the unknot. Flat diagrammatic moves and the virtualization move are allowed.

14. Does $\mathcal{E}_{i,j}$ extend to an invariant of free links? If so, prove the invariance. If not, give an example of two equivalent free diagrams L and K such that $\mathcal{E}_{i,j}(L) \neq \mathcal{E}_{i,j}(K)$.

15. Does $\mathcal{L}_{i,j}$ extend to an invariant of free links? If so, prove the invariance. If not, give an example of two equivalent free diagrams L and K such that $\mathcal{L}_{i,j}(L) \neq \mathcal{L}_{i,j}(K)$.

16. Construct an infinite family of non-trivial free links.

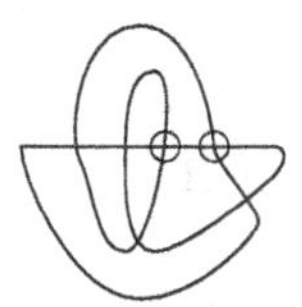

FIGURE 3.16: Free knot and links

3.2 WELDED, SINGULAR, AND PSEUDO KNOTS

In the previous section, we considered knot theories that were obtained by reducing the amount of information provided by the link diagram. In this section, we consider knot theories with a physical motivation.

Welded links were first introduced by Fenn, Rimanyi, and Rourke in the article "The braid-permutation group" published in the journal *Topology* in 1997. A **welded link diagram** is a underlying diagram with n components, decorated with two types of crossings: welded crossings and classical crossings. The welded crossings are indicated by circled, solid crossings and the classical crossings by over/under-markings. Welded links are defined on the same set of diagrams as virtual links but the equivalence relation is different. Two welded link diagrams are **welded equivalent** if they are related by a finite sequence of planar isotopy, Reidemeister moves, virtual Reidemeister moves and the welded move (see Figure 3.17). A **welded link** is an equivalence class of welded link diagrams.

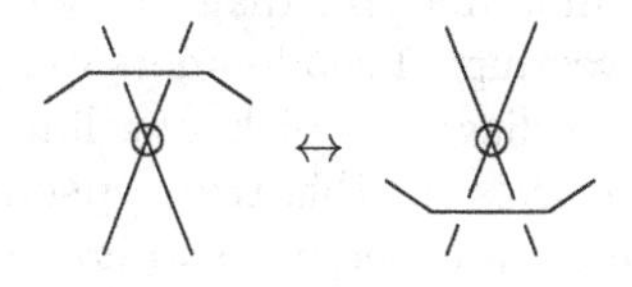

FIGURE 3.17: The welded move

We can visualize welded links as physical objects. Imagine that we've constructed the underlying diagram out of loops of string. At each classical crossing, the string is arranged to match the over/under-markings. At the welded crossings, we glue the crossing to the plane and render the crossing immobile. As a result, the string will float over the welded crossing in the welded move. The welded move does not affect linking numbers so all the linking invariants defined earlier are well defined on welded links.

The **forbidden move** is shown in Figure 3.18 and is frequently mistaken for the welded move. The forbidden move and the welded move can be used together with the other diagrammatic moves to unknot any virtual knot diagram. For this reason, we do not use the forbidden move and the welded move.

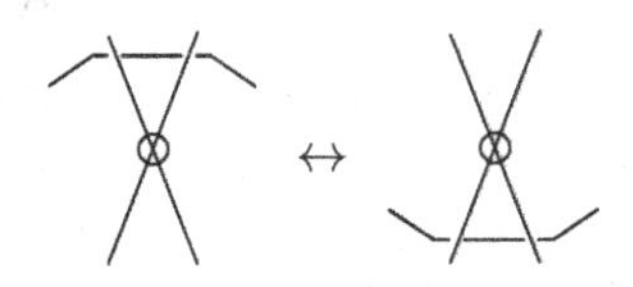

FIGURE 3.18: The forbidden move

A **singular link diagram** is an underlying diagram with n components that has two types of crossings: classical and singular crossings. **Singular crossings** are indicated by marking solid dots on the crossings. Two equivalent singular links are related by a sequence of Reidemeister moves and the **singular moves** (in Figure 3.19). **Singular links** are equivalence classes of equivalent singular link diagrams. We visualize these links as knotted and linked loops of string which are fused together at the singular crossings. Singular links and knots play an important role in the definition of the Vassiliev invariants.

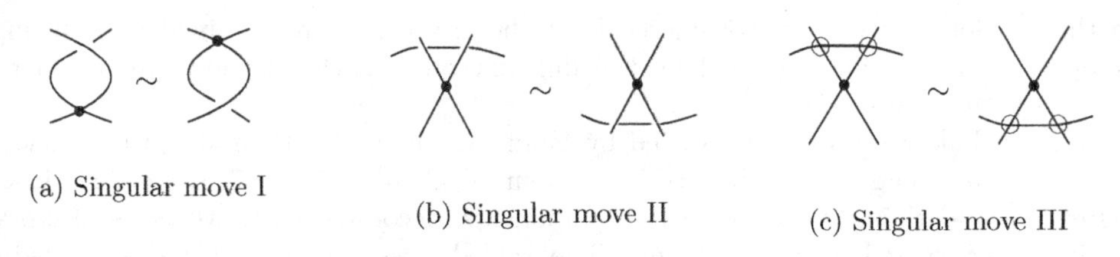

(a) Singular move I

(b) Singular move II

(c) Singular move III

FIGURE 3.19: Singular moves

Now, imagine taking a photograph of a classical link diagram. Suppose the focus on the photograph is off, and you can't determine the over/under-markings on some of the crossings. If we have two such blurred photographs, can we eliminate the possibility of two diagrams being related?

A **pseudo link diagram** is an underlying diagram with n components and two types of crossings: classical and pseudo crossings. Pseudo knots were first introduced by Ryo Hanaki in a 2010 paper entitled "Pseudo diagrams of knots, links and spatial graphs" published in the *Osaka Journal of Mathematics*. In this text, **pseudo crossings** are marked by a solid square to distinguish them from other types of crossings. Two pseudo link diagrams, L and K, are pseudo equivalent if there is a finite sequence of planar isotopy, Reidemeister moves, and **pseudo moves** (see Figure 3.20) relating the two diagrams. A **pseudo link** is an equivalence class of pseudo link diagrams.

The physical interpretation motivates the pseudo moves. Consider the move P_2. In Figure 3.21, we resolve the pseudo crossing on the left hand side as a classical crossing. There are two possible resolutions as a classical crossing, so we obtain two different diagrams (the upper and lower decks). We apply the Reidemeister II move to the upper deck twice, and reverse the overpassing strand. Comparing the diagrams in the upper and lower decks, we have two classical diagrams that differ only at a single crossing. The two possible diagrams

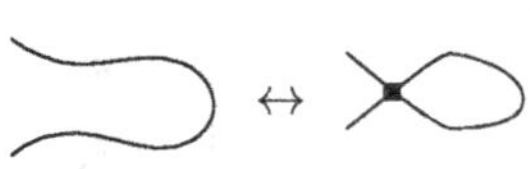

(a) Move P_1

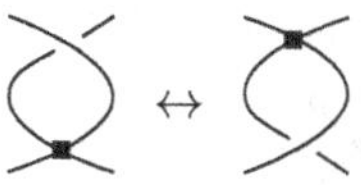

(b) Move P_2

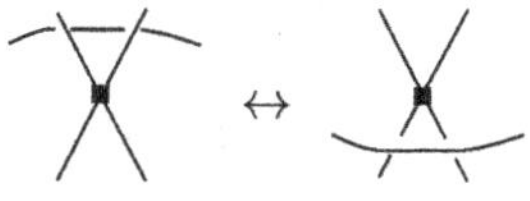

(c) Move P_{3a}

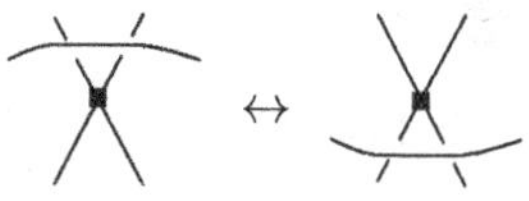

(d) Move P_{3b}

FIGURE 3.20: The pseudo moves

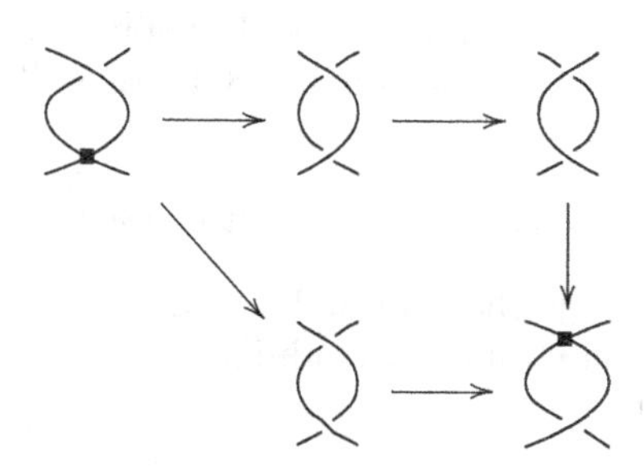

FIGURE 3.21: Examining P_2

are represented as the diagram with a single pseudo crossing on the right hand side of the P_2 move.

The **weighted resolution set** or the We-Re set of a pseudo knot P is a set of ordered pairs of the form $(K, p(K))$ where K is a classical knot and $p(K)$ the probability associated with K. The knot diagrams K are obtained by selecting over/under-markings for each pseudo crossing in the diagram P. A diagram obtained by selecting over/under-markings for each pseudo crossing is called a resolution. We consider two examples in Figure 3.22, a pseudo diagram with two classical crossings and a single pseudo crossing and a pseudo diagram with two pseudo crossings.

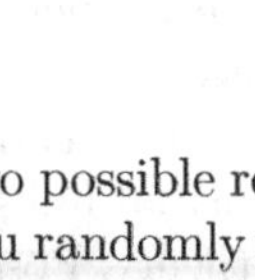

(a) A pseudo trefoil

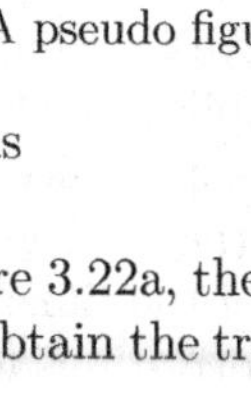

(b) A pseudo figure 8

FIGURE 3.22: Pseudo diagrams

There are two possible resolutions of the diagram in Figure 3.22a, the trefoil, T, and the unknot, U. If you randomly select a crossing resolution, you obtain the trefoil with probabil-

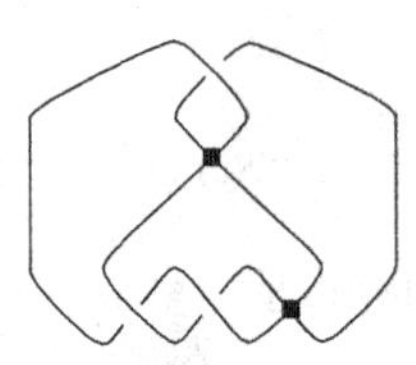

(a) Pseudo link: P_a

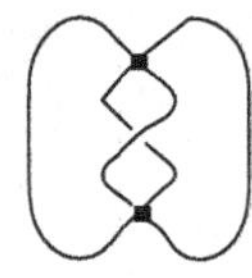

(b) Pseudo link: P_b

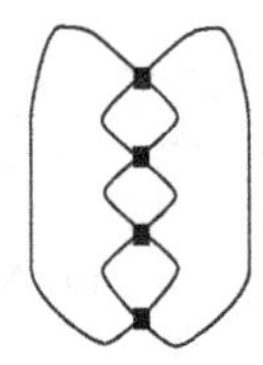

(c) Pseudo link: P_c

FIGURE 3.23: Compute the We-Re set

ity $\frac{1}{2}$ and the unknot with probability $\frac{1}{2}$. The We-Re set of this diagram is $\{(U, \frac{1}{2}), (T, \frac{1}{2})\}$ since there are only two possible resolutions: the unknot and the trefoil. In Figure 3.22b, there are two pseudo crossings and four possible resolutions. Of these four resolutions, three are the unknot and one is the classical figure 8 knot, K. The We-Re set of this diagram is $\{(U, \frac{3}{4}), (K, \frac{1}{4})\}$.

Theorem 3.6. *The We-Re set is an invariant of pseudo knots.*

We omit the proof of this theorem and encourage you to read the one of the original articles listed in the section on further reading.

Exercises

1. Find a sequence of moves, using any of the following; Reidemeister, virtual Reidemeister, forbidden, and welded moves, that transforms Kishino's knot into the unknot.

2. Find a sequence of moves, using any of the following; Reidemeister, virtual Reidemeister, forbidden, and welded moves, that transforms the virtual trefoil into the unknot.

3. Define singular virtual links by giving a set of diagrammatic moves that determine equivalence.

4. Construct sequences of diagrams illustrating the motivation for the moves: P_1, P_{3a}, and P_{3b}.

5. Compute the We-Re set of the pseudo knot shown in Figure 3.23.

6. For each pseudo link in Figure 3.23 determine the number of resolutions that are not equivalent to the unknot.

3.3 NEW KNOT THEORIES

We've detailed several different knot theories and their motivations in this section. An axiomatic approach to the construction of knot theories is laid out in Vladimir Turaev's paper (Vladimir Turaev, Topology of words, *Proc. Lond. Math. Soc, (3)*, 95(2):360–412). The key ideas involve 1) defining a set of underlying diagrams and 2) a set of crossing types. Then one can establish a set of diagrammatic moves that relate particular diagrams and from this set and equivalence relation construct equivalence classes.

In addition to the equivalence classes defined here, there are **twisted links** and various combinations of the links defined above (such as singular virtual links). There may be several different motivations for constructing a particular knot theory. Possible motivations include physical models: classical knots model the behavior of closed loops of string in three dimensional space while pseudo knots model the information loss potentially caused recording the diagram. In a paper from 2000, Shin Satoh used virtual knot diagrams to describe ribbon torus knots in four dimensional space. Other theories are inspired by the partitions determined by a particular invariant, $\mathcal{I}$. In this case, we define K to be equivalent to L if and only if $\mathcal{I}(K) = \mathcal{I}(L)$ (we will further explore this idea in subsequent chapters). We can then determine how to express equivalence in terms of diagrams. Other motivations arising include algebraic structures—such as link diagrams that model more abstract mathematical constructions.

3.4 OPEN PROBLEMS AND PROJECTS

Open problems

The goals below have given rise to a number of papers in the field of virtual knot theory.

1. Define a combinatorial knot theory and explain its topological or physical motivations.

2. Define the relationship (as a quotient or projection map) between two different knot theories.

Projects

The articles introducing many of the combinatorial knot theories are accessible to undergraduates.

1. Read about twisted knots and explain the diagrammatic conventions and moves.

2. Read about virtual pseudo knots and explain the diagrammatic conventions and moves.

3. Find pairs of flat virtual diagrams that are inequivalent but the diagrams are equivalent as free virtual link diagrams.

Further Reading

The suggested reading includes the original article on virtual knot theory, as well as the original articles on free knots, welded knots, and pseudo knots.

1. Louis H. Kauffman. Virtual knot theory. *European J. Combin.*, 20(7):663–690, 1999

2. O. V. Manturov and V. O. Manturov. Free knots and groups. *J. Knot Theory Ramifications*, 19(2):181–186, 2010

3. Roger Fenn, Richard Rimanyi, and Colin Rourke. The braid-permutation group. *Topology*, 36(1):123–135, January 1997

4. Ryo Hanaki. Pseudo diagrams of knots, links and spatial graphs. *Osaka J. Math.*, 47(3):863–883, 2010

5. A. Henrich, N. MacNaughton, S. Narayan, O. Pechenik, R. Silversmith, and J. Townsend. A midsummer knot's dream. *College Math. J.*, 42(2):126–134, 2011

6. A. Henrich, N. Macnaughton, S. Narayan, O. Pechenik, J. Townsend, Mikami Hirasawa, Naoko Kamada, and Seiichi Kamada. Classical and virtual pseudodiagram theory and new bounds on unknotting numbers and genus. *J. Knot Theory Ramifications*, 20(4):625–650, April 2011

7. Seiichi Kamada. Braid presentation of virtual knots and welded knots. *Osaka J. Math.*, 44(2):441–458, 2007

8. Taizo Kanenobu. Forbidden moves unknot a virtual knot. *J. Knot Theory Ramifications*, 10(1):89–96, 2001

CHAPTER 4

Crossing invariants

In Chapter 2, we counted sets of crossings (with sign) in order to compute different linking numbers. We return to the idea of counting crossings in order to define a link invariant. Here, we fix a diagram and count the number of crossings in a designated set. Each diagram in an equivalence class determines an integer, so that we obtain a set of integers for each link. The value of the invariant is the minimum value in this set of integers.

Given a set of real numbers, S, a **lower bound** on the set S is a number l such that $l \leq x$ for all $x \in \mathrm{S}$. For example, -1 is a lower bound on the set $\{0, 1, 2, 3\}$ since $-1 < x$ for all x in the set $\{0, 1, 2, 3\}$. Not all sets of integers have a lower bound. For example, the set of even integers, $\{\ldots - 4, -2, 0, 2, 4 \ldots\}$, does not have a lower bound. A number L is the **greatest lower bound** on the set S if L is a lower bound of S and $L \geq l$ for all lower bounds l of the set S. Returning to our example, 0 is the greatest lower bound for the set $\{0, 1, 2, 3\}$ since 0 is less than or equal to all elements of the set and greater than or equal to any lower bound. Similarly, an **upper bound** of a set S is a number u such that $u \geq x$ for all $x \in S$. The definition of the **least upper bound** parallels that of the greatest lower bound. The number U is a upper bound of S and $U \leq u$ for all upper bounds u of S. We conclude that 3 is the least upper bound for the set $\{0, 1, 2, 3\}$.

In this chapter, we construct sets containing non-negative integers and possibly the improper element ∞. If the set S has a greatest lower bound then the greatest lower bound is the minimum element in the set.

4.1 CROSSING NUMBERS

Let K denote a virtual link diagram. The **crossing number** of a virtual link diagram K is the number of classical crossings in the diagram K and is denoted as $\mathsf{c}_d(K)$. The crossing number of a link K, denoted $\mathsf{c}(K)$, is defined as the minimum:

$$\mathsf{c}(K) = min\{\mathsf{c}_d(K') | K' \sim K\}. \tag{4.1}$$

The crossing number of the unknot is zero, $\mathsf{c}(U) = 0$, since there is a diagram of U with zero crossings. In Figure 4.1, we illustrate some virtual links with their crossing number.

We can determine the crossing numbers of these knots because it is possible to construct all diagrams with two or fewer virtual crossings. In general, it may be hard to determine the crossing number of a diagram. It is much easier to determine an upper bound on $\mathsf{c}(K)$. To compute an upper bound, we simply choose a diagram K and compute $\mathsf{c}_d(K)$. We present this statement as a theorem.

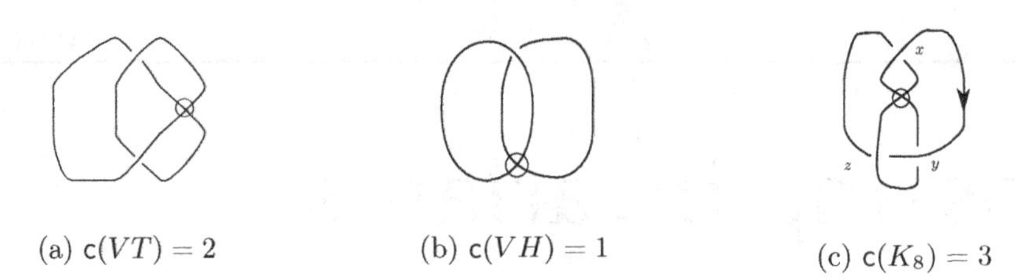

(a) $\mathsf{c}(VT) = 2$ (b) $\mathsf{c}(VH) = 1$ (c) $\mathsf{c}(K_8) = 3$

FIGURE 4.1: Crossing number examples

Theorem 4.1. *For all virtual links K, if D is a virtual link diagram and $D \sim K$ then $\mathsf{c}(K) \leq \mathsf{c}_d(D)$.*

Proof. Let D be a virtual link diagram equivalent to K. Suppose that D has n classical crossings. Then $n \in \{\mathsf{c}_d(D) | D \sim K\}$. Since $\mathsf{c}(K)$ is a lower bound of this set, then $n \geq \mathsf{c}(K)$. □

Proving that a number is the greatest lower bound of the set $\{\mathsf{c}_d(D)|D \sim K\}$ is much more challenging than computing an upper bound. Suppose that a diagram D has n crossings and we believe that $\mathsf{c}(K) = n$. One possible way to prove this fact is to check that all diagrams D' with fewer crossing are not equivalent to D. Another method involves using an invariant (such as a crossing weight number) to prove that no diagram of D can have fewer crossings. We give a lower bound on the crossing number of non-trivial links in Theorem 4.2.

Theorem 4.2. *For all non-trivial virtual links K, $\mathsf{c}(K) \geq 1$.*

Proof. Suppose that K is a non-trivial virtual link with $\mathsf{c}(K) = 0$. Then by definition, there exists a diagram L with zero classical crossings that is equivalent to K. However, a diagram with zero classical crossings (containing only virtual crossings) is equivalent to the unknot. □

Crossing number is a measure of complexity, so it is natural to ask if there is a knot with the most crossings. We need to rephrase this question to answer it. We ask if given a number N, can we always find a link K with $\mathsf{c}(K) = N$?

Theorem 4.3. *For all virtual knots K,*

$$\sum_{a \in \mathbb{Z} - \{0\}} |\mathcal{C}_a(K)| \leq \mathsf{c}(K).$$

Proof. Let K be a virtual link diagram. Recall that for $a \in \mathbb{Z} - \{0\}$, $\mathcal{C}_a(K)$ is link invariant, indicating that every diagram equivalent to K must contain at least $|\mathcal{C}_a(K)|$ crossings of weight a. Summing, we see that

$$\sum_{a \in \mathbb{Z} - 0} |\mathcal{C}_a(K)| \leq \mathsf{c}(K).$$

□

We apply Theorem 4.3 to the virtual trefoil, VT, shown in Figure 4.1a. The weight of both crossings in the diagram is 1, so that $\mathcal{C}_1(VT) = -2$. Then, we determine that $\mathsf{c}(VT) \geq 2$. The diagram of VT has two crossings indicating that $\mathsf{c}(VT) \leq 2$. As a result, we conclude that $\mathsf{c}(VT) = 2$. Notice that the virtual figure eight knot, K_8; shown in Figure 4.1c, has crossing number 3. But we can't prove this using crossing weights, at best we can prove that $2 \leq \mathsf{c}(K_8) \leq 3$ since $|C_1(K_8)| = 2$. Using Theorem 4.3, we have the following corollary.

Corollary 4.4. *For all $n \in \mathbb{N}$, $n \geq 2$, there exists a virtual knot diagram K_n such that $\mathsf{c}(K) \geq n$.*

Proof. We construct a sequence of virtual knot diagrams. For $n \geq 2$, the knot K_n has crossing number n. We construct a knot diagram with n classical crossings and $n-1$ virtual crossings positioned as in knot diagrams shown in Figure 4.2. In Figure 4.2a, the knot K_2

(a) K_2 (b) K_3

FIGURE 4.2: Constructing knots with $C_{n-1}(K) = n$

has 2 crossings of weight 1. The knot K_3 (shown in Figure 4.2b) has 3 crossings of weight 2. We add pairs of crossings (as shown in the gray box) until the diagram contains n classical crossings to form K_n. Computation shows that $C_{n-1}(K_n) = n$, so that $\mathsf{c}(K_n) \geq n$. The diagram that we constructed has n classical crossings with weight $n-1$ and $\mathsf{c}(K_n) = n$. □

The **connected sum** of two knot diagrams K_1 and K_2 is constructed by selecting a point on each knot diagram and splicing the two diagrams together. The connected sum of K_1 and K_2 is denoted as $K_1 \sharp K_2$. Pictorially, this is shown in Figure 4.3. The connected sum of two trefoils ($T \sharp T$) is shown in Figure 4.4a. We state (without proof) that $\mathsf{c}(T \sharp T) = 6$.

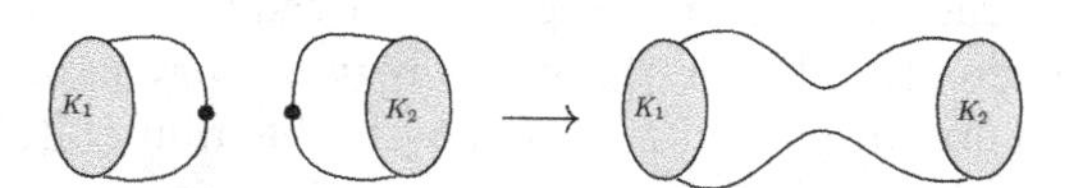

FIGURE 4.3: Schematic connected sum

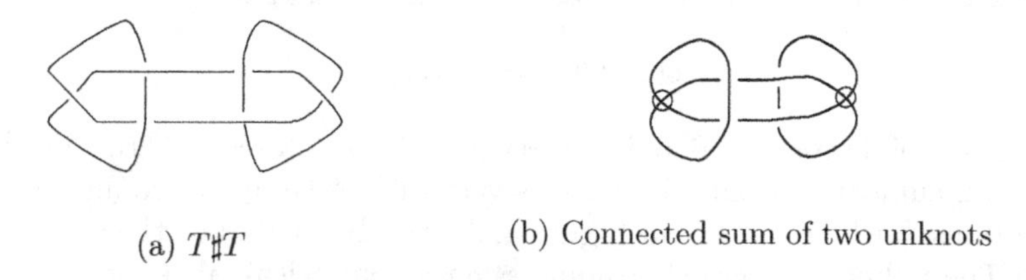

(a) $T \sharp T$ (b) Connected sum of two unknots

FIGURE 4.4: Connected sum examples

For virtual knot diagrams, the equivalence class of the connected sum is dependent on the points selected for the splice. For classical knot diagrams, the connected sum always results in the same knot regardless of the location of the sum.

Theorem 4.5. *For all classical knot diagrams K_1 and K_2, the equivalence class of the diagram $K_1 \sharp K_2$ is independent of the splicing points.*

Proof. We prove that the equivalence class of $K_1 \sharp K_2$ is independent of the location of the splicing point on K_2. We shrink the diagram K_1 until it is contained in a small disk in the plane. We can then slide the disk containing K_1 around K_2 via a sequence of classical Reidemeister II and III moves to any point on the knot diagram K_2. □

We restrict our attention to classical links. If we consider equivalence classes of classical link diagrams with no virtual crossings as determined by the classical Reidemeister moves and planar isoptopy, we define the **classical crossing number**. The classical crossing number of a classical link K is

$$\mathsf{cc}(K) = min\{\mathsf{c}_d(D) | D \text{ is a classical knot diagram and } D \sim K\}. \tag{4.2}$$

From the definition, it is clear that

$$\mathsf{c}(K) \leq \mathsf{cc}(K). \tag{4.3}$$

In fact, $\mathsf{c}(K) = \mathsf{cc}(K)$, but we do not currently have the tools to prove this fact.

For classical knots only, we have a theorem that relates $\mathsf{cc}(K_1), \mathsf{cc}(K_2)$, and $\mathsf{cc}(K_1 \sharp K_2)$.

Theorem 4.6. *For all classical knot diagrams K_1 and K_2 with zero virtual crossings then*

$$\mathsf{cc}(K_1 \sharp K_2) \leq \mathsf{cc}(K_1) + \mathsf{cc}(K_2). \tag{4.4}$$

Proof. Let K_1 (respectively K_2) be a classical knot diagram with no virtual crossings. We let $\mathsf{cc}(K_1) = n_1$ and $\mathsf{cc}(K_2) = n_2$. Notice that diagram K_i can be transformed via classical Reidemeister moves into a classical diagram D_i with $\mathsf{cc}_d(D_i) = n_i$. Form the knot diagram $K_1 \sharp K_2$. Then shrink the K_2 into a small circular region. There is a sequence of classical Reidemeister moves that transform K_1 in D_1. Since K_2 is contained in a small region of the plane, we can minimize the diagrammatic moves that interact with the region. We visualize the moves interacting with the region as a sequence of Reidemeister II and III moves that move a single strand over the small region. Now expand K_2 and shrink D_1, repeating the process to obtain D_1 and D_2. We have a diagram $D_1 \sharp D_2$ equivalent to $K_1 \sharp K_2$ with $\mathsf{cc}_d(D_1 \sharp D_2) = \mathsf{cc}(K_1) + \mathsf{cc}(K_2)$. □

In fact, people conjecture that the following statement is true for equivalence classes of classical knot diagrams related by the Reidemeister moves and planar isotopy.

Conjecture 4.1. *For all classical knot diagrams, K_1 and K_2,*

$$\mathsf{cc}(K_1 \sharp K_2) = \mathsf{cc}(K_1) + \mathsf{cc}(K_2). \tag{4.5}$$

The analog of Theorem 4.6 and Conjecture 4.1 for $\mathsf{c}(K)$ is not true for all virtual knots. The crossing number invariant, c, behaves very differently under connected sum for virtual knot diagrams. Kishino's knot (in Figure 4.4b) can be viewed as the connected sum of two unknots. The unknot has crossing number zero—but Kishino's knot has crossing number 4! The proof described above fails because the diagram of the connected sum may contain virtual crossings. The sequence of diagrammatic moves could contain a forbidden move.

Theorem 4.7. *For all two component virtual links K, $\mathsf{c}(K) \geq |\mathcal{L}_2^1(K)| + |\mathcal{L}_1^2(K)|$.*

Proof. The proof of this theorem is part of the exercises. □

By now, you've probably realized that the crossing number of a diagram is easy to compute, but computing the crossing number of a knot is not! Each diagram provides an upper bound on the knot's crossing number. In spite of this difficulty, even the bounds on the crossing number can be used to provide an estimate of how difficult it will be to compute the value of other link invariants.

We define a virtual analog of crossing numbers. Let K denote a virtual link diagram and let $\mathsf{v}_d(K)$ denote the number of virtual crossings in the diagram K. The **virtual crossing number** of the link K is:

$$\mathsf{v}(K) = min\{\mathsf{v}_d(K')|K' \sim K\}. \tag{4.6}$$

(a) $\mathsf{v}(K) = 2$

(b) $\mathsf{v}(VH) = 1$

(c) $\mathsf{v}(K) = 1$

FIGURE 4.5: Virtual crossing numbers

In the Figure 4.5, we give the virtual crossing number of several knots without proof. Then, we examine the virtual crossing number of oriented, two component links. We use linking numbers to determine a lower bound on the virtual crossing number. Recall the difference linking number $\mathcal{D}_{1,2}(K)$. For an oriented, two component link L, the value of $|\mathcal{D}_{1,2}(L)|$ does not depend on the ordering of the link.

Theorem 4.8. *For all oriented, two component links K, $\mathsf{v}(K) \geq |\mathcal{D}_{1,2}(K)|$.*

Proof. Let K be an oriented, two component link diagram. We consider the underlying diagram and realize that the two components intersect an even number of times. Following the orientation of component 1, we pair the closest virtual and classical crossings involving both components of the diagram and obtain one of the four types of pairings shown below. Component 1 is marked with a bold line. The pairs are classified as follows: the same component overpasses in both classical crossings in type 1, different components overpass in the classical crossings in type 2, both crossings are virtual in type 3, and in type 4, one crossing is virtual and one crossing is classical.

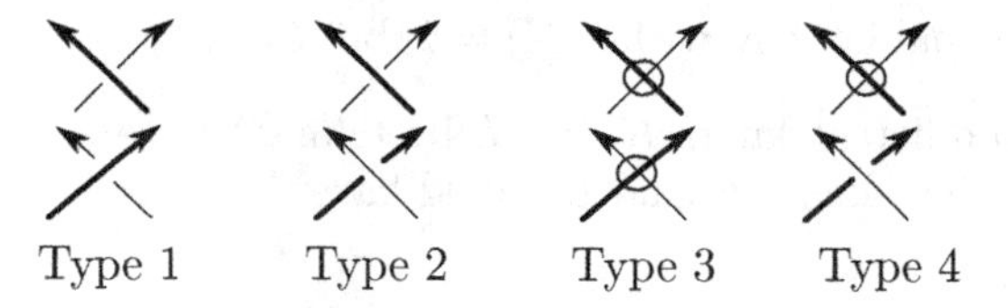

The net contribution to the difference linking number ($\mathcal{D}_{1,2}(K)$) from pairings of types 1–3 is zero. Only a pairing of type 4 makes a contribution to $\mathcal{D}_{1,2}(K)$, increasing or decreasing the value based on the sign of the classical crossing. The diagrammatic moves do not change $\mathcal{D}_{1,2}(K)$. The value of $|\mathcal{D}_{1,2}(K)|$ indicates that any diagram of K must have at least that many virtual crossings. □

We generalize this theorem to virtual knots using crossing weights.

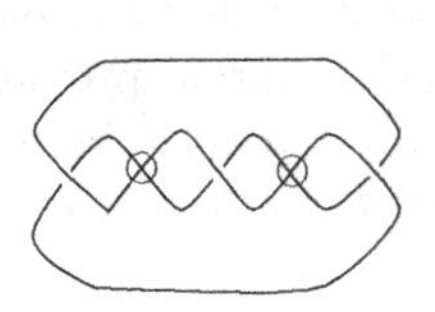

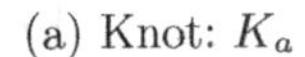

(a) Knot: K_a

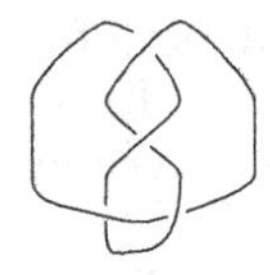

(b) Knot: K_b

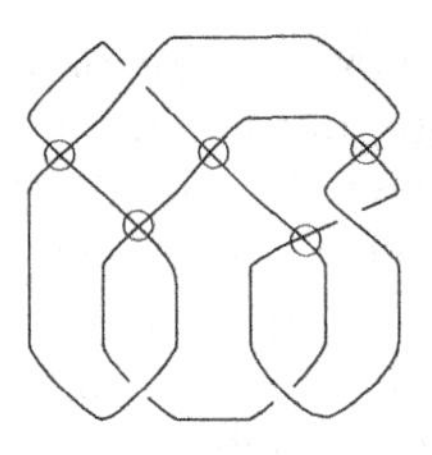

(c) Knot: K_c

FIGURE 4.6: Compute bounds on the crossing and virtual crossing number

Theorem 4.9. *For all virtual knots K if*

$$A = \max\{|a| \mid a \in \mathbb{Z} - 0 \text{ and } \mathcal{C}_a(K) \neq 0\}, \tag{4.7}$$

then

$$\mathsf{v}(K) \geq A.$$

Proof. Let K be a virtual knot diagram. By the hypothesis, $\mathcal{C}_a(K)$ is non-zero. Then for every virtual link diagram D equivalent to K, there is a crossing c in the oriented diagram D such that $\mathsf{w}(c) = a$. We form the two component diagram D_c. Now, $|\mathcal{D}_{1,2}(D_c)| = A$ by definition of $\mathcal{C}_a(K)$. Then by Theorem 4.8, D_c has at least A virtual crossings. Since D_c was formed by smoothing the classical crossing c in D, we see that $\mathsf{v}_d(D) = \mathsf{v}_d(D_c) \geq A$. We conclude that for all virtual knot diagrams $D \sim K$, $\mathsf{v}_d(D) \geq A$. Then $\mathsf{v}(K) \geq A$. □

Kishino's knot provides a counterexample to the conjecture that for all virtual knots K_1 and K_2, $\mathsf{v}(K_1 \sharp K_2) \leq \mathsf{v}(K_1) + \mathsf{v}(K_2)$. Kishino's knot, K_{Kish}, is the connected sum of two unknots and $\mathsf{v}(K_{Kish}) = 2$.

Exercises

1. Calculate upper and lower bounds for $\mathsf{c}(K)$ for the knots shown in Figure 4.6. Justify your bounds using evidence.
2. Calculate upper and lower bounds for $\mathsf{v}(K)$ for the knots shown in Figure 4.6.
3. Construct a family of 2 component virtual links L_n, such that $\mathsf{c}(L_n) = n$.
4. Construct a virtual knot K with $\mathsf{c}(K) = 5$, but $\mathcal{C}_4(K) \neq 5$.
5. Construct two different knots K and L by taking the connect sum of the same two knot diagrams. The knots K and L should have the following property: $\mathsf{c}(K) \neq \mathsf{c}(L)$.
6. Prove Theorem 4.7.
7. Show that the family of knots constructed in Corollary 4.4 has $\mathsf{v}(K_n) = n - 1$ for all $n \in \mathbb{N}$.
8. Given $n \in \mathbb{N}$, construct a family of virtual links, K_n, with $\mathsf{v}(K) = n$.
9. Construct a sequence of virtual links L_n where $\mathsf{c}(L) = \mathsf{v}(L) = n$.
10. Construct a sequence of virtual links L_n where $\mathsf{c}(L_n) - \mathsf{v}(L_n) = n$.

4.2 UNKNOTTING NUMBERS

Another measure of a link's complexity is unknotting numbers. We may be able to unknot a knot by performing an action on the crossings until we obtain a diagram of the unknot. There are several different choices for the action. We examine two cases: (1) changing the sign of a set of crossings and (2) changing a set of classical crossings into virtual crossings.

We define the **unknotting number** of a virtual link. Let D be a virtual link diagram. Select n classical crossings and switch the crossings' over/underpassing arcs so that a positive crossing becomes a negative crossing (or vice versa) to form the virtual knot diagram D'. The unknotting number of the diagram D ($\mathsf{u}_d(D)$) is the minimum number of crossings required to transform D into a diagram D' that is equivalent to the unknot. It may not be possible to transform D into the unknotted unlink via crossing changes. In this case, we define $\mathsf{u}_d(D) = \infty$. The unknotting number of a virtual link K is the minimum:

$$\mathsf{u}(K) = min\{\mathsf{u}_d(D) | D \sim K\}.$$

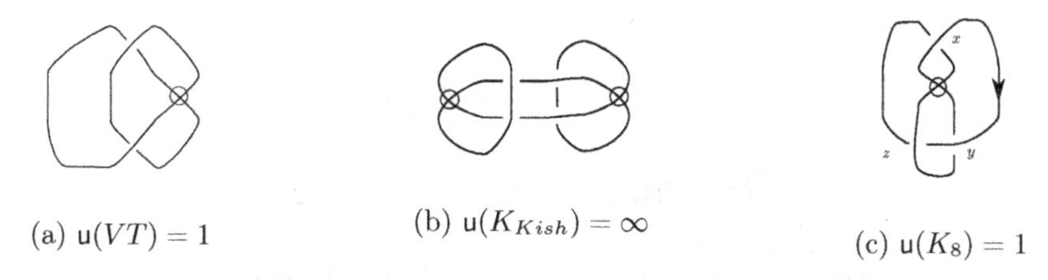

(a) $\mathsf{u}(VT) = 1$ (b) $\mathsf{u}(K_{Kish}) = \infty$ (c) $\mathsf{u}(K_8) = 1$

FIGURE 4.7: Unknotting number examples

In Figure 4.7, we show (without proof) several knots and their unknotting number. For most virtual knots, we can determine an upper bound on u from a diagram of the virtual knot. However, there are knots that cannot be unknotted through crossing change. Kishino's knot, K_{Kish}, has $\mathsf{u}(K_{Kish}) = \infty$. We may not always be able to find an upper bound for unknotting number. We can prove that if K is a non-trivial knot then $\mathsf{u}(K) \geq 1$.

Theorem 4.10. *For all non-trivial virtual knots K, $\mathsf{u}(K) \geq 1$.*

Proof. We use proof by contradiction. Suppose that K is a non-trivial virtual knot such that $\mathsf{u}(K) = 0$. Then by definition, K is equivalent to the unknot. But we assumed that K was not trivial. □

For classical knot diagrams, we define the **classical unknotting number** based on equivalent classical knot diagrams. We define the classical unknotting number of a classical knot diagram K with zero virtual crossings as

$$\mathsf{cu}(K) = min\{\mathsf{u}_d(D) | D \text{ is a classical diagram and } D \sim K\}. \tag{4.8}$$

For all classical knots K, $\mathsf{u}(K) \leq \mathsf{cu}(K)$.

We have the following theorem about unknotting number for classical knot diagrams.

Theorem 4.11. *For all classical knot diagrams (no virtual crossings), K_1 and K_2,*

$$\mathsf{cu}(K_1 \sharp K_2) \leq \mathsf{cu}(K_1) + \mathsf{cu}(K_2). \tag{4.9}$$

Proof. Let K_1 and K_2 be classical knot diagrams. The equivalence class of $K_1 \sharp K_2$ is independent of the splice points. Using the classical Reidemeister moves, we can transform $K_1 \sharp K_2$ into the diagram $D_1 \sharp D_2$ where $\mathsf{u}_d(D_1) = \mathsf{cu}(K_1)$ and $\mathsf{u}(D_2) = \mathsf{cu}(K_2)$. We can unknot $D_1 \sharp D_2$ with $\mathsf{cu}(K_1) + \mathsf{cu}(K_2)$ crossing changes. We conclude that $\mathsf{cu}(K_1 \sharp K_2) \leq \mathsf{cu}(K_1) + \mathsf{cu}(K_2)$. □

We immediately obtain the following corollary.

Corollary 4.12. *Let K_1 and K_2 be classical knot diagrams with no virtual crossings. Then* $\mathsf{u}(K_1 \sharp K_2) \leq \mathsf{cu}(K_1) + \mathsf{cu}(K_2)$.

In general, a version of Theorem 4.11 is not true for virtual knots. The diagram of Kishino's knot cannot be unknotted via crossing change, but it is formed by taking the connect sum of two unknots.

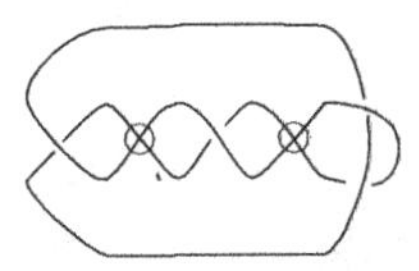

FIGURE 4.8: $\mathsf{u}(K) = 1, \mathsf{c}(K) \geq 2$

The lowest unknotting number does not always occur in the diagram with the lowest crossing number. The example in Figure 4.9 is due to S. Bleiler and Y. Nakanishi. The diagram K (in Figure 4.9a) is equivalent to diagram K' in Figure 4.9b. The knot diagram K has the lowest possible classical crossing number: $\mathsf{c}_d(K) = \mathsf{cc}(K) = 10$, while K' has 14 crossings. The unknotting number for the diagram is $\mathsf{u}_d(K) = 3$. The knot diagram K' (in Figure 4.9a) has the lowest possible unknotting number: $\mathsf{u}_d(K') = 2$. This demonstrates that the lowest unknotting number and the lowest crossing number do not necessarily occur in the same diagram.

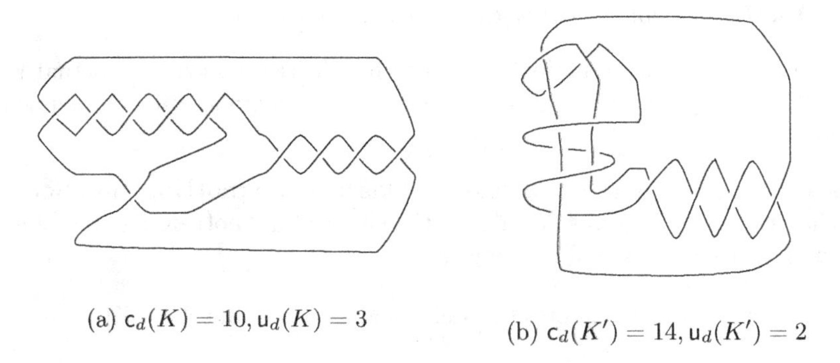

(a) $\mathsf{c}_d(K) = 10, \mathsf{u}_d(K) = 3$

(b) $\mathsf{c}_d(K') = 14, \mathsf{u}_d(K') = 2$

FIGURE 4.9: Bleiler-Nakanishi example

For classical knots, we know that the unknotting number is finite.

Theorem 4.13. *For all classical knots K, $\mathsf{u}(K) \leq \frac{1}{2}\mathsf{cc}(K)$.*

Proof. Let K be a classical knot with $\mathsf{cc}(K) = n$. Then K has a classical diagram D with zero virtual crossings and $\mathsf{c}_d(D) = n$. Our goal is to change the over/underpassing information at a set of crossings in D and construct either a descending diagram or an ascending diagram. (Either of these types of diagram are equivalent to the unknot.)

Assign an orientation to the diagram D and a basepoint. Traverse the diagram in the direction of the orientation. We travel across the underpassing strand first at j crossings and travel across the overpassing strand at the remaining $n-j$ crossings.

If $j \leq n-j$ then we switch the underpassing strand to an overpassing strand at the j crossings where the underpassing strand was traversed first. This creates a descending diagram.

If $n-j \leq j$ then we switch the overpassing strand to an underpassing strand at the $n-j$ crossings where the overpassing strand was traversed first. This creates an ascending diagram.

We compute that

$$min\{j, n-j\} \leq \frac{1}{2}n. \tag{4.10}$$

Then we can conclude that

$$\mathsf{u}(K) \leq \frac{1}{2}\mathsf{cc}(K). \tag{4.11}$$

□

A different approach to unknotting involves changing a set of classical crossings into virtual crossings to obtain a knot diagram equivalent to the unknotted unlink. Given a virtual knot diagram D, we define $\mathsf{vu}_d(D)$ to be the minimum number of classical crossings that need to be changed into virtual crossings to obtain an unknotted unlink. Then the **virtual unknotting number** is

$$\mathsf{vu}(K) = \min\{\mathsf{vu}_d(D) | D \sim K\}.$$

In Figure 4.10, we look at several examples of knot diagrams and state the virtual unknotting number without proof.

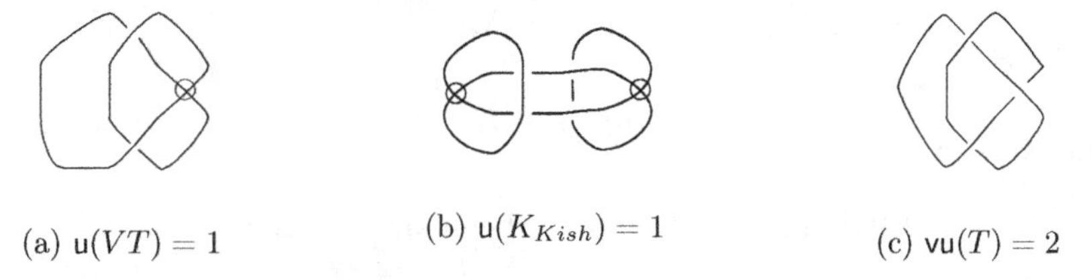

(a) $\mathsf{u}(VT) = 1$ (b) $\mathsf{u}(K_{Kish}) = 1$ (c) $\mathsf{vu}(T) = 2$

FIGURE 4.10: Virtual unknotting number examples

One inequality that immediately comes to mind is the following. For all virtual links K_1 and K_2,

$$\mathsf{vu}(K_1 \sharp K_2) \leq \mathsf{vu}(K_1) + \mathsf{vu}(K_2).$$

This inequality is not true for all virtual links. Kishino's knot immediately provides a counterexample.

We can also ask if the virtual unknotting number and the crossing number always occur in the same virtual link diagram.

For classical knots, we have the following conjecture due to Bernhard and Jablan. Although the unknotting number and the minimal crossing number are not linked (which makes it much more challenging to find the unknotting number), Bernhard and Jablan proposed the Bernhard-Jablan linking number, $\mathsf{BJ}(K)$, which is calculated using classical diagrams with the minimum possible number of crossings.

Conjecture 4.2. *Given a knot diagram L, we compute* $\mathsf{BJ}(L)$ *recursively as follows.*

1. *Define* $\mathsf{BJ}(L) = 0$ *for an unlink.*
2. *Define* $\mathsf{BJ}(L) = \mathsf{BJ}(L^*) + 1$ *where L^* is obtained from L by changing a single crossing and performing Reidemeister moves until the number of classical crossings is minimal.*

This process terminates for classical links and that $BJ(L) = \mathsf{cu}(L)$.

The crossing number of a virtual link imposes an upper bound on the virtual unknotting number.

Theorem 4.14. *For all virtual knots K,* $\mathsf{vu}(K) \leq \mathsf{c}_d(K)$.

Proof. Let D be a virtual knot diagram equivalent to K with $\mathsf{c}(K)$ crossings. From D, we construct the virtual link diagram D' by replacing each classical crossing in D with a virtual crossing. The diagram D' is equivalent to the unknot. As a result, $\mathsf{vu}_d(D) \leq \mathsf{c}(K)$. We conclude that $\mathsf{vu}(K) \leq \mathsf{vu}_d(D)$. □

Theorem 4.15. *For all non-trivial virtual knots K,* $\mathsf{vu}(K) \geq 1$.

Proof. If K is equivalent to a diagram D with $\mathsf{vu}_d(D) = 0$, then K is equivalent to the unknot. □

Exercises

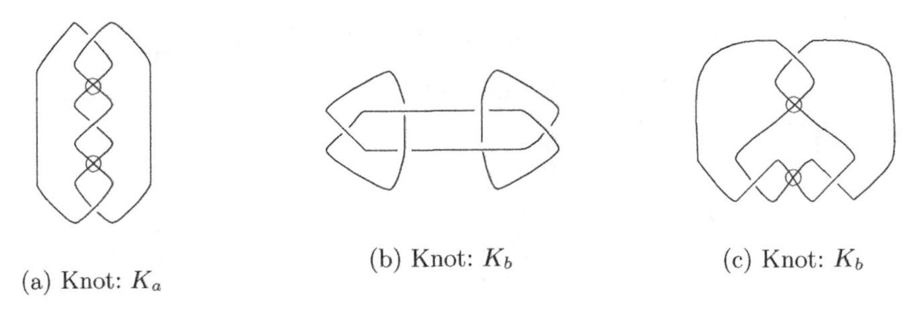

(a) Knot: K_a (b) Knot: K_b (c) Knot: K_b

FIGURE 4.11: Compute the unknotting numbers

1. Compute bounds on the unknotting number, u, for the knot diagrams shown in Figure 4.11.
2. Compute bounds on the virtual unknotting number, vu, for the knot diagrams shown in Figure 4.11.
3. Compute bounds on the unknotting number, u, for the knot diagrams shown in Figure 4.12.
4. Compute bounds on the virtual unknotting number, vu, for the knot diagrams shown in Figure 4.12.
5. Construct a knot diagram of the form $K_1 \sharp K_2$ such that $\mathsf{u}(K_1 \sharp K_2) \leq \mathsf{u}(K_1) + \mathsf{u}(K_2)$.
6. Construct a knot with $|\mathsf{u}(K) - \mathsf{c}(K)| = n$.

7. Find a family of virtual knots where $\mathsf{u}(K) = \infty$.

8. Prove that Kishino's knot (K_{Kish}) has $\mathsf{vu}(K_{Kish}) = 1$. Change one classical crossing to a virtual crossing and then find a sequence of diagrammatic moves that reduces the diagram to the unknot.

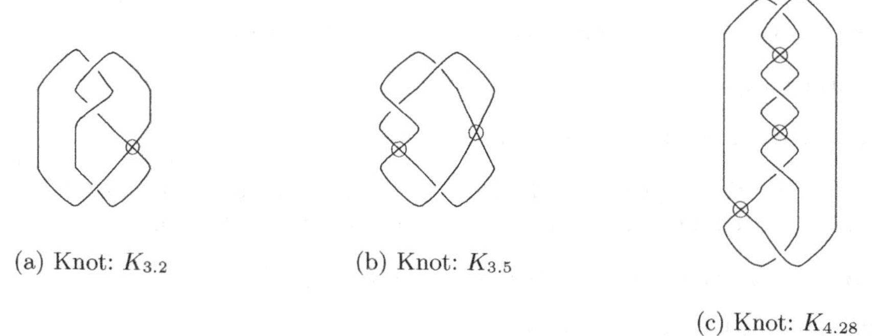

(a) Knot: $K_{3.2}$ (b) Knot: $K_{3.5}$ (c) Knot: $K_{4.28}$

FIGURE 4.12: Compute the unknotting numbers

4.3 UNKNOTTING SEQUENCE NUMBERS

We modify the definition of unknotting number and allow Reidemeister moves to be performed in between crossing changes. This leads to the definition of unknotting sequence numbers and several conjectures.

We define the **unknotting sequence number** as follows. Let K be a virtual knot diagram. We then

1. Perform a sequence of diagrammatic moves.
2. Make n crossing changes on the diagram.
3. Perform a sequence of diagrammatic moves.
4. Repeat steps 2–3 until the diagram is equivalent to the unknot.

We define $\mathsf{us}(K)$ to be the minimum number of crossing changes required to obtain the unknot from K.

Let us_R denote the unknotting sequence number when we restrict the allowed moves to the classical Reidemeister moves. Colin Adams proved the following theorem for classical knots and links.

Theorem 4.16. *For all classical link diagrams K, $\mathsf{us}_R(K) = \mathsf{u}(K)$.*

The proof of this theorem is in *The knot book* by Colin Adams, published in 2004 by the American Mathematical Society. The proof relies on the fact that only the classical Reidemeister moves are used—if we try to extend the proof to virtual knots and links, the argument is not valid. Forbidden moves can occur during the sequence of diagrammatic moves required. We do have a weaker theorem involving $\mathsf{u}(K)$ and $\mathsf{us}(K)$.

Theorem 4.17. *For all virtual knot diagrams K, $\mathsf{us}(K) \leq \mathsf{u}(K)$.*

Proof. The virtual unknotting number $\mathsf{u}(K)$ is a specific case of the unknotting sequence number. □

Similarly, we can define the **virtual unknotting sequence number** of a virtual knot K. On the diagram K,

1. Perform a sequence of diagrammatic moves.
2. Change n classical crossings in virtual crossings on the diagram.
3. Perform a sequence of diagrammatic moves.
4. Repeat steps 2–3 until the diagram is equivalent to the unknot.

The virtual unknotting sequence number, $\mathsf{vus}(K)$, is the minimum number of classical crossings that are changed into virtual crossings in order to obtain the unknot from K.

Theorem 4.18. *For all virtual knots K then $\mathsf{vus}(K) \leq \mathsf{vu}(K) \leq \mathsf{c}(K)$.*

Proof. The proof is left as an exercise. □

Using unknotting sequences, we define the **relative unknotting number**. Let K and K' be two inequivalent knots. We define $\mathsf{ru}(K_1, K_2)$ to be the minimum number of crossing changes required to transform K into K' via a sequence of diagrams: $K = K_1, K_2, \ldots, K_n = K'$.

1. Perform diagrammatic moves on K_i.
2. Switch j_i classical crossings to obtain diagram K_{i+1}.

The value $\mathsf{ru}(K, K')$ is defined to be the minimum of the possible sums of the form $\sum j_i$. If K and K' have finite unknotting numbers then we can prove that $\mathsf{ru}(K, K')$ is finite.

Theorem 4.19. *For all virtual knots K and K' such that $\mathsf{u}(K)$ and $\mathsf{u}(K')$ are finite, $\mathsf{ru}(K, K') \leq \mathsf{u}(K) + \mathsf{u}(K')$.*

Proof. The proof is left as an exercise. □

Exercises

1. Construct a relative unknotting sequence for the K_1 and K_2 in Figure 4.13.
2. Find a knot diagram K with seven classical crossings that can be transformed into Kishino's knot using diagrammatic moves and crossing changes.
3. Suppose that a knot K cannot be transformed into the unknot. Prove that K can not be transformed into the trefoil via a relative unknotting sequence.
4. Prove Theorem 4.18.
5. Prove Theorem 4.19.

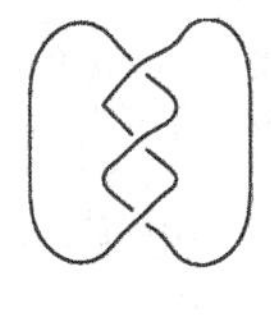

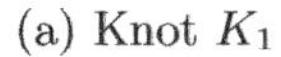

(a) Knot K_1

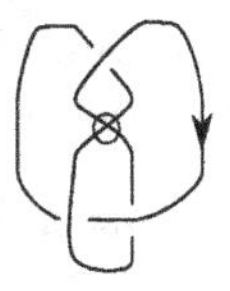

(b) Knot K_2

FIGURE 4.13: Transform K_1 into K_2 via a relative unknotting sequence

4.4 OPEN PROBLEMS AND PROJECTS

Open Problems

Researchers are interested in understanding the relationship between these invariants.

1. Given a virtual link diagram K, do $\mathsf{c}(K)$ and $\mathsf{v}(K)$ always occur in the same diagram?
2. For a particular subset of classical knots, we know that $\mathsf{c}(K) = \mathsf{cc}(K)$. However, we do not know if this is true for all classical knots. Can we prove that for all classical knot diagrams K that $\mathsf{c}(K) = \mathsf{cc}(K)$ and $\mathsf{u}(K) = \mathsf{cu}(K)$?
3. Given a knot K, can we find a mathematical equation relating $\mathsf{u}(K)$ and $\mathsf{vu}(K)$?
4. For all virtual knots K, is $\mathsf{u}(K) = \mathsf{us}(K)$?
5. For all virtual knots K, is $\mathsf{vu}(K) = \mathsf{vus}(K)$?
6. Find a Bleiler-Nalsanishki example for virtual links: two equivalent virtual knot diagrams K and K' such that $\mathsf{c}_d(K) < \mathsf{c}_d(K')$ but $\mathsf{v}_d(K) > \mathsf{v}_d(K')$.

Projects

These projects involve constructing families of knot diagrams with specific properties to better understand the relationship between the invariants. Another option is to investigate some of the invariants of this type that were not mentioned in the text.

1. Chose an invariant $(\mathsf{u}, \mathsf{vu}, \ldots)$ and construct an infinite family of knots K_n such that the value of this invariant is n (for example, $\mathsf{u}(K_n) = n$).
2. Find knots that are related to Kishino's knot by a relative unknotting sequence of length at least two.
3. Research ascending and descending numbers.
4. Research the bridge number of classical and virtual links.

Further Reading

1. A. Henrich, N. Macnaughton, S. Narayan, O. Pechenik, J. Townsend, Mikami Hirasawa, Naoko Kamada, and Seiichi Kamada. Classical and virtual pseudodiagram theory and new bounds on unknotting numbers and genus. *J. Knot Theory Ramifications*, 20(4):625–650, April 2011

2. Ryo Hanaki and Junsuke Kanadome. On an inequality between unknotting number and crossing number of links. *J. Knot Theory Ramifications*, 19(7):893–903, July 2010

3. Makoto Ozawa. Ascending number of knots and links. *J. Knot Theory Ramifications*, 19(1):15–25, January 2010

4. Slavik Jablan and Radmila Sazdanovic. Unlinking number and unlinking gap. *J. Knot Theory Ramifications*, 16(10):1331–1355, 2007

5. Denis Mikhailovich Afanasiev and Vassily Olegovich Manturov. On virtual crossing number estimates for virtual links. *J. Knot Theory Ramifications*, 18(6):1–16, 2009

6. Vassily Olegovich Manturov. Virtual crossing numbers for virtual knots. *J. Knot Theory Ramifications*, 21(13):13,1240009, 2012

7. Mikami Hirasawa, Naoko Kamada, and Seiichi Kamada. Bridge presentations of virtual knots. *J. Knot Theory Ramifications*, 20(06):881–893, June 2011

CHAPTER 5

Constructing knots

In this chapter, we use different techniques to construct knot and link diagrams. We use these techniques to create knots and links that meet specific criteria. In some cases, the diagrams are altered in a way that is not detectable by the invariants that have been defined so far in this book. First, we use symmetry to create a family of link diagrams. We then apply a linking invariant to these diagrams to explore whether or not these new diagrams are equivalent. These examples suggest several theorems about the crossing weight numbers and diagrams obtained using symmetry. This process illustrates how mathematicians move from examples and computation to hypothesis and then finally to proof and theorems. We also introduce some additional ways to construct knots.

5.1 SYMMETRY

Given an oriented knot or link diagram K, we construct new knot diagrams by exploiting the symmetry of K. We consider four specific methods of obtaining a new knot diagram from the diagram K. The **switch** of K, denoted K^S, is obtained by switching all the crossings in the knot. The **flip** of K, denoted K^F, is obtained by literally "flipping" the diagram along with the over/undercrossing information. The **mirror** of K is obtained by reflecting K across a line that does not intersect K. We denote the mirror of K as K^M. The **inverse** of K is obtained by reversing the orientation and is denoted K^I. The four symmetries are shown in Figure 5.1.

To study these different symmetries, we consider a virtual Hopf link, VH, as shown in Figure 5.2. We performed a Reidemeister I move on one of the components to aid in visualization. The diagram has two crossings: one involving two components and the crossing in a Reidemeister I twist. The crossing c in VH that involves both components has sign $+1$. The corresponding crossings in the other diagrams have the following signs.

$$VH^S : sgn(c) = -1 \qquad VH^F : sgn(c) = +1$$
$$VH^M : sgn(c) = -1 \qquad VH^I : sgn(c) = +1$$

The sign of the crossing immediately tells us that: $VH^S \nsim VH$ and $VH^M \nsim VH$. But we do not know whether or not VH is equivalent to VH^F. The sign of the crossing is correct, but now the overpassing strand is on the component with the twist.

We return to the more complicated knot K and its symmetries (Figure 5.1). In this case, a knot has only a single component and we can't use a linking number invariant to distinguish the diagrams. We consider the uppermost classical crossing in K and denote the crossing as x. In each of the diagrams obtained by symmetry, there is a corresponding

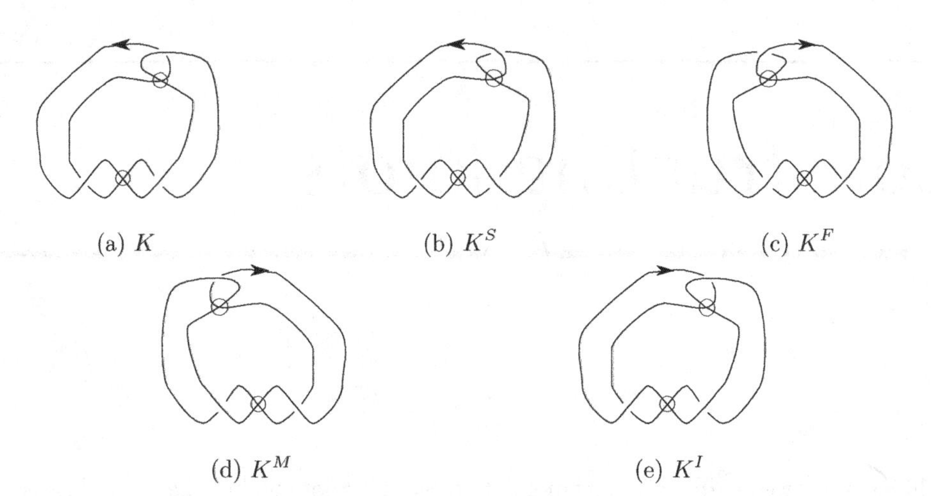

(a) K (b) K^S (c) K^F

(d) K^M (e) K^I

FIGURE 5.1: Symmetries of K

	$\mathcal{C}_{-2}(*)$	$\mathcal{C}_2(*)$
K	0	1
K^S	-1	0
K^F	0	1
K^M	-1	0
K^I	1	0

TABLE 5.1: $\mathcal{C}_2$ and $\mathcal{C}_{-2}$

crossing denoted x^S, x^F, and so on. Calculation shows that $\mathsf{w}(x) = 2$ and $\text{sgn}(x) = +1$. Recall that $\mathsf{w}(x) = \text{sgn}(x)(\mathcal{L}_2^1(K_x) - \mathcal{L}_1^2(K_x))$. For each diagram, we calculate the weight and sign of the crossing corresponding to x.

$$\mathsf{w}(x^S) = -2 \qquad \text{sgn}(x^S) = -1 \tag{5.1}$$

$$\mathsf{w}(x^F) = 2 \qquad \text{sgn}(x^F) = +1 \tag{5.2}$$

$$\mathsf{w}(x^M) = -2 \qquad \text{sgn}(x^M) = -1 \tag{5.3}$$

$$\mathsf{w}(x^I) = -2 \qquad \text{sgn}(x^I) = +1 \tag{5.4}$$

Further computation shows that none of the other crossings in K and its symmetries have weight ± 2. We indicate the value of $\mathcal{C}_2$ and $\mathcal{C}_{-2}$ for each knot in Table ??.

We conclude that K is not equivalent to K^S, K^M, and K^I. The only possible equivalent diagram pairs are: 1) K and K^F and 2) K^M and K^S. Are these pairs equivalent? Consider the example shown in Figure 5.3. In this case, the pairs L and L^F, and L^M and L^S do not seem to be equivalent. We cannot make a definite statement about whether or not these pairs are equivalent.

The following terminology is used when a knot diagram is equivalent to one of its symmetries. A knot K is **achiral** or **amphichiral** if $K \sim K^M$. A knot K is said to be **invertible** if $K^I \sim K$.

In our example, the crossing weight invariant $\mathcal{C}_j$ indicates an obstruction to some equivalences. We formalize this in the following theorem.

Theorem 5.1. *For all virtual knot diagrams K,*

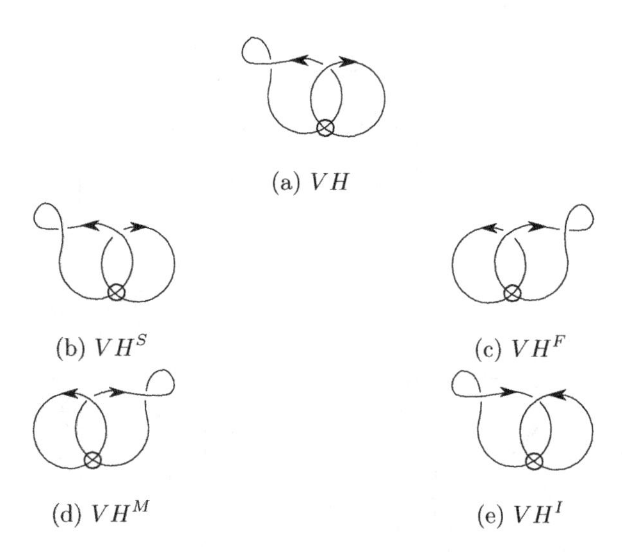

(a) VH

(b) VH^S

(c) VH^F

(d) VH^M

(e) VH^I

FIGURE 5.2: Symmetries of VH

1. $\mathcal{C}_a(K) = -\mathcal{C}_{-a}(K^S)$
2. $\mathcal{C}_a(K) = \mathcal{C}_a(K^F)$
3. $\mathcal{C}_a(K) = -\mathcal{C}_{-a}(K^M)$
4. $\mathcal{C}_a(K) = \mathcal{C}_{-a}(K^I)$

We prove that $\mathcal{C}_a(K) = -\mathcal{C}_{-a}(K^S)$ and $\mathcal{C}_a(K) = \mathcal{C}_{-a}(K^I)$. The remaining two cases are left as exercises.

Proof. Let K be a knot diagram. We recall that the weight of a crossing c is given by

$$\mathsf{w}(c) = \text{sgn}(c)\left(\mathcal{L}_2^1(K_c) - \mathcal{L}_1^2(K)\right) \tag{5.5}$$

where K_c is the two component diagram obtained by vertically smoothing the crossing c in K. Then

$$\mathcal{C}_a(K) = \sum_{c \in \mathcal{W}_a(K)} \text{sgn}(c). \tag{5.6}$$

Let x denote a crossing in K and let x^s denote the corresponding crossing in K^S. Switching changes the sign of a crossing, so that $\text{sgn}(x) = -\text{sgn}(x^s)$. At each crossing, the overpassing and underpassing components have exchanged places and the sign has changed so that $\mathcal{L}_j^i(K^S_{x^s}) = -\mathcal{L}_i^j(K_x)$. Rewriting, we find that

$$\begin{aligned}
\mathsf{w}(x^s) &= \text{sgn}(x^s)\left(\mathcal{L}_2^1(K^S_{x^s}) - \mathcal{L}_1^2(K^S_{x^s})\right) && (5.7)\\
&= -\text{sgn}(x)\left(-\mathcal{L}_1^2(K_x) - (-\mathcal{L}_2^1)(K_x)\right) && (5.8)\\
&= -\text{sgn}(x)\left(\mathcal{L}_2^1(K_x) - \mathcal{L}_1^2(K_x)\right) && (5.9)\\
&= -\mathsf{w}(x). && (5.10)
\end{aligned}$$

Since $\text{sgn}(x) = -\text{sgn}(x^s)$, $\mathcal{C}_a(K) = -\mathcal{C}_{-a}(K^S)$.

We consider the inverse diagram, K^I. Let x be a crossing in K and let x^I be the

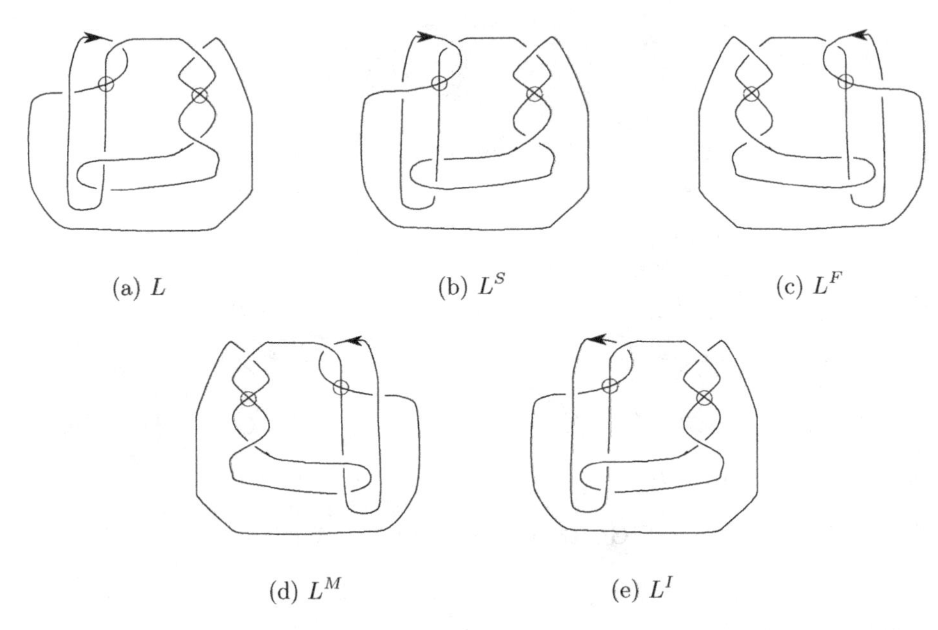

(a) L (b) L^S (c) L^F

(d) L^M (e) L^I

FIGURE 5.3: Symmetries of L

corresponding crossing in K^I. If we diagram the two crossings, we see that $\text{sgn}(x) = \text{sgn}(x^I)$, but the left and right components in K_c and $K^I_{x^I}$ have been interchanged. (The bottom of the crossing is now the top and vice versa.) Hence, $\mathcal{L}^1_2(K^I_x) = \mathcal{L}^2_1(K_x)$ and $\mathcal{L}^2_1(K^I_{x^I}) = \mathcal{L}^1_2(K_x)$.

$$\mathsf{w}(x^I) = \text{sgn}(x^I)\left(\mathcal{L}^1_2(K^I_{x^I}) - \mathcal{L}^2_1(K^I_{x^I})\right) \tag{5.11}$$

$$= \text{sgn}(x)\left(\mathcal{L}^2_1(K_x) - \mathcal{L}^1_2(K_x)\right) \tag{5.12}$$

$$= -\text{sgn}(x)\left(\mathcal{L}^1_2(K_x) - \mathcal{L}^2_1(K_x)\right) \tag{5.13}$$

$$= -\mathsf{w}(x). \tag{5.14}$$

We observe that $\mathcal{C}_a(K) = \mathcal{C}_{-a}(K^I)$. □

Based on our theorem, we quickly obtain the following corollaries.

Corollary 5.2. *For all knot diagrams K, if $K \sim K^S$ or $K \sim K^M$ then for $a \neq 0$, $\mathcal{C}_a(K) = -\mathcal{C}_{-a}(K)$.*

Corollary 5.3. *For all knot diagrams K, if $K \sim K^I$ then for $a \neq 0$, $\mathcal{C}_a(K) = \mathcal{C}_{-a}(K)$.*

Exercises

1. Construct K^s, K^M, K^F, and K^I for each diagram in Figure 5.4.
2. Prove that the figure 8 knot is achiral by finding a sequence of diagrammatic moves relating the figure 8 knot and its mirror.
3. Compute the invariants $\mathcal{C}_a$ for each knot diagram in Figure 5.4.
4. Compute the invariants $\mathcal{C}_a$ for each diagram obtained via symmetry from the diagrams in Figure 5.4.

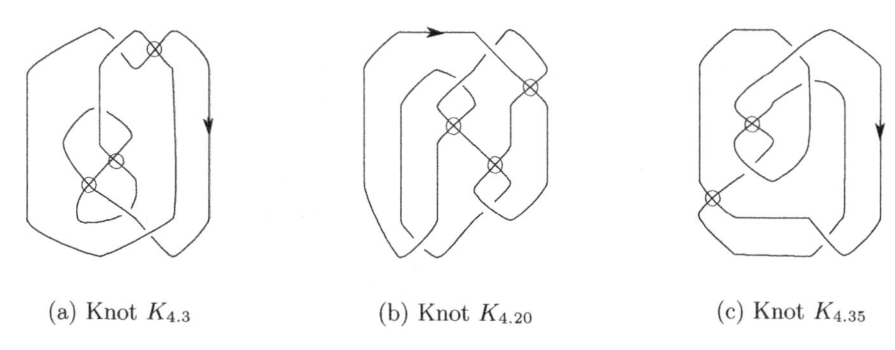

(a) Knot $K_{4.3}$ (b) Knot $K_{4.20}$ (c) Knot $K_{4.35}$

FIGURE 5.4: Construct diagrams using symmetry

5. Prove that $\mathcal{C}_a(K) = -\mathcal{C}_{-a}(K^M)$.
6. Prove that $\mathcal{C}_a(K) = \mathcal{C}_a(K^F)$.
7. State the contrapositive of Corollary 5.2.
8. State the contrapositive of Corollary 5.3.
9. Construct a knot K where $\mathcal{C}_a(K) = \mathcal{C}_{-a}(K)$.

5.2 TANGLES, MUTATION, AND PERIODIC LINKS

A $2-2$ tangle is a small "piece" of a link diagram. A $2-2$ **tangle** in a link diagram can be enclosed by a simple closed curve that intersects the link diagram in exactly four points. A $2-2$ tangle is typically drawn as a ball (where the knotting occurs) and 4 arcs that project to the northeast (NE), southeast (SE), northwest (NW), and southwest (SW) map directions. The tangle is referred to as a $2-2$ tangle because two arcs project from the top and two arcs project from the bottom. A sample $2-2$ tangle and a schematic drawing are shown in Figure 5.5. We can construct families of links based on tangles using a simple notation. We can also modify a link diagram by identifying a $2-2$ tangle in the diagram and changing only the $2-2$ tangle.

(a) Sample tangle (b) Schematic tangle

FIGURE 5.5: Tangles

To form a link (or knot) diagram from a $2-2$ tangle, we take the numerator closure as shown in Figure 5.6a. We can also produce a knot or link from a $2-2$ tangle by taking the denominator closure, shown in Figure 5.6b.

The simplest possible $2-2$ tangles are shown in Figure 5.7. These are the elementary tangles. The tangles $+1$, -1, and i are connected together to form more complicated tangles. We give a version of Conway notation for virtual tangles that was developed by Slavik Jablan for use with the Mathematica program LinKnot. There are three binary operations on tangles: addition, multiplication, and ramification and a unary operator called negation.

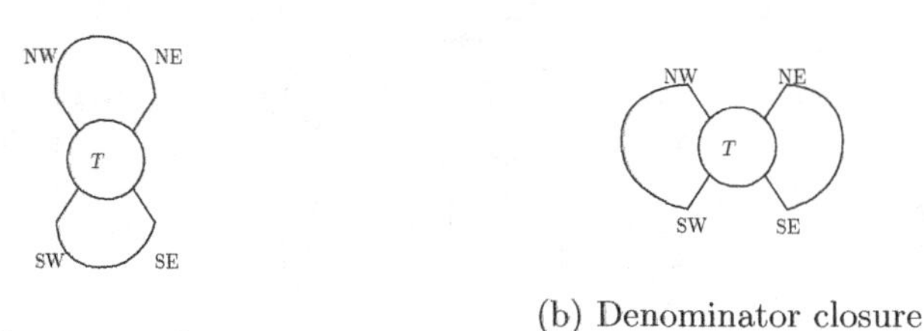

(a) Numerator closure

(b) Denominator closure

FIGURE 5.6: Tangle closures

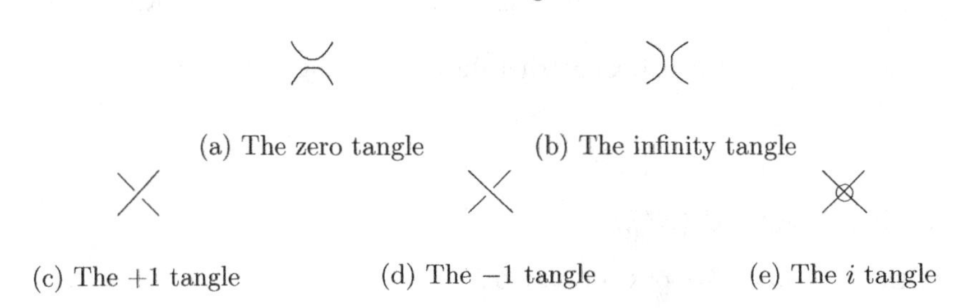

(a) The zero tangle

(b) The infinity tangle

(c) The +1 tangle

(d) The −1 tangle

(e) The i tangle

FIGURE 5.7: Elementary tangles

Tangle **addition** is performed by concatenating two tangles as shown in Figure 5.8. We denote the result of adding the tangles T_1 and T_2 as $T_1 + T_2$. We use a shorthand notation

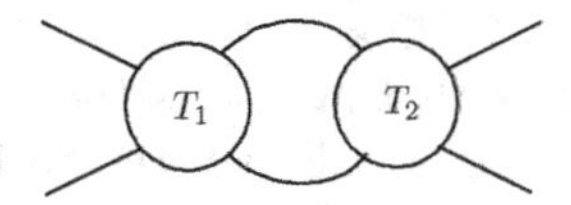

FIGURE 5.8: Tangle addition: $T_1 + T_2$

for addition for a finite sum of elementary tangles. We write $i + 1 + 1 + i$ as $(i, 1, 1, i)$ and $1 + 1 + i + -1$ as $(1, 1, i, -1)$. The tangle sums $(i, 1, 1, i)$ and $(1, 1, i, -1)$ are shown in Figure 5.9. Tangle addition does not commute, so the order of the elementary tangles in the parenthesized list is important.

Given a tangle L, we can also form the tangle $\sim L$ (which is also sometimes denoted as $-L$). The process of forming $\sim L$ from L is shown in Figure 5.10. The tangle L is reflected and then rotated 90 degrees in the counterclockwise direction. This unary operator is called the negation operator.

The second binary operation is tangle **multiplication**. The tangle $L \times K$ is shown in Figure 5.11a. In the parenthesized notation, we denote the product $L \times K$ as (LK). Tangle multiplication is not commutative or associative. We use parentheses to indicate order of operation. Two examples of tangle multiplication are shown in Figure 5.11. Changing the position of the parentheses in a tangle expression changes the resulting tangle since tangle multiplication is non-associative. In Figure 5.12, we see how the parentheses (and order of operations) affect the resulting tangle.

The third binary operation on the set of tangles is **ramification**. The ramification of L and K is denoted as L, K. In ramification, the tangles are rotated into vertical position and then concatenated as shown in Figure 5.13a. We see an example of ramification in Figure 5.13.

(a) Horizontal $(i, 1, 1, i)$ tangle

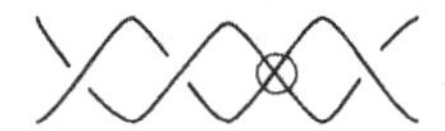

(b) Horizontal $(1, 1, i, -1)$ tangle

FIGURE 5.9: Tangles

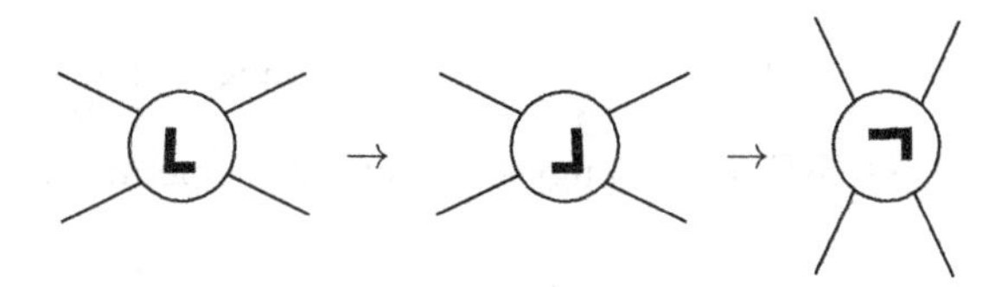

FIGURE 5.10: Forming $\sim L$

The notation $L_1, L_2, \ldots L_n$ indicates the link pictured in Figure 5.14a. We show another example of ramification in Figure 5.14b.

Algebraic tangles are formed by performing sum and product operations on elementary tangles. Rational links are formed by taking the numerator closure of a finite product of elementary tangles. The tangle operations, sum, product, and ramification, limit us to specific families of knots and links. There are also multiple descriptions of the same virtual link. For example, $\sim (1, 1)+ \sim (1, 1)$ and $(1, 1), (1, 1)$ are equivalent tangles. In the further reading, Schubert's 1956 paper on classical rational knots describes additional equivalences. For example, the links obtained from the tangles $(1, i)$ and $(1, i, 1)(-1)$ are equivalent links.

If we flip or mirror a $2-2$ tangle contained in a knot or link diagram, then we obtain a **mutation** of the diagram.

Let L be a virtual link diagram and let C be a simple closed curve in the plane that intersects the diagram L at exactly 4 points. The interior of the simple closed curve separates a $2-2$ tangle from the remainder of the diagram. We schematically indicate a tangle as a disk labeled with a stylized R and 4 projecting arcs as shown in Figure 5.15. We can flip or mirror the tangle along the horizontal or vertical axis as shown in Figure 5.15, using the stylized R to track how the symmetry mapping affects the tangle.

The possible symmetry mappings are shown in Figure 5.16. The tangle can be flipped horizontally (F_H) or flipped vertically (F_V), as well as mirrored horizontally M_H or mirrored

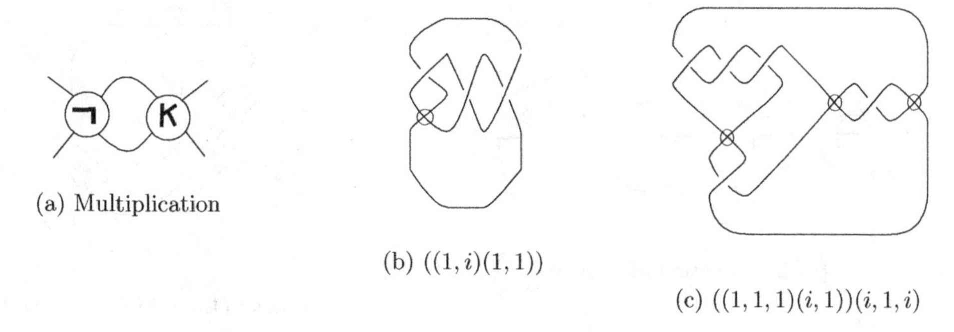

(a) Multiplication

(b) $((1, i)(1, 1))$

(c) $((1, 1, 1)(i, 1))(i, 1, i)$

FIGURE 5.11: Tangle multiplication

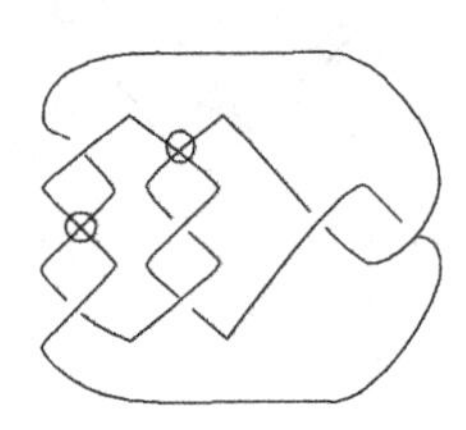

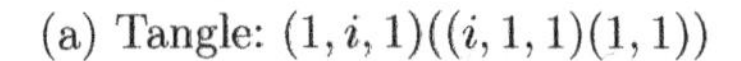

(a) Tangle: $(1, i, 1)((i, 1, 1)(1, 1))$

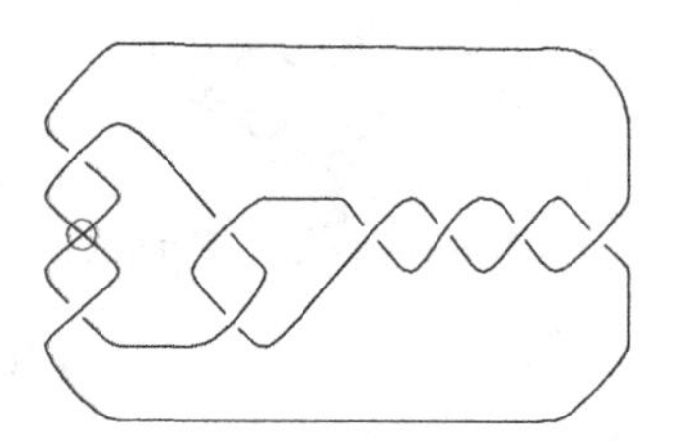

(b) Tangle: $(1, i, 1)((1, 1)(1, 1, 1, 1))$

FIGURE 5.12: Tangles and order of operation

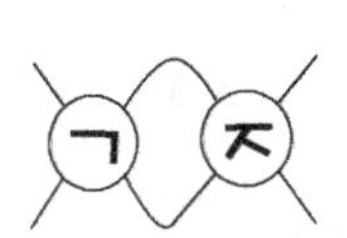

(a) Tangle ramification

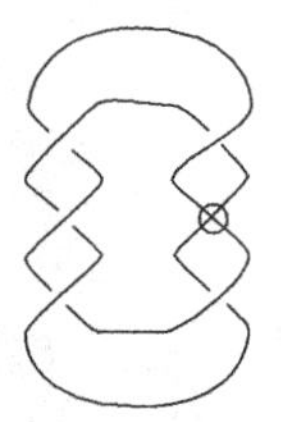

(b) Tangle: $(1, 1, 1), (1, i, 1)$

FIGURE 5.13: Ramification

vertically M_V. To form the **mutant** of a knot, we remove a tangle, perform a sequence of symmetry mappings, and then replace the (mutated) tangle in the diagram. For a link diagram K, we denote the mutants as $F_H(K)$ and so on. We construct two specific examples in Figure 5.17. Mutation results in a new diagram that may (or may not be equivalent to the old diagram).

Exercises

1. Construct link diagrams by taking the numerator closure of the tangles:

   (a) $(1, 1, i, 1, 1)$

   (b) $(1, i, 1)(1, 1)$

   (c) $(i, 1, 1, 1)(i, 1)$

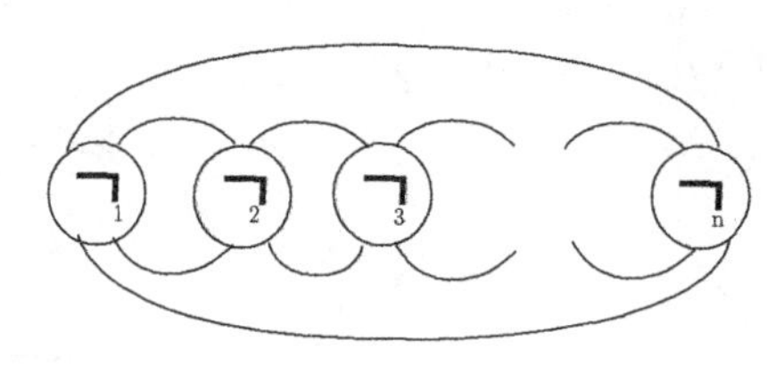

(a) Ramification of n tangles

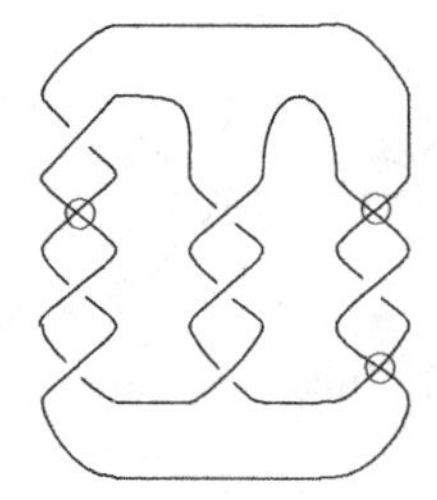

(b) Tangle: $(1, i, 1, 1), (1, 1, 1), (i, 1, i)$

FIGURE 5.14: Ramification

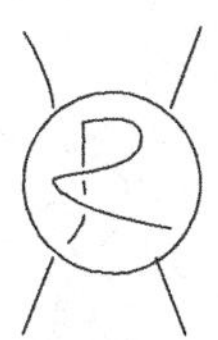

FIGURE 5.15: Schematic of tangle

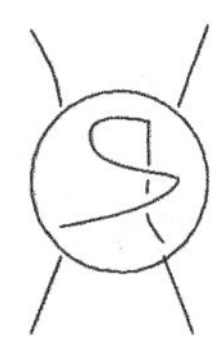

(a) Vertical mirror mutation: M_V

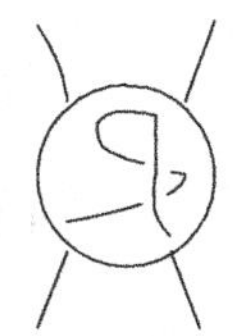

(b) Vertical flip mutation: F_V

(c) Horizontal mirror mutation: M_H

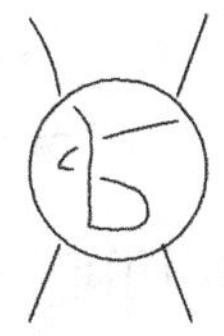

(d) Horizontal flip mutation: F_H

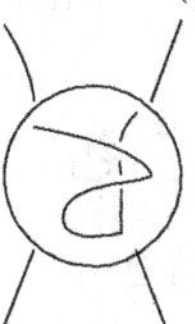

(e) Rotation mutation: R

FIGURE 5.16: Five mutation possibilities for virtual tangles

(d) $(1, 1, 1)(1, 1)(1, i, -1)(1, i)$
(e) $(1, i)(1, i)(-1, i)(1, 1)$
(f) $(1, i, -1)i(1, i, -1)$
(g) $(1, 1), (1, i), (1, i, -1)$
(h) $(1, 1)((1, i)(1, i, -1))$
(i) $((1, 1, 1)(1, i)), (1, i, -1)$
(j) $(1, 1, 1), ((1, i)(1, i, -1))$

2. Prove that the diagrams in Figure 5.18 are equivalent.
3. Give the Conway notation for the tangle in Figure 5.19.
4. For the knot diagram K and the indicated tangle in Figure 5.17, construct the mutations $F_V(K)$ and $M_H(K)$.

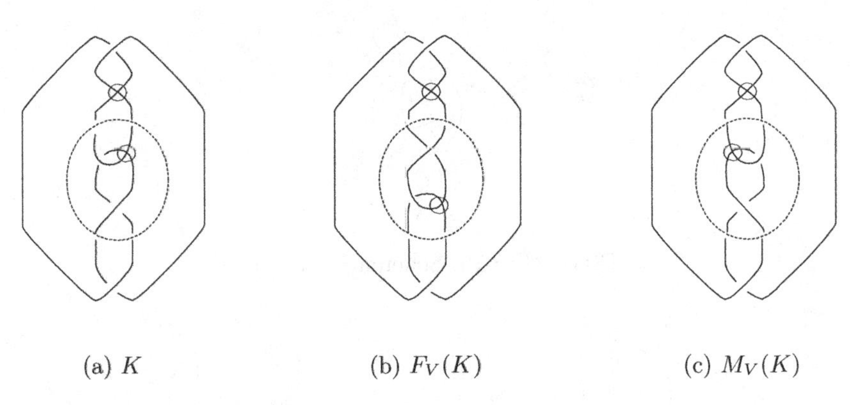

(a) K (b) $F_V(K)$ (c) $M_V(K)$

FIGURE 5.17: Mutation example

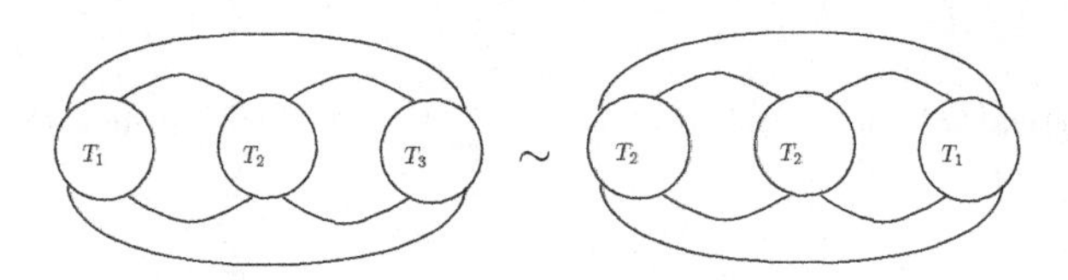

FIGURE 5.18: Prove these diagrams are equivalent

5. For all virtual knot diagrams K with tangle T, prove that $F_V(F_H(K)) = M_V(M_H(K))$.
6. Prove that mutation does not change the number of components in the diagram.
7. Construct a knot diagram K with a $2-2$ tangle T such that for some a, the value of $\mathcal{C}_a(K)$ is changed by mutation.

5.3 PERIODIC LINKS AND SATELLITE KNOTS

Recall the definition of the connected sum of two virtual knot diagrams. Let K_1 and K_2 be two virtual knot diagrams and with splice points d_1 and d_2, respectively (see Figure 5.20). To emphasize the splice points, we denoted the connected sum of K_1 and K_2 as $(K_1, d_1)\sharp(K_2, d_2)$. Previously, we demonstrated that it is possible to obtain different knots by

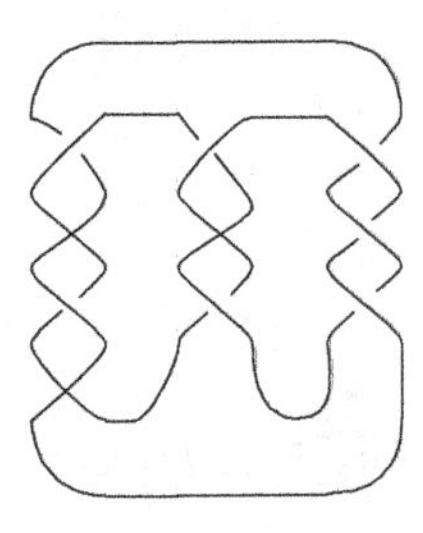

FIGURE 5.19: Describe the link diagram using tangle notation

choosing different connection points at which to take the connected sum. We use connected sum to construct some specific types of diagrams.

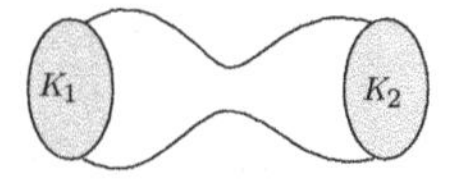

FIGURE 5.20: Connected sum example

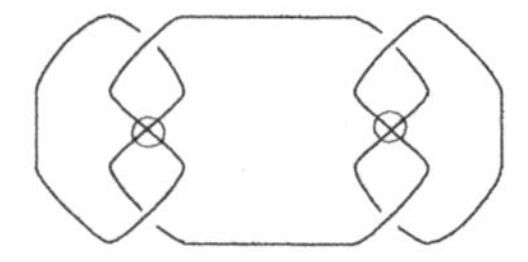

FIGURE 5.21: Ribbon connected sum

We form the **ribbon connected sum** of K by taking K and its mirror K^M and forming the ribbon connected sum $(K,d)\sharp(K^M,d^M)$ as in Figure 5.21. A knot K is **periodic** with period n if K has a diagram that misses the origin and is invariant under rotations of order n. We show several examples of periodic knots in Figure 5.22.

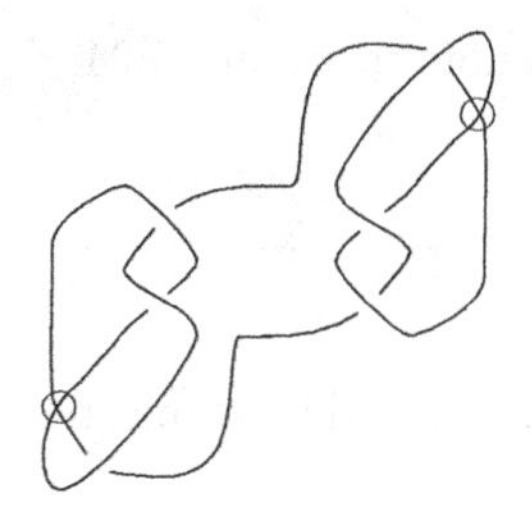

(a) A knot with period 2

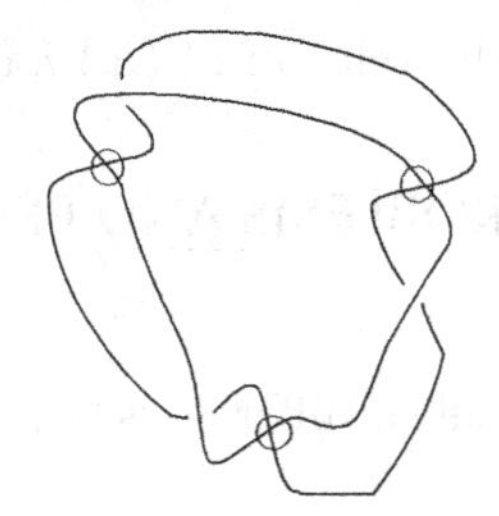

(b) A virtual link with period 3

FIGURE 5.22: Periodic knots

A **satellite knot** (or link) is constructed from two knots: a link diagram P that is called the **pattern** and a **companion** knot Q. The pattern link diagram is contained in the neighborhood of a circle. The pattern diagram is inserted into the companion diagram; the pattern diagram is arranged as the companion knot to form the satellite knot. We insert the patterns in Figure 5.23 into the virtual trefoil knot in Figure 5.24.

Exercises

1. Construct a virtual link of period 3 using the numerator closure of tangle $(1, i, 1, 1)$.
2. Construct a family of periodic links K_n where K_n has period n. Be able to prove each link in the family is inequivalent to the others.
3. Compute the value of $C_a(K)$ for an oriented version of the knot diagram in Figure 5.22a.

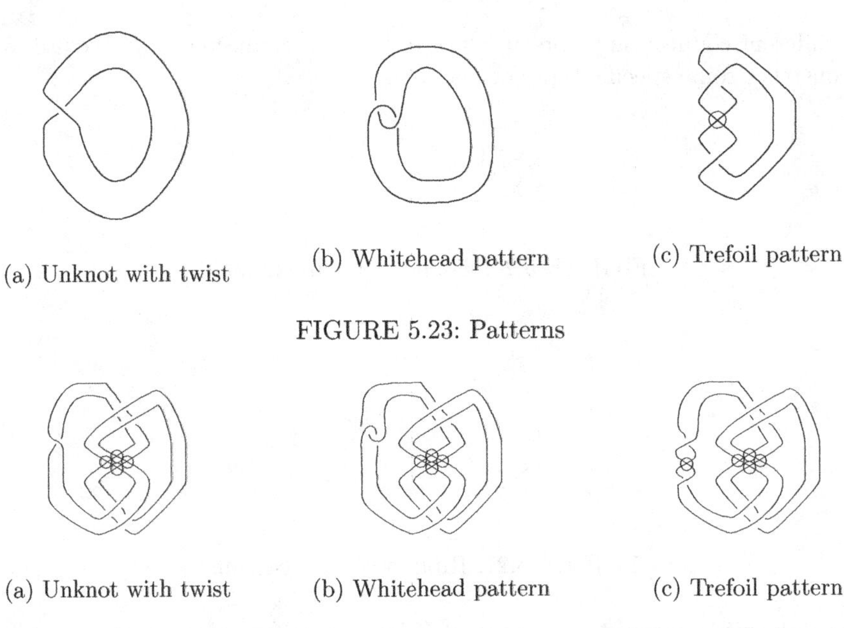

(a) Unknot with twist (b) Whitehead pattern (c) Trefoil pattern

FIGURE 5.23: Patterns

(a) Unknot with twist (b) Whitehead pattern (c) Trefoil pattern

FIGURE 5.24: Satellite knots

4. Construct satellites using the pattern and companions in Figure 5.25.
5. Compute the value of $C_a(K)$ for the satellite knots in Figure 5.24.

5.4 OPEN PROBLEMS AND PROJECTS

Open problems

We focus on the subtle differences between diagrams produced by connected sums, mutation, and symmetries.

1. Is there only one way to form a connected sum? That is, if $L \sim K_1 \sharp K_2$ and $L \sim P_1 \sharp P_2$ then is $K_1 \sim P_1$ and $K_2 \sim P_2$?
2. Can satellites be used to detect mutation and differentiate between the symmetries of a knot diagram?
3. Can we determine how many different virtual links can be generated by a pattern knot

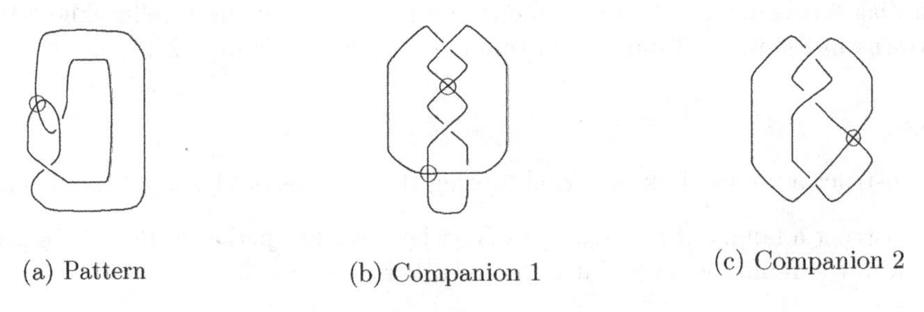

(a) Pattern (b) Companion 1 (c) Companion 2

FIGURE 5.25: Constructing satellites

and a companion? Changing the position of the pattern knot may result in forbidden moves.

Projects

We focus on using our link invariants to see if we can distinguish pairs of virtual knot diagrams.

1. Write the Conway notation each knot in Jeremy Green's knot table.
2. Let (w_i) denote a simple tangle. Compute the virtual unknotting number for $(w_1), (w_2), \ldots (w_n)$.
3. Suppose $K = (w_1), (w_2), \ldots, (w_n)$ is a knot. What is the value of $\mathcal{C}_a(K)$?
4. Let K be a non-classical knot. Let $m(K)$ denote a mutation of K. Construct satellites of K and $m(K)$ using a 2-strand cable of the unknot as the pattern. Compare values of $\mathcal{C}_a(K)$.
5. Let K_n be a family of periodic knots such that K_n has period n. Compare values of $\mathcal{C}_a(K)$. Compute $\mathsf{u}(K)$ and $\mathsf{vu}(K)$.
6. Construct two virtual knot diagrams related by mutation. Then, construct a pair of satellite knots using the mutants as companions. Compute one of the invariants for each member of this family of diagrams. Can you differentiate between the mutants?

Further Reading

1. Colin C. Adams. *The knot book.* American Mathematical Society, Providence, RI, 2004
2. Allison Henrich and Slavik Jablan. On the coloring of pseudoknots. *J. Knot Theory Ramifications*, 23(12):22,1450061, 2014
3. J. H. Conway. An enumeration of knots and links, and some of their algebraic properties. In *Computational problems in abstract algebra (Proc. Conf., Oxford, 1967)*, pages 329–358. Pergamon, Oxford, 1970
4. Louis H. Kauffman and Sofia Lambropoulou. Tangles, rational knots and –DNA". In *Lectures on topological fluid mechanics*, volume 1973 of *Lecture Notes in Math.*, pages 99–138. Springer, Berlin, 2009
5. Peter R. Cromwell. *Knots and links.* Cambridge University Press, Cambridge, 2004
6. Louis H. Kauffman, Slavik Jablan, Ljiljana Radovic, and Radmila Sazdanovic. Reduced relative Tutte, Kauffman bracket and Jones polynomials of virtual link families. *J. Knot Theory Ramifications*, 22(4):1340003, April 2013
7. Louis H. Kauffman and Sofia Lambropoulou. On the classification of rational tangles. *Advances in Applied Mathematics*, 33(2):199—-237, November 2004

II

Knot polynomials

CHAPTER 6

The bracket polynomial

In this section, we introduce the normalized Kauffman bracket polynomial, which is also sometimes referred to as the bracket polynomial or the f-polynomial. As we study the bracket polynomial, we focus on theorems that relate the bracket polynomial and properties of the links. In these theorems, quantifiers play an important role.

There are two types of quantifiers that appear in theorems, the universal quantifier and the existential quantifier. The universal quantifier is "for all". Theorems using this quantifier are stated as "For all x, $P(x)$." The existential quantifier statement is "there exists". We use this quantifier as "There exists an x, $P(x)$." Theorems also use the phrase "if and only if" or the phrase "necessary and sufficient conditions". Necessary and sufficient conditions are an implied reference to the biconditional ($\Leftrightarrow$) and a pair of if-then statements. Consider possible versions of the statement $P(X)$. We want to prove theorems that relate properties of the link with properties of a link invariant. A theorem of the form "for all links with property X, the links have property Y" increases the value of a link invariant.

We will also use the negation of a quantified statement. The negation of the statement "For all x, $P(x)$" is "There exists an x, $\neg P(x)$". The negation of the statement "There exists an x, $P(x)$" is "For all x, $\neg P(x)$".

6.1 THE NORMALIZED KAUFFMAN BRACKET POLYNOMIAL

The normalized Kauffman bracket polynomial (or f-polynomial) is an invariant of oriented links that was defined by Louis Kauffman in 1987. The f-polynomial is a combinatorial reformulation of the Jones polynomial that is known for its recursive computational formula. The formula is based on the skein relation and its computation is straightforward. The Jones polynomial was defined by Vaughn F. R. Jones in the 1984 paper "A polynomial invariant for knots via von Neumann algebras" which was published in the *Bulletin of the American Mathematical Society* in 1985. Vaughn F. R. Jones was awarded the Fields medal for this work in 1990.

The f-polynomial was originally defined as a map on the set of classical link and knot diagrams. Here, we define an extension of the f-polynomial as a map from the set of virtual knot and link diagrams to a set of polynomial functions in the variables A and A^{-1}. These polynomials are **Laurent polynomials** which can be expressed as a finite sum of the form

$$\sum_{i=-n}^{m} c_i A^i \tag{6.1}$$

where the coefficients c_n are elements of $\mathbb{Z}$.

To compute the f-polynomial, we begin by computing the bracket polynomial. The bracket polynomial is not a link invariant. Like writhe, the bracket polynomial is only invariant under the Reidemeister II and III moves and the virtual Reidemeister moves. The key to the computation of the bracket polynomial is the **skein relation**, one of the three axioms that define how to calculate the **bracket polynomial**.

1. The skein relation:

$$\langle \times \rangle = A\langle)(\rangle + A^{-1}\langle \asymp \rangle.$$

2. Disjoint union with an unknot:

$$\langle O \cup K \rangle = -(A^2 + A^{-2})\langle K \rangle.$$

3. Evaluating a unknot:

$$\langle O \rangle = 1.$$

The variable d is used to denote $-(A^2 + A^{-2})$.

The skein relation is applied recursively to a virtual link diagram. Each application produces two new link diagrams, each with one less classical crossing. Applying the skein relation to a virtual link diagram with n classical crossings results in a set of 2^n diagrams, each of which is called a **state** of the diagram. Each state is a collection of closed curves that only intersect at virtual crossings. (This means that each closed curve is equivalent to an unlinked unknot.) The states are then evaluated using the other two axioms.

We apply these rules to compute the bracket polynomial of the virtual trefoil. We expand the first crossing in Equation 6.2 using the skein relation.

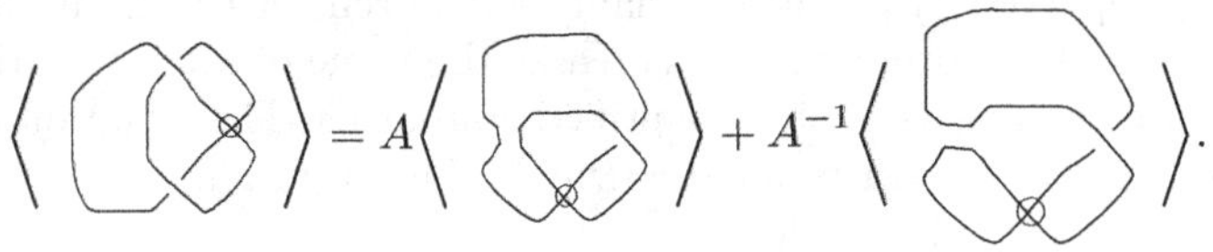

$$\langle \ \rangle = A\langle \ \rangle + A^{-1}\langle \ \rangle. \tag{6.2}$$

Then expand the remaining crossing in Equation 6.3.

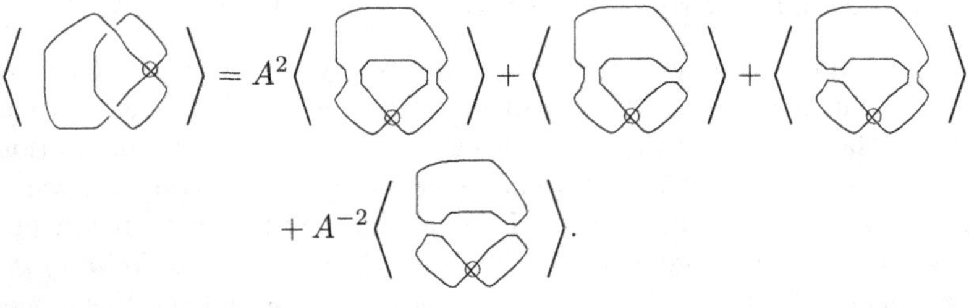

$$\langle \ \rangle = A^2\langle \ \rangle + \langle \ \rangle + \langle \ \rangle + A^{-2}\langle \ \rangle. \tag{6.3}$$

The only remaining crossings in the diagrams are virtual, and we evaluate the diagrams using the last two axioms.

$$\langle \ \rangle = 1, \qquad \langle \ \rangle = 1, \tag{6.4}$$

$$\langle \ \rangle = 1, \qquad \langle \ \rangle = -A^2 - A^{-2}. \tag{6.5}$$

Then, substituting the evaluations into Equation 6.3, we see that

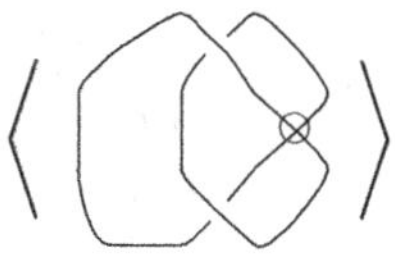

$$\langle \; \rangle = A^2 + 1 + 1 + A^{-2}d$$

$$= A^2 + 1 + A^{-4}. \quad (6.6)$$

We compute the bracket of the Hopf link following the same process.

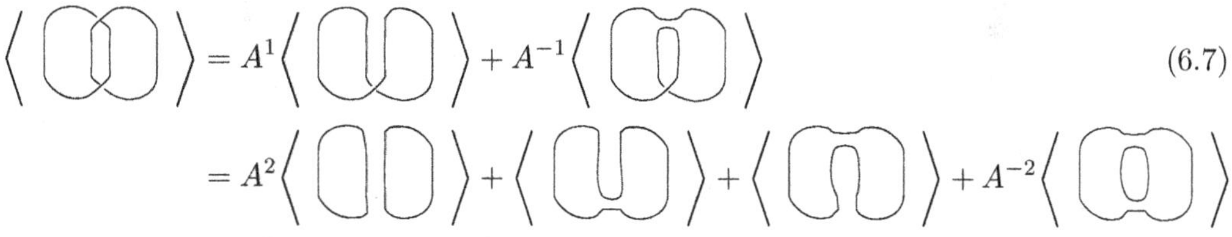

$$\langle \; \rangle = A^1 \langle \; \rangle + A^{-1} \langle \; \rangle \quad (6.7)$$

$$= A^2 \langle \; \rangle + \langle \; \rangle + \langle \; \rangle + A^{-2} \langle \; \rangle$$

$$= A^2 d + 1 + 1 + A^{-2}d$$

$$= -A^4 - A^{-4}.$$

The bracket polynomial is invariant under the Reidemeister II and III moves and the virtual Reidemeister moves. We prove that the value of the bracket polynomial is unchanged by these moves.

Theorem 6.1. *For all virtual link diagrams K, $\langle K \rangle$ is unchanged by the Reidemeister II and III moves, and the virtual Reidemeister moves.*

Proof. We prove that $\langle K \rangle$ is unchanged by the Reidemeister moves. If K_1 and K_2 are related by a single diagrammatic move then the diagrams are identical outside the region containing the diagrammatic move. To prove that both the left hand and right hand sides of a diagrammatic move result in the same contributions to the recursive equations, we apply the skein relation to the classical crossings in the diagrammatic move and compare the results.

We first expand the Reidemeister II move.

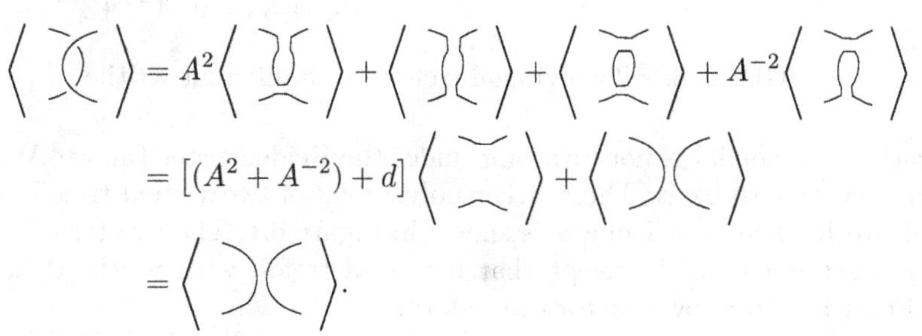

$$\langle \; \rangle = A^2 \langle \; \rangle + \langle \; \rangle + \langle \; \rangle + A^{-2} \langle \; \rangle$$

$$= [(A^2 + A^{-2}) + d] \langle \; \rangle + \langle \; \rangle$$

$$= \langle \; \rangle. \quad (6.8)$$

We need only partially expand the Reidmeister III move to show its invariance.

$$\langle\ \rangle = A\langle\ \rangle + A^{-1}\langle\ \rangle = +A\langle\ \rangle + A^{-1}\langle\ \rangle = \langle\ \rangle. \tag{6.9}$$

The virtual Reidemeister moves I – III do not involve classical crossings. We only need to check the virtual Reidemeister IV move. The equation that we need to verify is shown in Equation 6.10.

$$\langle\ \rangle = \langle\ \rangle. \tag{6.10}$$

We leave this step as an exercise. □

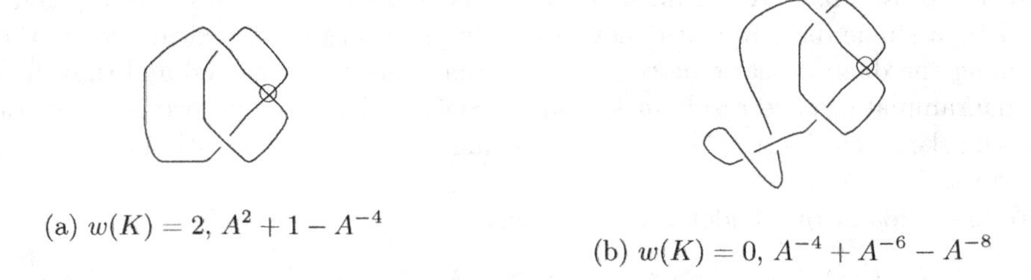

(a) $w(K) = 2$, $A^2 + 1 - A^{-4}$

(b) $w(K) = 0$, $A^{-4} + A^{-6} - A^{-8}$

FIGURE 6.1: Two virtual trefoils with different writhe

The bracket polynomial is not invariant under the Reidemeister I move. We investigate what occurs and how to fix it. The bracket polynomial of two virtual trefoils related by a sequence of two Reidemeister I moves is shown in Figure 6.1. The two trefoils have almost the same bracket polynomial, except that a virtual trefoil with writhe 0 has a bracket polynomial that is shifted by a factor of the form A^{-6}.

We apply the skein relation to a Reidemeister I move with a single positive crossing in Equation 6.11.

$$\langle\ \rangle = A\langle\ \rangle + A^{-1}\langle\ \rangle = -A^3\langle\ \rangle. \tag{6.11}$$

The calculation shows that a Reidemeister I move that inserts a positive twist shifts the bracket polynomial by a factor of $-A^3$. Each crossing in a diagram has the potential to be part of a Reidemeister I move, so we want to normalize the impact of the crossing signs. We do this by calculating the writhe of the link (the sum of all crossing signs in the diagram) and using writhe to determine the factor that we need to shift the bracket polynomial. Let K be an oriented link diagram with writhe $w(K)$. The **f-polynomial** (which is also called the **normalized bracket polynomial**) is defined as

$$f_K(A) = (-A^{-3})^{w(K)}\langle K\rangle. \tag{6.12}$$

We revisit our two virtual trefoils with different writhe in Figure 6.2.

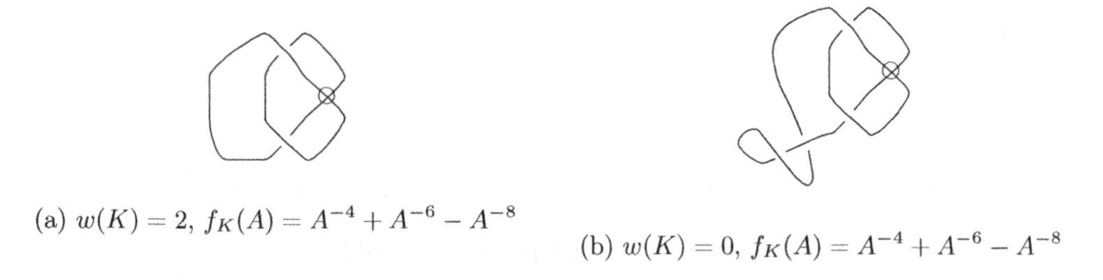

(a) $w(K) = 2$, $f_K(A) = A^{-4} + A^{-6} - A^{-8}$

(b) $w(K) = 0$, $f_K(A) = A^{-4} + A^{-6} - A^{-8}$

FIGURE 6.2: Two virtual trefoils with different writhe

We have proved the following theorem.

Theorem 6.2. *The Laurent polynomial $f_K(A) = (-A^{-3})^{w(K)}\langle K\rangle$ is a link invariant.*

Notice that if we reverse the orientation of a knot diagram, the sign of the crossings does not change. As a result, reversing orientation does not change the writhe of a knot diagram. For all virtual knot diagrams, the f-polynomial is independent of the orientation of the knot. For link diagrams, reversing the orientation of a component can change the sign of crossings and the writhe of the diagram.

The Jones polynomial of a link K is obtained by evaluating $f_K(A)$ at $A = t^{\frac{-1}{2}}$. The f-polynomial is a specialization of the HOMFLY-PT polynomial. The HOMFLY-PT polynomial is an invariant of oriented classical link diagrams that is defined by a two variable skein relation. The HOMFLY-PT polynomial was introduced in the article "A new polynomial invariant of knots and links" published in the *Bulletin of the American Mathematical Society* in 1985. The article has six authors: P. Freyd, D. Yetter, J. Hoste, W. B. R. Lickorish, K. Millett, and A. Oceanu.

The article is accompanied by a note from the editor. The note states that four articles were independently submitted to the journal between late September and early October. All four papers contained the same result from different approaches. The resolution of this situation was that the journal editor requested that the authors write a combined article presenting each of the four approaches. The "PT" part of the polynomial's name refers to J. Pryztycki and P. Traczyk who also obtained a version of the same polynomial, during approximately the same time period. Pryztycki and Traczyk's article "Invariants of link s of Conway type" was independently published in the the *Kobe Journal of Mathematics* in 1987.

Exercises

1. Compute the bracket polynomial of the knot diagrams in Figure 6.3.

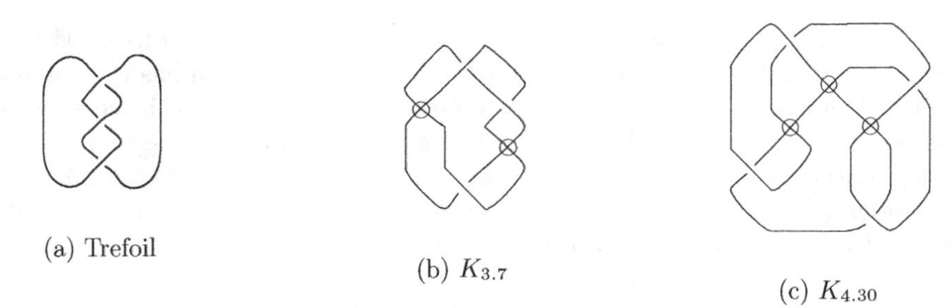

(a) Trefoil (b) $K_{3.7}$ (c) $K_{4.30}$

FIGURE 6.3: Compute the bracket polynomial

(a) Negative virtual trefoil (b) Hopf link

FIGURE 6.4: Compute the $f_K(A)$ for all possible orientations

2. Compute the f-polynomial of the knot diagrams in Figure 6.3.
3. Compute the f-polynomial for all possible orientations of the virtual trefoil in Figure 6.4a. Then compute the f-polynomial for all possible orientations of the virtual Hopf link in Figure 6.4b. How do changes in orientation affect the f-polynomial of a knot versus the f-polynomial of a link diagram? Explain.
4. Verify that the bracket polynomial is invariant under the virtual Reidemeister IV move (Equation 6.10).
5. Construct two equivalent knot diagrams, T_3 and T_0, such that $w(T_3) = 3$ and $w(T_0) = 0$. Then compute the bracket of both diagrams.
6. Calculate $\langle \quad \rangle$ in terms of $\langle \quad \rangle$.

6.2 THE STATE SUM

The skein relation is not the only way to define the bracket polynomial. Another approach involves writing the bracket polynomial as a weighted sum. Let L be a link diagram with n classical crossings. A **smoothing** of a crossing is obtained by a two step process. First, remove a small neighborhood of a crossing to obtain 4 endpoints. Then, glue together the endpoints so that there is no crossing. The outcome is designated as a type A or type B smoothing based on the relative orientation of the crossing as shown in Figure 6.5.

A **state** s of the knot diagram is obtained by choosing a type A or B smoothing for each classical crossing. Smoothing every classical crossing in a diagram results in a collection of closed curves that intersect only at virtual crossings. If a diagram has n classical crossings then there are 2^n states associated with the diagram. We indicate states by labeling the classical crossings with numbers. Then we associate a vector of length n, $\overline{x} = (x_1, x_2, \dots x_n)$, with entries from the set $\{A, B\}$ to each state. The notation $s(x_1, x_2, \dots x_n)$ indicates that crossing i has smoothing type x_i. We frequently refer to the all A state and the all B state,

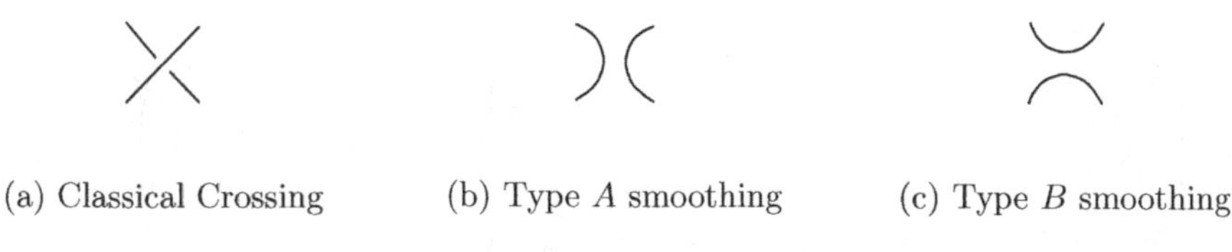

(a) Classical Crossing (b) Type A smoothing (c) Type B smoothing

FIGURE 6.5: Smoothings

so we denote them as s_α and s_β, respectively. A state $\overline{s}$ is the **dual** of a state s if all the smoothing choices are opposite. The states $s(A, A)$ and $s(B, B)$ are dual states, as are $s(A, B)$ and $s(B, A)$.

The number of curves in state s is denoted as $|s|$. We use $\alpha(s)$ to denote the number of type A smoothings in the state s. Similarly, $\beta(s)$ denotes the number of type B smoothings in the state s. Let S denote the set of all states of L. The variable d again denotes $-(A^2+A^{-2})$. Then the bracket polynomial is a weighted sum over all possible states:

$$\langle L \rangle = \sum_{s \in \mathrm{S}} A^{\alpha(s)-\beta(s)} d^{|s|-1}.$$

This expression is referred to as the **state sum**.

We compute the bracket polynomial of the positive virtual trefoil as a state sum. The diagram of the virtual trefoil has two positively oriented classical crossings (Figure 6.6a). The knot diagram has four states: $s_\alpha = s(A, A), s(A, B), s(B, A)$, and $s_\beta = S(B, B)$ as shown in Figure 6.6.

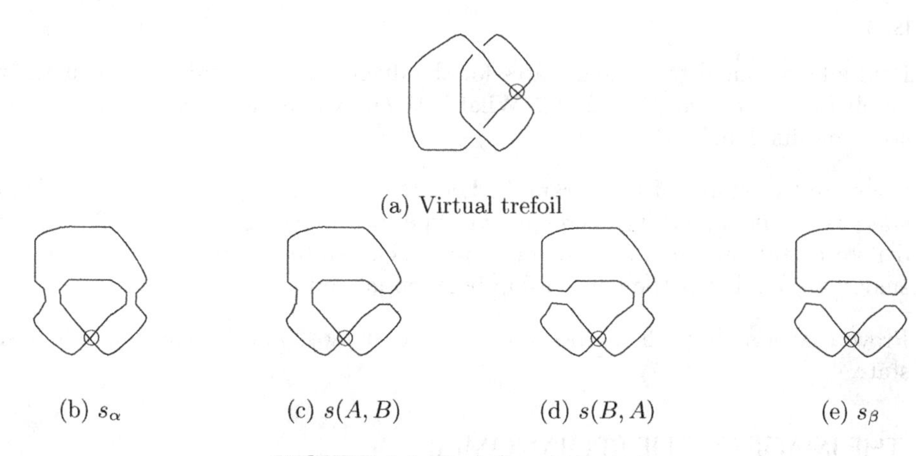

(a) Virtual trefoil

(b) s_α (c) $s(A, B)$ (d) $s(B, A)$ (e) s_β

FIGURE 6.6: Trefoil and its states

We compute the number of closed curves in each state using Figure 6.6:

$$|s_\alpha| = 1, \qquad |s(A, B)| = 1,$$
$$|s(B, A)| = 1, \qquad |s_\beta| = 2. \tag{6.13}$$

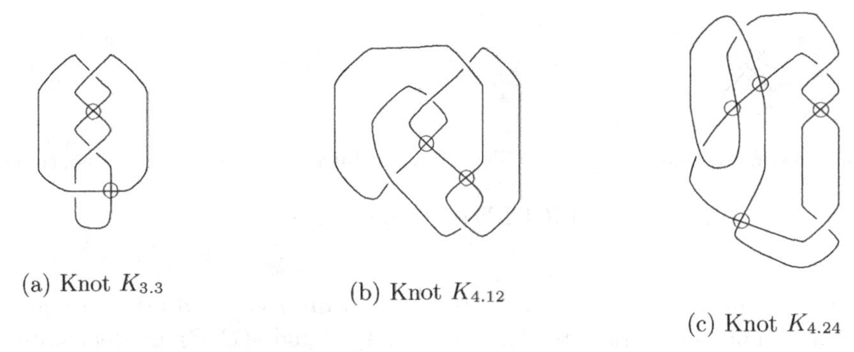

(a) Knot $K_{3.3}$

(b) Knot $K_{4.12}$

(c) Knot $K_{4.24}$

FIGURE 6.7: Find the state with the most components

Next, we use this information to compute our weighted sum.

$$\begin{aligned}\langle K\rangle &= \sum_{s\in S} A^{\alpha_s-\beta_s} d^{|s|-1}\\ &= A^2 d^{|S_\alpha|-1} + 1d^{|S(A,B)|-1} + 1d^{|S(B,A)|-1} + A^{-2}d^{|S_\beta|-1}\\ &= A^2 d^{1-1} + 1d^{1-1} + 1d^{1-1} + A^{-2}d^{2-1}\\ &= A^2 + 1 + 1 + A^{-2}(-A^2 - A^{-2})\\ &= A^2 + 1 - A^{-2}.\end{aligned}$$

Exercises

1. Find all states and their dual states for the diagrams in Figure 6.7. Count the total number of components in s and $\overline{s}$. What is the maximum number of components in a state and its dual?

2. Compare two states of a diagram D. Let $s(x_1, x_2, \ldots, x_{i-1}, A, x_{i+1}, \ldots x_n)$ denote a state where the smoothing of crossing i is type A. Let $s'(y_1, y_2, \ldots, y_{i-1}, B, y_{i+1}, \ldots y_n)$ denote a state where $x_j = y_j$ for $j \neq i$. (The states s and s' differ at only one smoothing.) What is the relationship between $|s|$ and $|s'|$?

3. Prove that a state with n crossings can have at most $n+1$ components in a single state.

6.3 THE IMAGE OF THE f-POLYNOMIAL

The properties of the f-polynomial are investigated in this section. First, we analyze the behavior of the bracket polynomial under the symmetries: flip, mirror, switch, and inverse. Recall the knots related by symmetries: K, K^I, K^F, K^S, and K^M. Crossing invariants could not differentiate the pairs (K^M, K^S) and (K, K^F). Since $K^F = (K^M)^S$, if the f-polynomial can distinguishes the link diagrams K^M and K^S, then the f-polynomial will also distinguish the second pair of link diagrams, K and K^F.

Theorem 6.3. *For all virtual link diagrams K,*

$$f_K(A) = f_{K^I}(A), \tag{6.14}$$
$$f_K(A^{-1}) = f_{K^M}(A), \tag{6.15}$$
$$f_K(A) = f_{K^F}(A), \tag{6.16}$$
$$f_K(A^{-1}) = f_{K^S}(A). \tag{6.17}$$

Proof. Apply the state sum definition to K to prove Equations 6.14 and 6.15. If K has n classical crossings, then any link diagram obtained by symmetry has the same number of classical crossings.

Notice that $w(K) = w(K^I)$. Next, comparing the state $s(x_1, x_2, \ldots x_n)$ of K to the corresponding state $s_I(x_1, x_2, \ldots x_n)$ of K^I, we see that they are identical diagrams. Hence, $|s| = |s_I|$, $\alpha(s) = \alpha(s_I)$, and $\beta(s) = \beta(s_I)$. As a result, $f_K(A) = f_{K^I}(A)$.

Consider the diagrams K and K^M. Note that $w(K) = -w(K^M)$. Define

$$\overline{x_i} = \begin{cases} A & \text{if } x_i = B, \\ B & \text{if } x_i = A. \end{cases} \tag{6.18}$$

Now, the state $s(x_1, x_2, \ldots, x_n)$ of K is the mirror image of the state $s_M(\overline{x_1}, \overline{x_2}, \ldots \overline{x_n})$ of K^M. Then $|s(x_1, x_2, \ldots, x_n)| = |s_M(\overline{x_1}, \overline{x_2}, \ldots \overline{x_n})|$. Additionally, $\alpha(s) = \beta(s_M)$ and $\beta(s) = \alpha(s_M)$. Then

$$f_{K^M}(A) = (-A^{-3})^{w(K^M)} \sum_{s_M \in S_M} A^{\alpha(s_M)-\beta(s_M)} d^{|s_M|-1} \tag{6.19}$$
$$= (-A^{-3})^{-w(K)} \sum_{s \in S} A^{\beta(s)-\alpha(s)} d^{|s|-1} \tag{6.20}$$
$$= f_K(A^{-1}). \tag{6.21}$$

□

The f-polynomial can not detect the difference between the pairs: K and K^F or K^M and K^S. A natural question is: can we improve the bracket polynomial to detect these symmetries? Or do we need an entirely new invariant?

Virtualization involves taking a single classical crossing and flanking the crossing with two virtual crossings. There are two possible methods of virtualization, depending on whether one wants to preserve the over passing strand or the sign of the crossing. In terms of $2-2$ tangles, the two methods of virtualization are:

Way virtualization: $(1) \to (i, 1, i), (-1) \to (i, -1, i)$

Sign virtualization: $(1) \to (i, -1, i), (-1) \to (i, 1, i)$.

Diagrammatically, the two types of virtualization are illustrated in Figure 6.8. **Way vir-**

(a) Way virtualization

(b) Sign virtualization

FIGURE 6.8: Virtualization moves

tualization changes the overpassing strand and preserves the sign of the crossing. Way virtualization is also sometimes referred to as the **z-move.** **Sign virtualization** changes the sign of the center crossing. When reading papers that discuss virtualization, be careful to distinguish what the authors mean when they use the term virtualization. We apply the skein relation to the result of way virtualization in Equation 6.22.

$$\langle \quad \rangle = A\langle \quad \rangle + A^{-1}\langle \quad \rangle$$
$$= \langle \quad \rangle. \qquad (6.22)$$

In the case of way virtualization, the bracket polynomial only "sees" a single classical crossing. We can exploit this fact to construct virtual knot diagrams K with $f_K(A) = 1$. See Figure 6.9 for an example. This technique was introduced by Louis Kauffman in his 1999 paper on virtual knot theory. Begin with a non-trivial classical knot diagram K. Select a set of classical crossings S such that switching overpassing information in the diagram turns K into the unknot. Apply way virtualization to the crossings in S. Then the resulting diagram has f-polynomial 1. (These diagrams are also not equivalent to the unknot, but proving this fact requires a different invariant.)

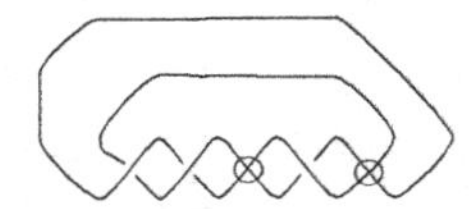

FIGURE 6.9: K with $f_K(A) = 1$

The bracket polynomial of the connected sum of $K_1 \sharp K_2$ can be computed as the product of the bracket polynomial of the two diagrams, K_1 and K_2. Using the skein relation, we smooth some of the crossings in $K_1 \sharp K_2$, and then express $\langle K_1 \sharp K_2 \rangle$ as the product of $\langle K_1 \rangle$ and $\langle K_2 \rangle$. Remember that the bracket polynomial of the disjoint union of the virtual links K and the unknot is given by the equation: $\langle K \sqcup U \rangle = d\langle K \rangle$. Using the skein relation and the formula for $\langle K \sqcup U \rangle$, we prove the following theorem.

Theorem 6.4. *For all knot diagrams* K_1 *and* K_2, $\langle K_1 \sharp K_2 \rangle = \langle K_1 \rangle \langle K_2 \rangle$.

Proof. Let K_1 and K_2 denote link diagrams. Denote the set of states of K_2 as $\mathrm{S}(K_2)$ and let c_s denote the coefficients of the Laurent polynomial $\langle K_2 \rangle$. Then

$$\langle K_2 \rangle = \sum_{s \in S(K_2)} c_s d^{|s|-1}. \qquad (6.23)$$

Using the skein relation, expand all classical crossings in the diagram K_2 and ignore the crossings in K_1. In a partially expanded state of $K_1 \sharp K_2$, where all the classical crossings in K_2 have been expanded, one of the components of $s \in \mathrm{S}(K_2)$ forms a connected sum with K_1. Using Equation 6.23 with the partially expanded state, evaluate the states associated with K_2,

$$\langle K_1 \sharp K_2 \rangle = \sum_{s \in S(K_2)} \langle K_1 \rangle c_s d^{|s|-1} \tag{6.24}$$

$$= \langle K_1 \rangle \sum_{s \in S(K_2)} C_s d^{|s|-1} \tag{6.25}$$

$$= \langle K_1 \rangle \langle K_2 \rangle. \tag{6.26}$$

□

As a corollary, the f-polynomial of $K_1 \sharp K_2$ can be computed.

Corollary 6.5. *For all virtual knots K_1 and K_2,*

$$f_{K_1 \sharp K_2}(A) = f_{K_1}(A) f_{K_2}(A). \tag{6.27}$$

Kishino's knot (shown in Figure 6.10) is the connect sum of two unknots and is not distinguished from the unknot by the f-polynomial. For this reason, Kishino's knot is frequently used to test the robustness of a link invariant.

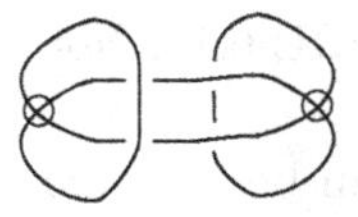

FIGURE 6.10: Kishino's knot, K_{Kish}

We are interested in the following features of the f-polynomial.

1. **Given a virtual link diagram, what are the maximum and minimum possible degrees of the bracket polynomial?** Another way to phrase this question is to ask what are possible spans of the f-polynomial. Since we can compute the polynomial using the skein relation, does the span only depend on the crossing number of the virtual link?

2. **What are the possible coefficients of the bracket polynomial?** We could rephrase this question as follows: are any restrictions placed on the coefficients? Perhaps the coefficients are bounded by some factor of the crossing number.

3. **What knots are contained in the pre-image of a polynomial (and how are they related)?** The answer to this question would give us an idea of how good the f-polynomial is at differentiating knots and links. For example, if we knew only three link types map to a distinct Laurent polynomial, we could focus on methods of differentiating those types. Currently, we have the following conjecture about classical knots:

   Conjecture 6.1. *Let K be a classical knot that is not equivalent to the unknot. Then $f_K(A) \neq 1$.*

   Many researchers have attempted to prove this conjecture.

4. **What topological information does the f-polynomial give us about the knot? Or is some of the information solely based on the diagram structure?** We don't yet know much about the topological information that is indicated by the link diagrams, so this question is difficult for us to consider right now. In classical knot theory, link diagrams are essentially recipes for three dimensional objects. So an invariant of a classical knot should (and many do) give us information about the topological structure of the three dimensional object. How this information extends to non-classical knots is unknown.

We spend the majority of Part II reflecting on these questions. We begin with an examination of the span of the bracket polynomial.

We introduce some definitions. A knot diagram is **connected**. A link diagram with more than one component is **connected** if every component intersects at least one other component in a classical crossing. A link diagram is **reduced** if it does not contain any crossings of the form shown in Figure 6.11. Notice that if T and T' contain only classical

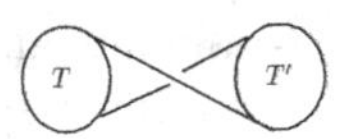

FIGURE 6.11: Not reduced

crossings then the central crossing can be removed by a sequence of classical Reidemeister moves. The crossing may not be removable if both T and T' contain virtual crossings. A virtual knot diagram is **prime** if the knot diagram cannot be expressed as a connected sum.

For classical knots, we have the following theorem due to Kauffman, Murasurgi, and Thistlethwaite:

Theorem 6.6. *Let D be a connected diagram of a classical link K with n classical crossings. Then*

1. $\text{Span}(f_K(A)) \leq 4n$.
2. *If D is a reduced, alternating diagram then* $\text{Span}(f_K(A)) = 4n$.
3. *If D is a non-alternating and prime diagram then* $\text{Span}(f_K(A)) < 4n$.

This theorem establishes a lower bound on the classical crossing number of a knot diagram based on the span of its f-polynomial. If we have a diagram D with n crossings and $\text{Span}(f_D(A)) = 4n$ then $\mathsf{c}(D) = n$.

We can also determine if a link does not have a reduced, alternating diagram. Suppose that a classical link K has a reduced, alternating diagram and $\text{Span} f_K(A) = 4n$. Then there is classical link diagram D with n classical crossings that is equivalent to K. If no diagram with n classical crossings is equivalent to K, then K does not have a reduced, alternating diagram.

For a classical link diagram K, we can also find a lower bound for $\mathsf{c}(K)$ using this theorem. If K is a classical knot and $20 \leq \text{Span}(f_K(A))$ then $20 \leq 4n$ where n is the number of classical crossings in any diagram of K. Any diagram of K has 5 or more classical crossings.

This theorem is not true for virtual knots. The proof of the theorem relies on the fact that the maximum number of components in a state and its dual is $n + 2$ for a classical knot diagram with n crossings. We would like a version of this theorem for virtual knots because it is so useful.

We construct a weak version of Theorem 6.6 for virtual knot diagrams. Given a connected link diagram K, let $n = \mathsf{c}(K)$. Let D be a diagram equivalent to K with n crossings. Since $K \sim D$ then $\pm 1 A^i \langle K \rangle = \langle D \rangle$ for some integer i. For the diagram D, denote a state of D as $s(D)$. First, bound the maximum number of curves in any state $s(D)$.

Lemma 6.7. *For all connected virtual link diagrams D with n classical crossings, $|s(D)| \leq n$.*

Proof. Let D be a connected virtual link diagram with 1 crossing. Then a state of D has at most two components. Now, suppose that a connected link diagram D has $n-1$ crossings. By the induction hypothesis, $s(D)$ has at most n components ($|s(D)| \leq n$).

Let D be a diagram with n classical crossings. We select a classical crossing c and smooth the crossing horizontally to form a diagram D' with $n-1$ classical crossings. Mark the position of crossing c on $s(D')$. For any state $s(D')$, $|s(D')| \leq n$ since D' has only $n-1$ crossings. Notice that every state of D' corresponds to a state of D.

We construct the remaining states of D from the state $s(D')$. We re-smooth the curves in $s(D')$ at the position where c was located. If the markings indicating the position of c connect the same component then $|s(D)| = |S(D)| + 1$ or $|s(D)| = |S(D')|$. If the markings from c connect two different components, then $|s(D)| = |s(D)| - 1$. This argument holds for all states of D', and we see that

$$|s(D)| \leq |s(D')| + 1 \leq n + 1. \tag{6.28}$$

□

Theorem 6.8. *For a connected virtual link diagram D with n classical crossings, the maximum (respectively minimum) degree of $\langle D \rangle$ is $3n$ (respectively $-(3n)$). Consequently,* $\mathrm{Span}(f_D(A)) \leq 6n$.

Proof. Let D be a connected virtual link diagram with n classical crossings. The all A state s_α has a coefficient A^n in the state sum. The value of $|s_\alpha| \leq n+1$. The state s_α contributes the polynomial $A^n d^i$ where $i \leq n$ to the state sum. The maximum possible exponent in this polynomial is $3n$. Hence, the maximum possible exponent of a term in the f-polynomial is $3n$. A similar argument applies to the all B state, which has a minimum possible exponent of $-3n$. Then $\mathrm{Span}(f_D(A)) \leq 6n$. □

The argument above bounds the span of the polynomial by $6n$. But we can improve this bound by examining the relationship between a state s and its dual state $\overline{s}$. In the next chapter, we introduce abstract link diagrams and use these to further refine our bound on the span of the bracket polynomial.

Exercises

1. Compute the f-polynomial for the connected sum of the knots in Figure 6.12a and Figure 6.12b.

2. Compute the f-polynomial for the connected sum of the knot diagrams in Figure 6.12c and Figure 6.12b.

3. Prove Equation 6.16, $f_K(A) = f_{K^F}(A)$.

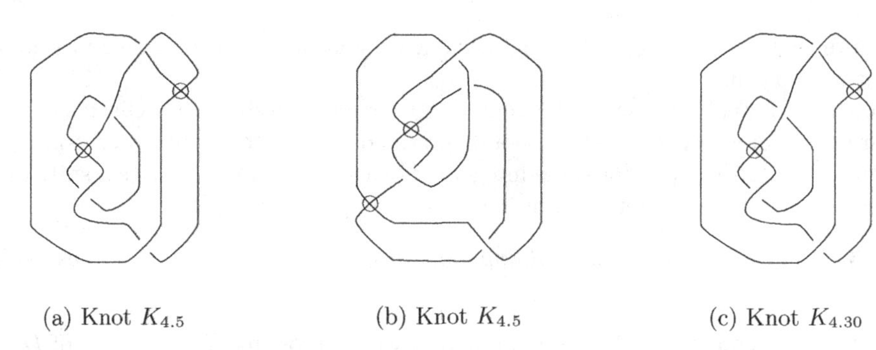

(a) Knot $K_{4.5}$ (b) Knot $K_{4.5}$ (c) Knot $K_{4.30}$

FIGURE 6.12: Knot diagrams for Exercises

4. Prove Equation 6.17, $f_K(A) = f_{K^S}(A^{-1})$.
5. Prove the bracket polynomial is not invariant under sign virtualization.
6. Construct a family of link diagrams K_n such that $f_{K_n}(A) = 1$.
7. For the knot diagram K shown in Figure 6.9, compute $f_K(A)$ and $\mathcal{C}_2(K)$.
8. Prove Corollary 6.5.
9. For connected link diagrams K_1 and K_2, compute $\langle K_1 \sqcup K_2 \rangle$. (The bracket polynomial of two disjoint, connected link diagrams.)
10. Compute the f-polynomial of the link diagrams shown in Figure 6.13 in terms of $\langle K_1 \rangle$ and $\langle K_2 \rangle$.

6.4 OPEN PROBLEMS AND PROJECTS

Open problems

1. Does the f-polynomial distinguish all non-trivial classical knots from the unknot? This is a problem from classical knot theory that is unsolved.
2. Do virtual knots have a minimal crossing number in particular projections? For example, classical knots with reduced alternating projections have a minimal number of classical crossings. Is there a similar conjecture for non-classical virtual links and knots?
3. Is there a virtual knot diagram K with trivial f-polynomial that is not detected by constructing a satellite of K with the triple banded unknot as the pattern?

Projects

1. Select a family of rational virtual knots. Compute the bracket polynomial for this family.
2. Construct a family of virtual link diagrams that are "almost" connected sums and compute a formula for the f-polynomial of each element in the family.
3. Define an operation similar to virtualization and apply the skein relation to determine how the bracket polynomial changes.

4. Find two virtual knot diagrams with the same f-polynomial that are not related by a sequence of virtualization moves and the Reidemeister moves. Note that a knot diagram K_1 that is related by virtualization by the unknot is not a solution to this problem.

5. We compute the f-polynomial of knot diagrams that are formed using a modification of connected sum as shown in Figure 6.13

(a) Two crossings (b) Two crossings

FIGURE 6.13: Compute the f-polynomial

Further Reading

1. Toshimasa Kishino and Shin Satoh. A note on non-classical virtual knots. *J. Knot Theory Ramifications*, 13(7):845–856, November 2004

2. L. H. Kauffman. New invariants in the theory of knots. *Amer. Math. Monthly*, 95(3):195–242, 1988

3. Heather A. Dye. Virtual knots undetected by 1-and 2-strand bracket polynomials. *Topology and its Applications*, 153:141–160, 2005

4. P. Freyd, D. Yetter, J. Hoste, W. B. R. Lickorish, K. Millett, and A. Ocneanu. A new polynomial invariant of knots and links. *Bull. Amer. Math. Soc. (N.S.)*, 12(2):239–246, April 1985

5. Morwen B. Thistlethwaite. A spanning tree expansion of the Jones polynomial. *Topology*, 26(3):297–309, 1987

CHAPTER 7

Surfaces

To extend the proof bounding the span of the bracket polynomial to virtual links, we need to know about two-dimensional surfaces. Fortunately, compact, oriented, two-dimensional surfaces without boundary are completely classified. There is a complete list of oriented two-dimensional surfaces with identifying characteristics. In this section, we learn about two-dimensional surfaces, the Euler characteristic, and genus. Compact, closed, two-dimensional, oriented surfaces are classified by calculating the genus.

7.1 SURFACES

A **two-dimensional surface** S is a set of points in an ambient space (such as $\mathbb{R}^3$ or $\mathbb{R}^4$). Informally, we can describe a two-dimensional surface S as follows. For a point p in S, the surface "looks like" $\mathbb{R}^2$ near p. Visualize the surfaces as being constructed from a thin, flexible rubber sheet.

We give a technical definition of a two-dimensional surface in an ambient Euclidean space such as $\mathbb{R}^3$ or $\mathbb{R}^4$. For two sets X and Y in an ambient Euclidean space, a map $f: X \to Y$ is a **homeomorphism** if f is injective and surjective, and both f and f^{-1} are continuous maps. Now, let B denote the set of points (x, y) in $\mathbb{R}^2$ such that $x^2 + y^2 < 1$. For each p in S, fix $\delta > 0$. Then a **neighborhood of** p is the set of points in S within distance δ from p. We denote the neighborhood of p as N_p. Then for all p in S, there is a homeomorphism from B to N_p. We introduce some examples of two-dimensional surfaces.

A **sphere** (shown in Figure 7.1a) meets this criterion as does the **torus** (shown in Figure 7.1b). Note that when mathematicians say "sphere", they are referring to the surface of a

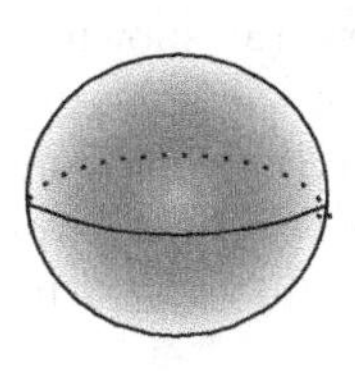

(a) Sphere

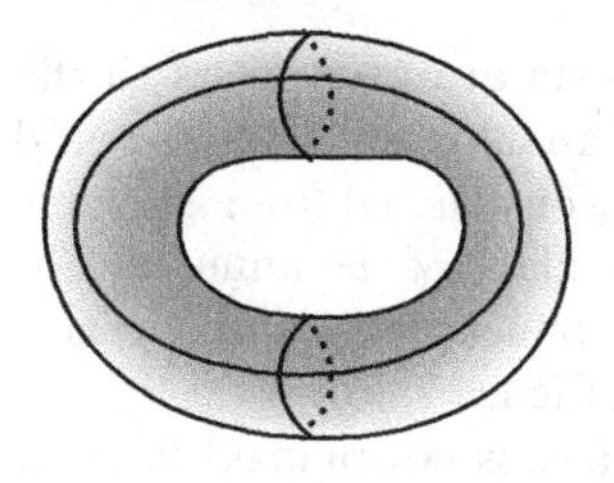

(b) Torus

FIGURE 7.1: Surfaces

ball. To clarify, the unit sphere in $\mathbb{R}^3$ is the set of points (x, y, z) that satisfy

$$x^2 + y^2 + z^2 = 1. \tag{7.1}$$

The unit ball (which contains the unit sphere) is the set of points (x, y, z) that satisfy

$$x^2 + y^2 + z^2 \leq 1. \tag{7.2}$$

The **torus** is the surface of a donut shape.

The sphere and torus are closed and compact. The technical definition of compact is beyond the scope of this book (it involves finite open subcovers). However, since the ambient space is $\mathbb{R}^n$, the important implication of the term compact is that the surface is bounded. There exists a finite number N such that for any pair of points p_1 and p_2 in the surface S, the distance between p_1 and p_2 is less than N.

A surface with boundary almost satisfies the condition that a surface "looks like" $\mathbb{R}^2$ at every point p. The boundary of a surface S is a collection of closed curves that we will denote as ∂S. The half disk is the set of points that satisfies the equation $\{(x, y) | x^2 + y^2 \leq 1; x \geq 0\}$. For a point p on the boundary of the surface S, the set N_p is homeomorphic to the half disk. A **closed** two-dimensional surface has no boundary components.

When we discuss surfaces with boundary, we often refer to a **puncture** or **punctured surface**. A puncture in a surface is formed by selecting a point p and then removing all points within some distance δ from the surface. The set of points in the surface at distance δ remains in the surface. This set of points forms a boundary curve. A sphere with a single puncture is shown in Figure 7.2a. The punctured sphere has a single closed curve as a boundary component. Since we visualize our surfaces as being constructed from rubber sheets, we can deform the punctured sphere by enlarging the puncture until the boundary of the puncture forms the edge of a bowl shape as shown in Figure 7.2b. Then we flatten the surface to form a disk as shown in Figure 7.2c.

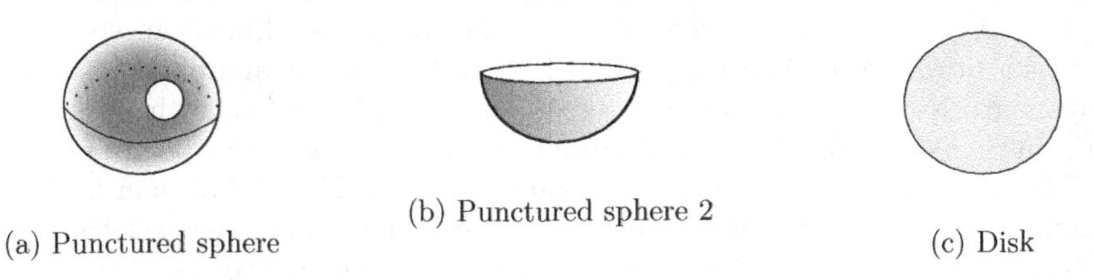

(a) Punctured sphere (b) Punctured sphere 2 (c) Disk

FIGURE 7.2: Deforming a punctured sphere

The term **annulus** refers to the twice punctured sphere (as shown in Figure 7.3a). To form the annulus, we remove two disks from the surface of the sphere. Since we view the sphere as constructed from a flexible rubber sheet, we stretch out one punctured to form the outer boundary of the annulus. The other puncture forms the inner boundary. The sphere with four punctures is a disk with three holes; the fourth puncture forms the boundary of the disk (Figure 7.3b).

A surface is **connected** if given any pair of points p_1 and p_2 in the surface, the points are connected by a curve contained in the surface. (If you placed an ant on the surface at point p_1 then it could crawl to point p_2 without leaving the surface.)

In our descriptions, we visualized the surfaces as constructed from a thin, flexible rubber material. We describe and give a technical definition of this notion of flexibility. Two equivalent surfaces S_1 and S_2 are related by a finite sequence of ambient isotopies and homeomorphisms. Two surfaces S_1 and S_2 in the ambient space X are **ambient isotopic**

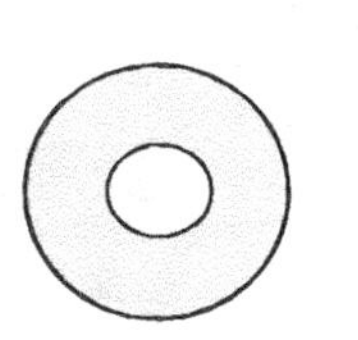

(a) Annulus

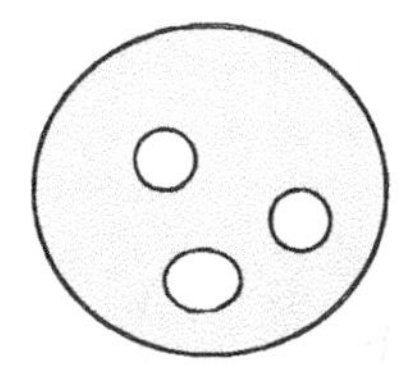

(b) Four times punctured sphere

FIGURE 7.3: Punctured spheres

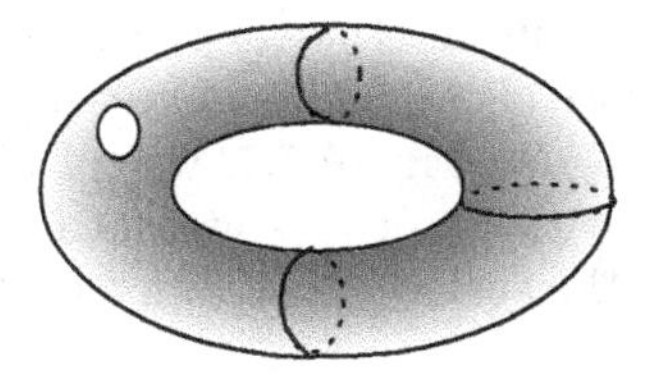

(a) Punctured torus

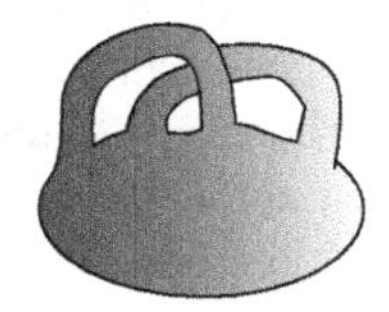

(b) Punctured torus—view 2

FIGURE 7.4: Torus with one puncture

if the surface S_1 can be deformed into a second surface S_2. Technically, this means that there is a map $F : X \times [0, 1] \to X$ where

1. Fixing an element of $[0, 1]$, we define $f_t : X \to X$ where $f_t(x) = f(x, t)$. For all t in $[0, 1]$, the map $f_t(x)$ is a homeomorphism.
2. The map $f_0(x)$ is the identify map.
3. The map $f_1(x)$ sends S_1 to S_2, $f_1(S_1) = S_2$.

Two pairs of isotopic surfaces are shown in Figures 7.5 and 7.6. The ambient space is important. For example, the unknot is not ambient isotopic to the trefoil in $\mathbb{R}^3$ but the unknot is ambient isotopic to the trefoil in $\mathbb{R}^4$.

However, equivalent surfaces S_1 and S_2 can be placed in the ambient space in a way that can't be accounted for by ambient isotopy. Let C be a simple closed curve in the surface

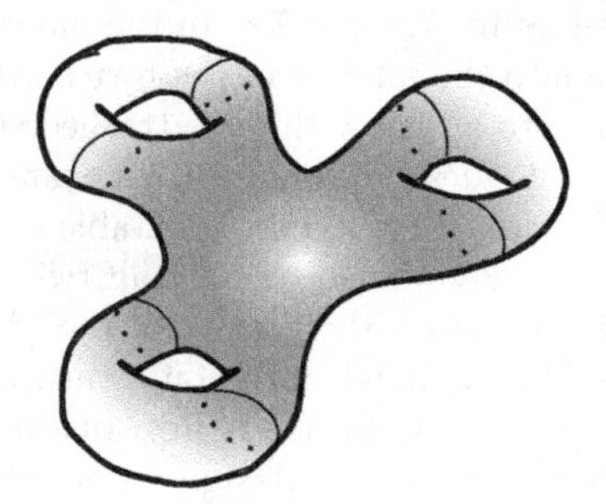

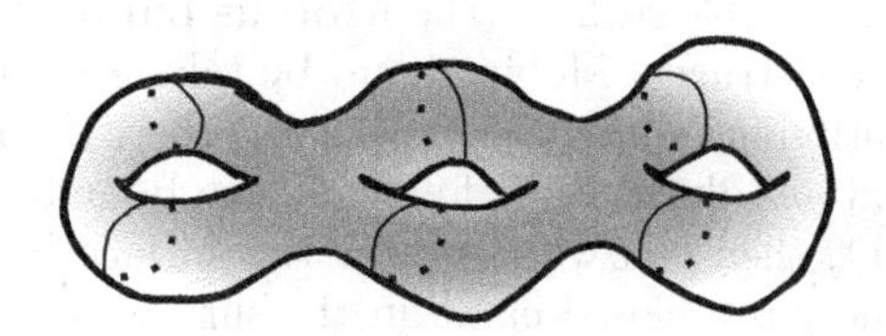

FIGURE 7.5: Example 1: A pair of isotopic surfaces

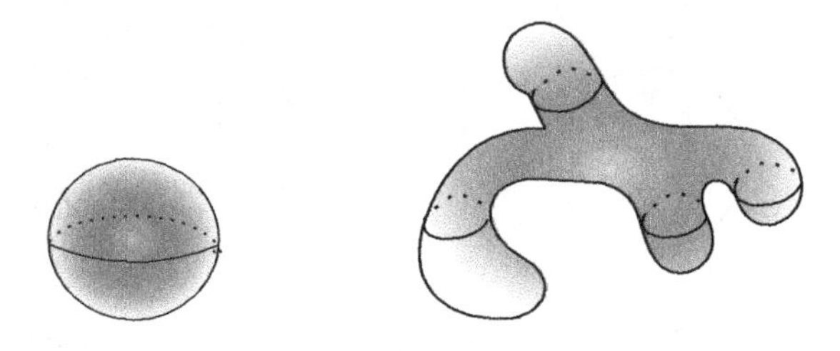

FIGURE 7.6: Example 2: A pair of isotopic surfaces

S_1 such that C does not bound a disk in S_1. Then we can cut the surface S_1 along the curve C, move the cut surface, and then reglue the surface S_2 along the curve. If we follow this process then there is a homeomorphism $f : S_1 \to S_2$. In Figure 7.7, we cut along the surface along the indicated curve producing a surface with two tubes attached. We then push the tubes inside the surface and reglue. Although the surfaces look different, they are homeomorphic..

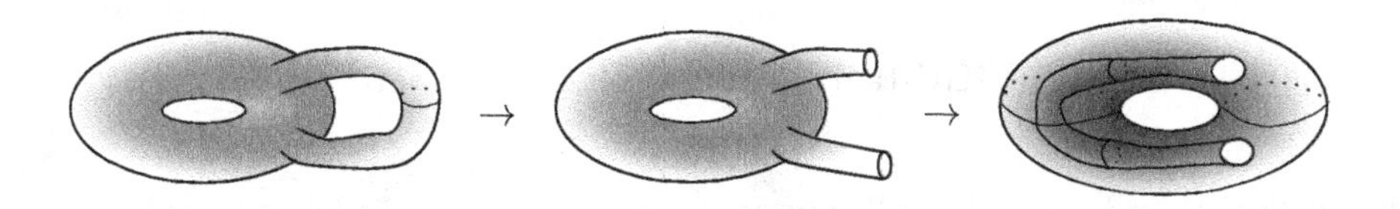

FIGURE 7.7: Homeomorphism

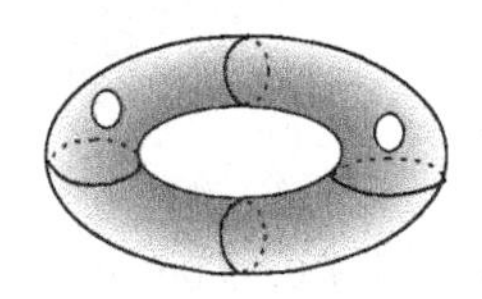

(a) Twice punctured torus

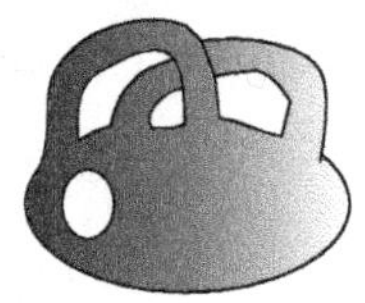

(b) Twice punctured torus—view 2

FIGURE 7.8: Tori with 2 puncture

Punctured tori (the plural of torus) are shown in Figures 7.4 and 7.8. In the punctured torus (see Figure 7.8a), the punctures allow us to see into the interior of the torus. We can stretch the puncture and deform the punctured torus into a disk with two attached strips.

A surface is **orientable** if the surface has two distinct sides. The sphere is an example of an orientable surface. The **Mobius band** (shown in Figure 7.9) is not an orientable surface. We construct a Mobius band by taking a strip of paper, giving the strip a half twist, and identifying (gluing) the edges together. If you select a point on the surface of the Mobius band and draw a line following the boundary of the Mobius band until you return to the point, then you will see that the surface has only one side. A surface is **non-orientable** if there is a closed curve in the surface with the following property. Assign an orientation to the curve and choose a point on the curve. Attach a normal vector at the point on the curve. Remember, a normal vector is perpendicular to the tangent vector points away from the surface. Push a copy of the normal vector along the curve until you return to the point.

If the two copies of the normal vector are pointing in opposite directions, then the surface is non-orientable.

We identify or glue the boundaries of two-dimensional surfaces with boundary to create new surfaces. The **projective plane** is a surface formed by gluing the boundary of a disk to the boundary of the Mobius band. The **Klein bottle** (Figure 7.9b) is a non-orientable surface obtained by gluing together two Mobius bands along the boundary. In our sketch of the Klein bottle, the neck of the bottle appears to pass through the body of the bottle. The Klein bottle cannot be contained in the ambient space $\mathbb{R}^3$.

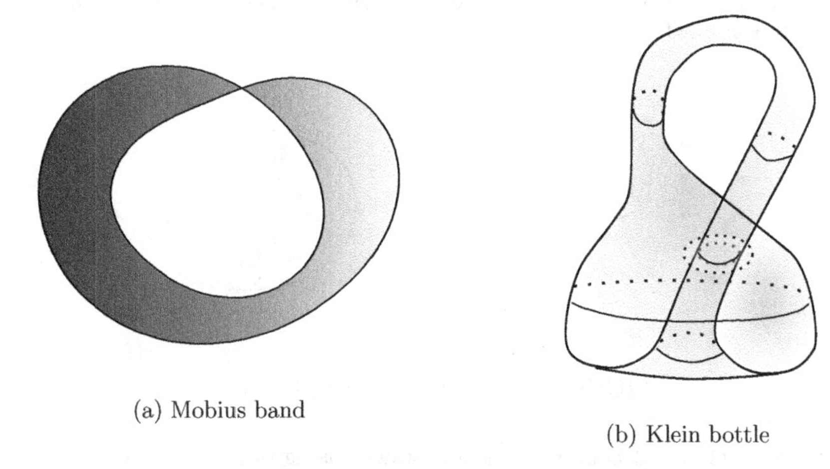

(a) Mobius band

(b) Klein bottle

FIGURE 7.9: Non-orientable surfaces

We use two-dimensional, orientable surfaces in the next several chapters. Our immediate goal is to identify and classify all possible connected, oriented, two-dimensional surfaces and surfaces with boundary.

Another characteristic of surfaces is the genus. Loosely speaking, the number of "donut holes" in a surface is the **genus** of the surface S, denoted as $g(S)$. A surface of genus g contains g non-intersecting, non-separating, simple closed curves that do not bound a disk. Cutting along the entire set of curves produces a punctured disk. We see from the diagram in Figure 7.1 that the genus of a sphere is zero since every simple closed curve bounds a disk. The torus contains only one non-intersecting closed curve that does not bound a disk and has genus one.

Puncturing a surface does not change the genus. Consider a torus with one puncture. We can place two non-intersecting curves that do not bound a disk in the punctured torus. However, one of these curves must contain the puncture and separates the surface into two components. The genus of a torus with one puncture is one. Technically, we define the genus of a surface as follows. Draw a simple closed curve that does not bound a disk on the surface. Cut the surface along the curve to produce a connected surface with boundary. Repeat this process until cutting along any simple closed curve (that does not bound a disk) in the surface produces two disconnected surfaces. The number of curves n required is the genus of the surface.

Cellular decomposition

To identify (or construct) a two dimensional surface, our first goal is to divide the surface into standardized components. These components are: points, line segments, and (possibly deformed) disks. We use the term **cell** to refer to each component. A 0-cell is a point, a 1-cell is a line segment, and a 2-cell is a disk. Our goal is to build a "wire frame" for the surface to which we can attach sheets of material and form the surface. This construction is called a **cell complex**. Conversely, we can decompose a surface S into cells forming a cell complex for S. This is called a **cellular decomposition**.

The torus is constructed using one 0-cell, two 1-cells, and one 2-cell in Figure 7.10. Begin with a single point. Next, the two 1-cells are attached to the 0-cell, forming two loops. (Notice that as the 1-cells wrap around the surface we mark the arcs on the back of the surface as dotted lines.) Finally the 2-cell is attached. The lines shown in the shaded torus are exactly the 1-cells.

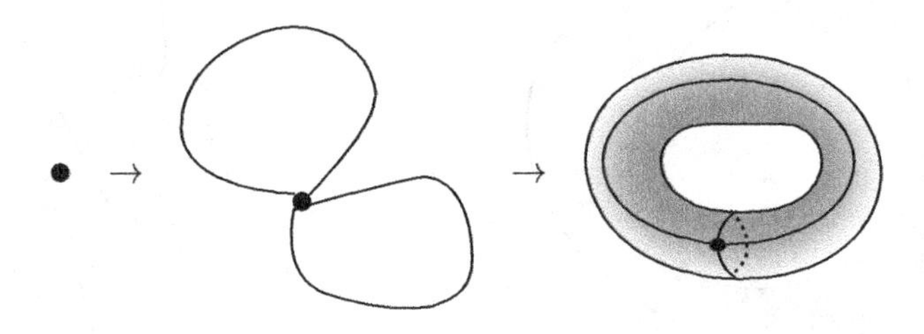

FIGURE 7.10: Constructing a torus

We construct a cellular decomposition of a sphere using one 0-cell, one 1-cell, and two 2-cells. We attach the 1-cell to the 0-cell forming a loop. The 2-cells are then glued to either side of the loop.

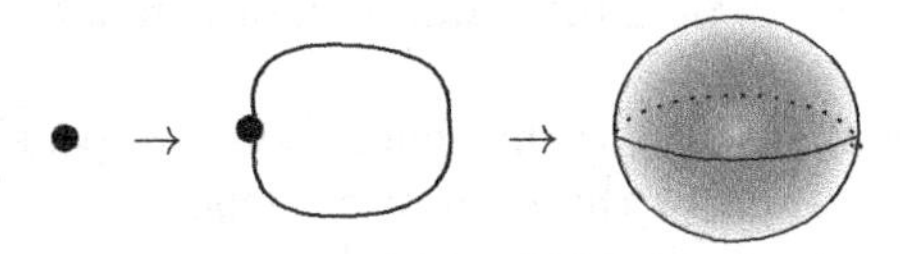

FIGURE 7.11: Constructing a sphere

A cellular decomposition of a connected, 2-dimensional surface satisfies the following rules.

1. The endpoints of each 1-cell are 0-cells.
2. Any pair of 0-cells is connected by a sequence of 1-cells.
3. Every 1-cell is adjacent to at most two 2-cells.
4. The boundary of every 2-cell is a finite collection of 1-cells.

There are many different cellular decompositions of a surface. The main restriction is that in a cellular decomposition of a surface, a simple closed curve that does not intersect the 1-cells (of the decomposition) bounds a disk. We show alternate decompositions of the torus in Figure 7.12, along with decompositions of a surface with boundary and another surface.

New two-dimensional surfaces can be formed from two existing surfaces, S_1 and S_2.

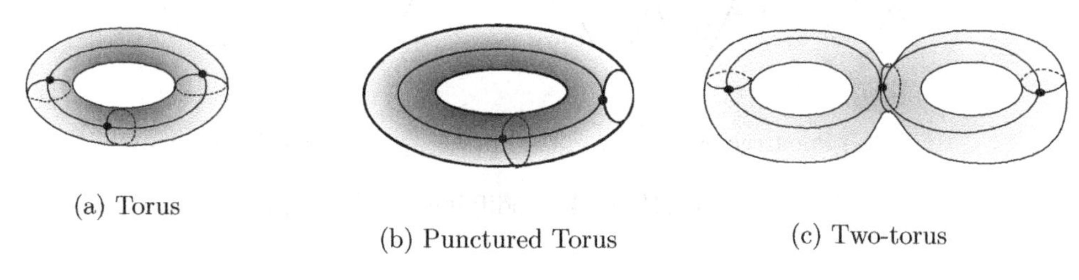

(a) Torus

(b) Punctured Torus

(c) Two-torus

FIGURE 7.12: Cellular decompositions

Puncture the two surfaces as shown in Figure 7.13. Then, we glue the boundaries of these punctures together to form a new surface, the **connected sum** of the two surfaces. The connected sum of two surfaces, S_1 and S_2, is denoted $S_1 \sharp S_2$. The example in Figure 7.13 involves two tori and resulting surface is denoted $T \sharp T$.

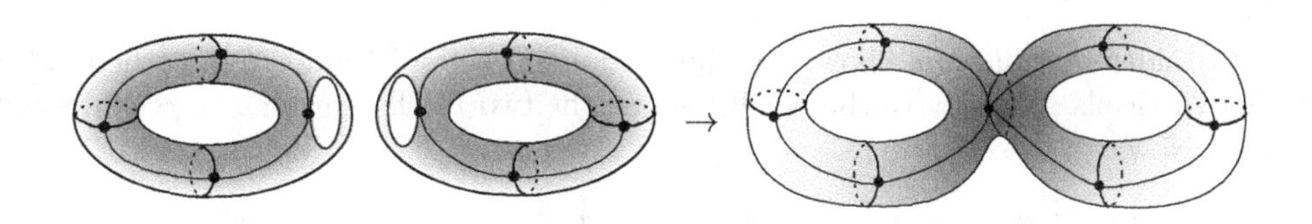

FIGURE 7.13: Forming a connected sum

We now describe all possible closed, compact, 2-dimensional surfaces. The description is called the classification theorem.

Theorem 7.1. *For all connected, closed, compact, 2-dimensional surfaces S, S is one of the following surfaces.*

1. *S is a sphere.*
2. *S is the connected sum of n tori.*
3. *S is the connected sum of p projective planes.*

Genus and the Euler characteristic

Given a cellular decomposition of a surface F, we let C_i denote the collection of i-cells in the cellular decomposition of F. We let $|C_i|$ denote the number of i-cells in the cellular decomposition. Then the **Euler characteristic** of the surface F is defined as

$$\chi(F) = \sum_{i=0}^{2} (-1)^i |C_i|. \tag{7.3}$$

We've defined the Euler characteristic of a surface, not the Euler characteristic of a surface and a fixed cellular decomposition. We prove that Euler characteristic is well defined; that it does not depend on the particular cellular decomposition. We have two lemmas (technical theorems) that demonstrate how we can modify the cellular decomposition and still have a cellular decomposition. These changes do not affect the Euler characteristic.

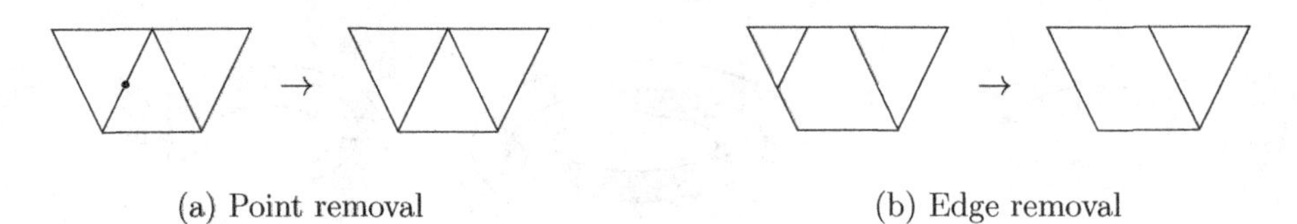

(a) Point removal (b) Edge removal

FIGURE 7.14: Modifying a cell complex

Lemma 7.2 (0-cell). *Let $C = \{C_2, C_1, C_0\}$ be a cellular decomposition of a surface F. If the cellular decomposition has two 1-cells meeting at a 0-cell as shown in Figure 7.14a, then we can modify C_1 and C_0 to obtain new cellular decomposition: $C' = \{C'_2, C'_1, C'_0\}$. Further,*

$$\sum_{i=0}^{2}(-1)^i|C_i| = \sum_{i=0}^{2}(-1)^i|C'_i|. \tag{7.4}$$

This process can also be reversed to add a 0-cell.

Proof. Consider the diagram shown on the right hand side of Figure 7.14a where a 0-cell bounds two 1-cells. We remove the 0-cell, joining the two 1-cells, forming the cell decomposition C'. Now

$$|C_2| = |C'_2|, \qquad |C_1| = |C'_1| + 1, \qquad |C_0| = |C'_0| + 1. \tag{7.5}$$

Summing, we see that both cell decompositions result in the same Euler characteristic. □

Next, we can remove or add a 1-cell to a cell complex under certain conditions.

Lemma 7.3 (1-cell). *Let $C = \{C_2, C_1, C_0\}$ be a cellular decomposition of a surface F. If we observe the configuration shown in Figure 7.14b, then we can modify C_2, C_1, and C_0 to obtain new cellular decomposition: $C' = \{C'_2, C'_1, C'_0\}$ or vice versa. Further,*

$$\sum_{i=0}^{n}(-1)^i|C_i| = \sum_{i=0}^{2}(-1)^i|C'_i|. \tag{7.6}$$

This process can be reversed, inserting a new 1 cell.

Proof. Consider the diagram shown on the right hand side of Figure 7.14b where a 1-cell divides a 2-cell. (Note that there are other possible configurations, based on whether or not the 1-cells share a 0-cell with another 1-cell. We remove the 1-cell, resulting in the removal of two 0-cells and joining two 1-cells. The removal also joins two 2-cells, forming the cell decomposition C'. Now

$$|C_2| = |C'_2| + 1, \qquad |C_1| = |C'_1| + 3, \qquad |C_0| = |C'_0| + 2. \tag{7.7}$$

Both cell decompositions result in the same Euler characteristic. □

Theorem 7.4. *For all cell decompositions, $C = \{C_0, C_1, C_2\}$ and $C' = \{C'_2, C'_1, C'_0\}$, of the connected, closed, compact surface F,*

$$\sum_{i=0}^{2}(-1)i|C_i| = \sum_{i=0}^{2}(-1)^i|C'_i|. \tag{7.8}$$

Proof. Construct a new cell complex $K = \{K_0, K_1, K_2\}$.

We begin with the complex C and insert 0-cells and 1-cells to form the complex K as in Lemmas 7.2 and 7.3 where we can identify a subset of K_1 with each element of C_1'. Two adjacent 1-cells in K are identified with adjacent 1-cells in C_1' or are identified with the same 1-cell. (We are subdividing the graph formed by the 0-cells and 1-cells of C and adding additional 1-cells until a subgraph can be identified with the graph formed by the 0-cells and 1-cells of C'.)

Delete 1-cells and 0-cells until we have the complex C'. □

The Euler characteristic of a surface is independent of cell decomposition and the following corollary immediately follows.

Corollary 7.5. *Let F be a connected, closed, compact, two-dimensional surface. Then $\chi(F)$ is an invariant of the surface.*

We have the following theorems.

Theorem 7.6. *For the sphere S^2, $\chi(S^2) = 2$.*

Proof. See Figure 7.11. □

Theorem 7.7. *For the torus T, $\chi(T) = 0$.*

Proof. See Figure 7.10. □

Theorem 7.8. *For all connected, compact, two-dimensional surfaces, S_1 and S_2,*

$$\chi(S_1 \sharp S_2) = \chi(S_1) + \chi(S_2) - 2.$$

Proof. Let S_1 be surface with a cell decomposition C. We assume that this surface has a 1-cell bounding a 2-cell as shown Figure 7.13. We remove the 2-cell bounded by the loop to obtain a surface with boundary S_1'. We note that $\chi(S_1') = \chi(S_1) - 1$. If this loop is part of a cell decomposition, the disk is bounded by a 0-cell and a 1-cell with a net contribution of 0 to the Euler characteristic. We repeat this process with the surface S_2, observing that $\chi(S_2') = \chi(S_2) - 1$. We glue S_1' and S_2' together along the loop. Hence, we obtain a cell complex for $S_1 \sharp S_2$ and

$$\chi(S_1 \sharp S_2) = \chi(S_1) + \chi(S_2) - 2. \tag{7.9}$$

□

In the remainder of this chapter, we focus on the connected sum of oriented surfaces. We end this section with a theorem about the Euler characteristic of a connected, closed, compact two-dimensional surface F.

Theorem 7.9. *For all connected, closed, compact, orientable two-dimensional surfaces F.*

1. *If F is the connected sum of n torii then $\chi(F) = -2n + 2$.*
2. *The genus of a surface can be computed directly from the Euler characteristic.*

$$g(F) = \frac{1}{2}(2 - \chi(F)).$$

3. *The genus of F, $g(F) \geq 0$ and*

$$0 \leq \frac{1}{2}(2 - \chi(F))$$

and $\chi(F) \leq 2$.

Proof. The proof is left as an exercise. □

Exercises

1. Construct a Mobius band. Take a strip of paper, give it a half twist, and tape the ends together. How many colors do you need to completely color the paper, if you color each side a different color?

2. Find different cellular decompositions for the surfaces shown in Figure 7.12.

3. Construct the connected sum of three tori. Sketch the diagram and a cellular decomposition.

4. Prove the remaining cases in Lemma 7.3.

5. Using the cell complex of the torus given in Figure 7.11, verify that the Euler characteristic of the torus is zero.

6. Construct a cell complex for a sphere with six 2-cells and compute the Euler characteristic of the sphere.

7. Prove (using induction) the following fact. If F is the connected sum of n torii then $\chi(F) = -2n + 2$.

8. Prove that if F is a connected, closed, and compact, oriented 2-dimensional surface then $g(F) = \frac{1}{2}(2 - \chi(F))$.

9. Prove that for a connected, closed, and compact surface F, $\chi(X) \leq 2$.

7.2 CONSTRUCTIONS OF VIRTUAL LINKS

Classical link diagrams represent links in three-dimensional space. The deformations that are possible in three dimensional space are captured exactly by the Reidemeister moves. Classical links are visualized by drawing diagrams on the Cartesian co-ordinate plane and applying the Reidemeister moves. However, instead of drawing the diagram on the Cartesian co-ordinate plane, we could draw the diagram on the sphere. We could also draw link diagrams on other orientable surfaces. We study how virtual links can be viewed as equivalence classes of classical link diagrams on oriented, two-dimensional surfaces.

Construction 1: Abstract links

Our first goal is to associate a surface with boundary containing a link diagram to a connected virtual link diagram D. We assume that any component of the diagram meets at least one other component at a classical crossing. To associate a link diagram with a surface with boundary, we define a **retract**. Let X be a subspace of Y. A retract of Y onto X is a continuous map $f : Y \rightarrow X$ such that f restricted to X is the identity map. If there is a retract of Y onto X, we can visualize Y shrinking into X.

We define an **abstract link diagram** as a pair (S, D) of a compact, oriented surface S and a link diagram D on S where S retracts onto D. We apply a thickening map to a connected virtual link diagram as shown in Figure 7.15 to produce an abstract link diagram that we denote as (S_D, D). The surface S_D has a non-empty boundary. We abbreviate abstract link diagram as ALD.

These surface-diagram pairs were first introduced by Naoko Kamada in a preprint from 1995, "Alternating link diagrams on compact oriented surfaces" (available at www.arxiv.org). This preprint was followed by a joint article with Seiichi Kamada, "Abstract link diagrams and virtual knots" that was published in 2000 (see the further reading list for complete bibliographic information).

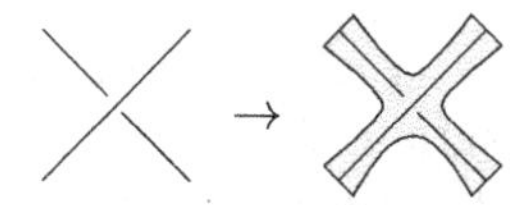

(a) Classical crossing to surface.

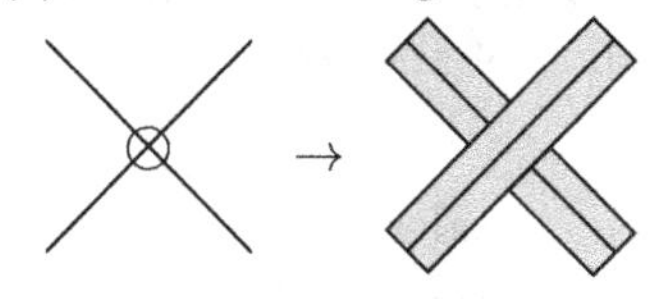

(b) Virtual crossing to surface.

FIGURE 7.15: Thickening map

We see in Figure 7.15b that the choice of which edge is contained in the upper surface does not matter. We can cut the surface and re-glue the thickened edge underneath, creating a homeomorphic surface that contains the same link diagram. Analogs of the diagrammatic moves for ALDs can be constructed using the thickening map.

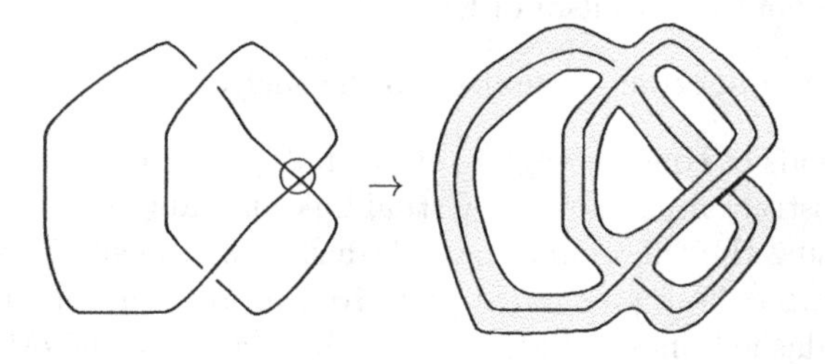

FIGURE 7.16: Surface link pair

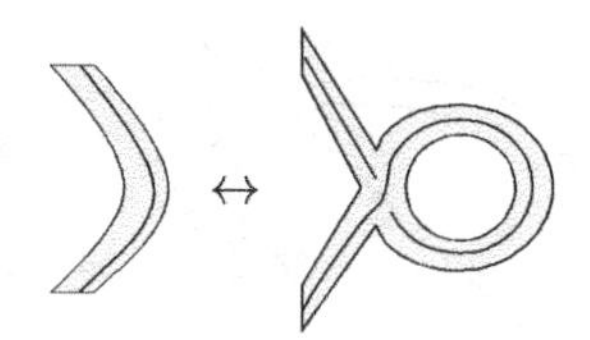

(a) Abstract link Reidemeiester I

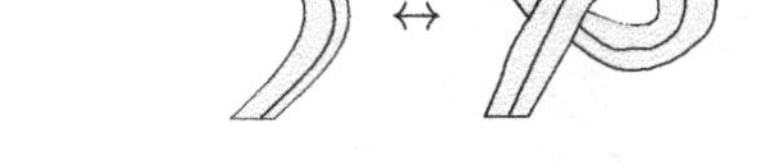

(b) Abstract link virtual Reidemeiester I

FIGURE 7.17: Abstract link moves

In an ALD, the virtual Reidemeister moves are given by deformations of the surface (such as the virtual Reidemeister IV move) or homeomorphisms of the surface. The classical Reidemeister moves change the surface S_D in very specific ways. Two abstract link diagrams are equivalent if they are related by a finite sequence of analogs of the Reidemeister moves, deformations of the surface, and homeomorphisms. An **abstract link** is an equivalence class of abstract link diagrams.

Theorem 7.10. *Abstract links are in one-to-one correspondence with virtual links.*

Proof. Construct a map T from the set of virtual link diagrams to the set of abstract link diagrams. For any virtual link diagram, L, we can construct an abstract link diagram, $(S_L, \hat{L})$, from the thickening map. The thickening map is surjective, since we can recover a virtual link diagram from an abstract link diagram. If L and L' are equivalent link diagrams, then $(S_L, \hat{L})$ is equivalent to $(S_{L'}, \hat{L}')$. If there is a sequence of diagrammatic moves relating L and L' we can apply the analog of the sequence to $(S_L, \hat{L})$ and $(S_{L'}, \hat{L}')$. Hence, the thickening map is injective. □

In the next construction, we consider closed, oriented two-dimensional surfaces. We examine what additional information these surfaces provide in the next section.

Construction 2: Surface link pairs

Given a closed, oriented, surface S, we can draw a link diagram D on the surface. We call this a **surface link pair**, denoted (S, D). We define a series of moves that define equivalence between surface link pairs.

1. **Reidemeister moves:** Planar isotopy and Reidemeister moves on the surface
2. **Dehn twist:** Cut the surface along a curve C, twist one of the boundary components, and re-glue (a homeomorphism of the surface)
3. **Handles:** Add or subtract "handles" to the surface

There are two methods of constructing a surface link pair from a virtual link diagram. The first method uses abstract links. For any virtual link diagram D, construct the abstract link diagram (S_D, D) using the thickening map. Attach disks to each boundary components of S_D to obtain the surface link pair $(S(D), D)$. To indicate that the diagram wraps around the surface, we use dashed lines to indicate that the edge is on the other side of the surface. The second method of construction is to draw the virtual link diagram on the surface of the sphere and then to insert a "handle" at each virtual crossing (Figure 7.18).

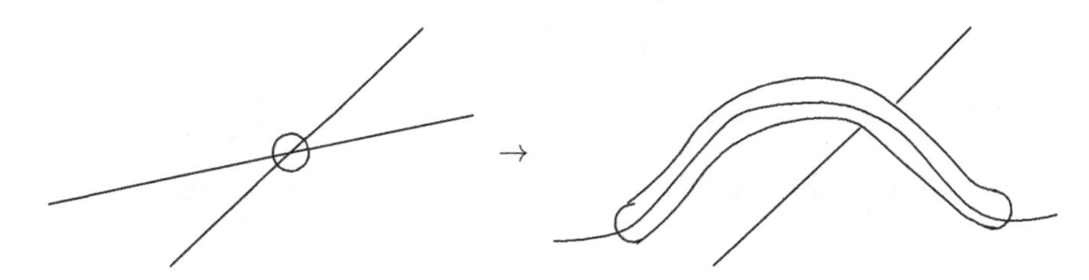

FIGURE 7.18: Diagram and surface link

A surface link pair corresponding to a virtual trefoil is shown in Figure 7.19. We can recover a virtual link diagram from the surface link pair by projecting the link diagram onto the plane. For classical crossings, the over/undermarkings are also projected onto the plane. The virtual crossings are the extra crossings that appear in the projection as the link components wrap around the surface.

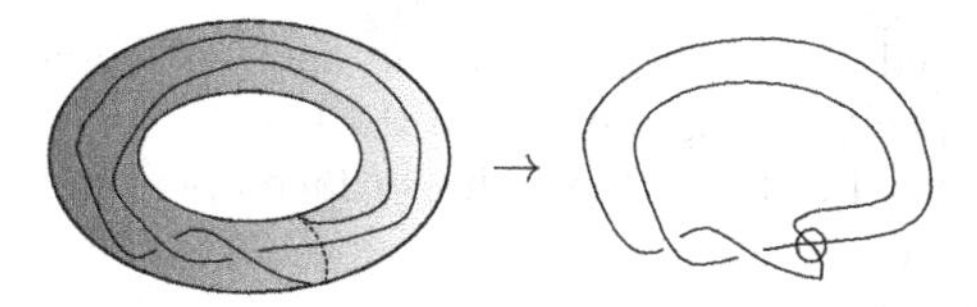

FIGURE 7.19: Recovering a diagram from a surface link

We describe the moves on surface link pairs in more detail. A **Dehn twist move** consists of four steps. First, specify a simple, closed curve on the surface that does not bound a disk. Second, cut the surface along the curve. Third, twist one of the boundaries 360 degrees and then finally re-glue the boundaries. A Dehn twist is a homeomorphism of the surface—a continuous, one-to-one and invertible map. This process is illustrated in Figure 7.20. If the surface does not contain a link, the surface looks exactly the same. However, the Dehn twist move twists the link in a surface link pair around the selected simple closed curve in the surface.

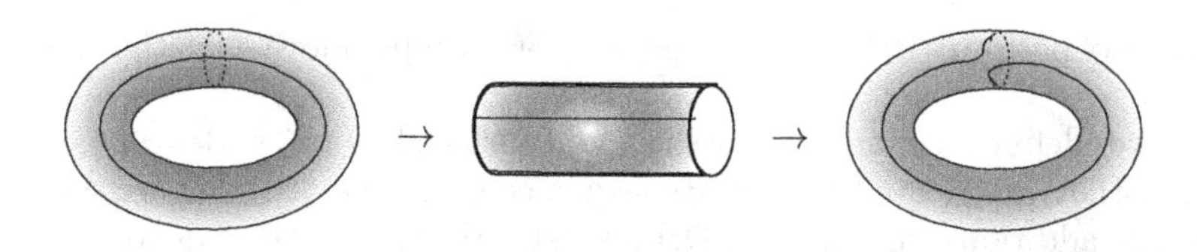

FIGURE 7.20: A Dehn twist

An example of a Dehn twist on a tube containing two strands is shown in Figure 7.21. The strands twist around the tube and appear to cross in our sketch of the tube. Any orientation-preserving homeomorphism of the surface can be expressed as a sequence of Dehn twists.

We can modify the surface S by adding or removing a **handle** to the surface. A handle is a tube (technically an annulus). The boundary of the tube consists of two circles. To add a handle to the surface, perform the following steps.

1. Locate two disks on the surface that do not intersect the diagram
2. Remove the interior of these disks
3. Glue the ends of the tube to the circles

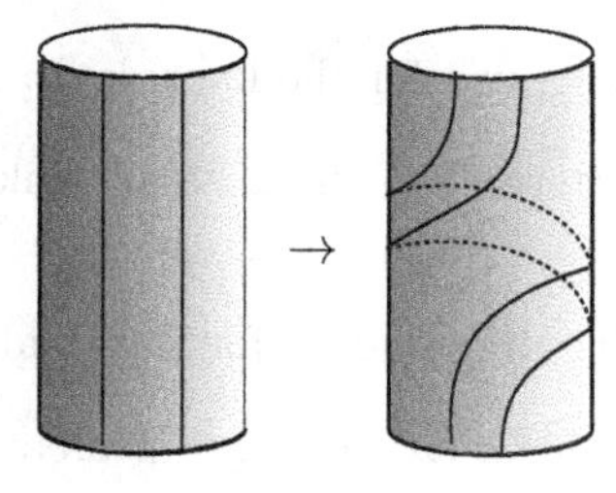

FIGURE 7.21: Dehn twist and a virtual Reidemeister II move

Removing a handle reduces the genus of the surface. Handles are removed by the following procedure: 1) specify a closed curve in the surface that does not intersect the link diagram and does not bound a disk, 2) cut the surface open along this curve to produce a surface with two boundary components, and 3) glue a disk to each boundary component, reducing the genus of the surface by one. An example of handle addition is shown in Figure 7.22. Adding or removing a handle from a surface is also sometimes called **stabilization**.

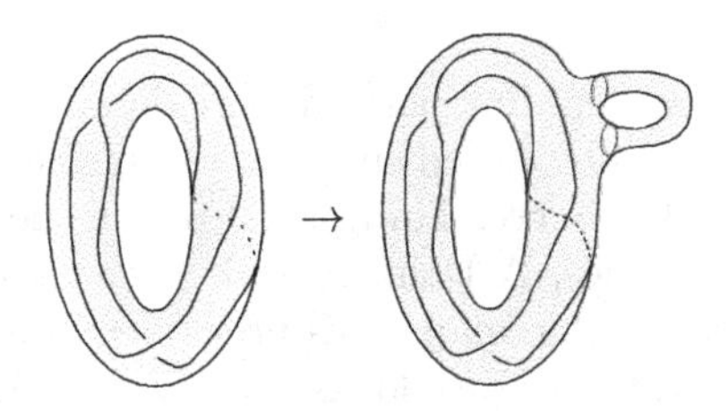

FIGURE 7.22: Adding and removing handles

We can also perform Reidemeister moves (or isotopy moves) of the diagram on the surface.

These moves define an equivalence relation on surface link pairs. Hence, $(S, D) \sim (S', D')$ if and only if they are related by a finite sequence of moves of Reidemeister moves in the surface, handle addition/removal, and Dehn twists. We use $[S, D]$ to denote the equivalence class of (S, D). To summarize, for a link diagram D, we can obtain a surface link pair $(S(D), \hat{D}))$ using either the handle map or by attaching disks to the abstract link diagram. The equivalence class of $(S(D), \hat{D})$ projects to the equivalence class of the virtual link diagram D.

Theorem 7.11. *Virtual links are in one-to-one correspondence with surface link pairs modulo handle addition/cancellation, Dehn twists, and Reidemeister moves.*

Proof. From $[S, D]$, take a representative (S, D). Project (S, D) onto the plane and recover a virtual link diagram. If two surface link pairs are equivalent, we can project the link diagrams onto the plane. (Remember that virtual crossings occur in the projection as the link diagram wraps around the surface.) If the surface link pairs are related by a sequence of moves, then projection results in a sequence of diagrammatic moves relating the projections. The projection map is onto since a surface link pair can be recovered by drawing a virtual link diagram on a sphere and then inserting a handle for each virtual crossing. □

Exercises

1. Construct abstract link diagrams for the knot diagrams in Figure 7.23.
2. Compute the Euler characteristic of the abstract link diagrams that you constructed from Figure 7.23.
3. Construct abstract link diagram analogs for the Reidemeister II and III moves.
4. Construct a surface link pair for a diagram equivalent to a virtual trefoil that requires surface of genus two.
5. Construct surface link pairs for the knot diagrams in Figure 7.23.

6. Apply a Dehn twist to the torus that contains a two component unlink (shown in Figure 7.24a).
7. Apply a Dehn twist along the dotted line to the surface link pair in Figure 7.24b.
8. Recover virtual link diagrams from the surface link pairs in Figure 7.25.

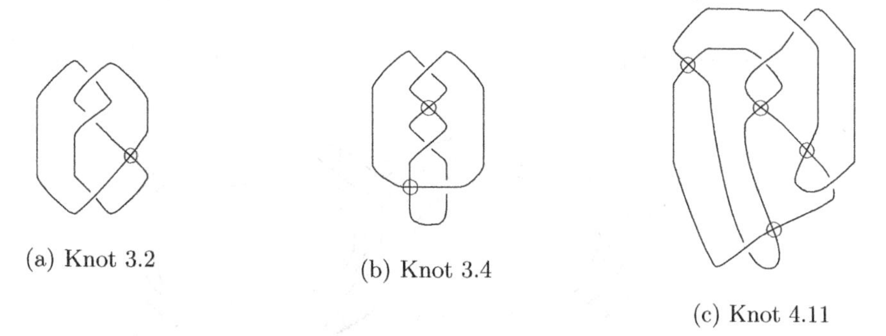

(a) Knot 3.2 (b) Knot 3.4 (c) Knot 4.11

FIGURE 7.23: Construct ALDs and surface link pairs

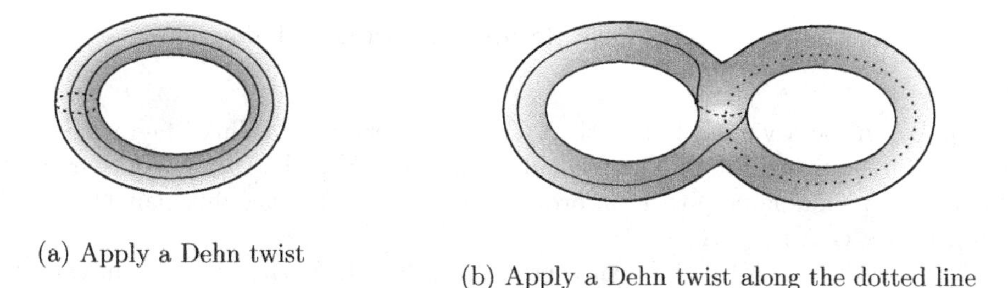

(a) Apply a Dehn twist (b) Apply a Dehn twist along the dotted line

FIGURE 7.24: Dehn twist exercises

7.3 GENUS OF A VIRTUAL LINK

The virtual genus of a virtual link K (denoted $G(K)$) is the minimum possible genus of a surface link pair $(S, \hat{K})$ that corresponds to K. Using set notation, we define the genus of K:

$$G(K) = min\{g(S')|(S', D) \in [S, \hat{K}]\}. \tag{7.10}$$

Greg Kuperberg proved that any surface link pair (S, D) in the equivalence class $[S, D]$ can be reduced by a sequence of Reidemeister moves and stabilization to a surface link pair with a surface that has minimum genus. His paper containing this result, "What is a virtual link?," was published in *Algebraic Geometric Topology* in 2003. Here, we investigate some theorems about virtual genus and how to compute a bound on virtual genus using abstract link diagrams and surface link pairs.

Theorem 7.12. *For all virtual links K with* $\mathsf{v}(K) > 0$, *K is non-classical and* $1 \leq G(K) \leq \mathsf{v}(K)$.

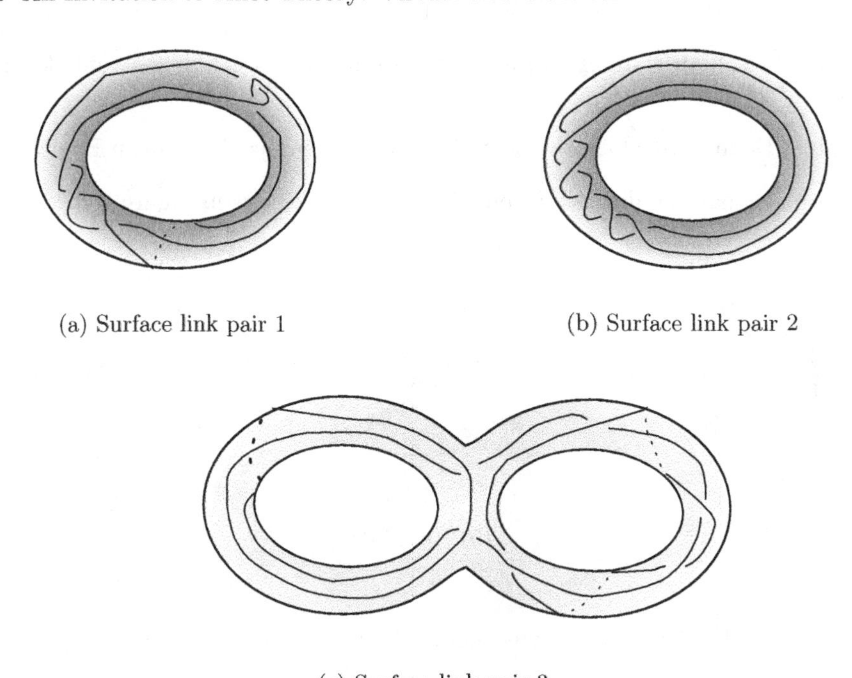

(a) Surface link pair 1

(b) Surface link pair 2

(c) Surface link pair 3

FIGURE 7.25: Recover the virtual link diagram

Proof. Let K be a virtual link with $\mathsf{v}(K) > 0$. Every virtual link diagram D of K has at least $\mathsf{v}(K)$ virtual crossings. There is a diagram D of K with exactly $\mathsf{v}(K)$ virtual crossings. We apply the handle map to D in order to construct a surface link pair $(S, \hat{D})$ with genus $\mathsf{v}(K)$. Hence $G(K) \leq \mathsf{v}(K)$.

Suppose that $G(K) = 0$. Then there is a surface link pair $(S', \hat{K})$ equivalent to $(S, \hat{D})$ where S' is the sphere. Then under the projection map, K is equivalent to a diagram with zero virtual crossings. This is a contradiction, so that $G(K) \geq 1$. □

We have the following partial converse.

Theorem 7.13. *For all virtual links K with $G(K) > 0$, K is non-classical.*

Proof. The proof is left as an exercise. □

For a virtual link diagram D and the surface link pair obtained from the abstract link diagram $(S(D), \hat{D})$, we use the link diagram D to construct a cell decomposition of S_D. Then we compute the genus of $S(D)$ from the cell decomposition. Remember that $S(D)$ is obtained by gluing disks to the boundary components of S_D. We begin by analyzing a cell decomposition of S_D. We analyze how the classical crossings, boundary components, and edges interact. At each classical crossing, we have: four 0-cells, four 1-cells and one 2-cell. Each thickened edge consists of two 1-cells and one 2-cell. If a connected diagram has n classical crossings then the corresponding cell complex C of the surface S_D with b boundary components satisfies:

$$|C_0| = 4n, \qquad |C_1| = 4n + 2(2n), \qquad |C_2| = n + n + b. \tag{7.11}$$

Summing, we see that the Euler characteristic is

$$\chi(S_D) = 2n + b - (8n) + 4n \tag{7.12}$$
$$\chi(S_D) = b - n. \tag{7.13}$$

We can then compute the genus of the surface S_D associated to diagram D in terms of the number of classical crossings and the number of boundary components

$$g(S_D) = \frac{1}{2}(2 - b + n). \tag{7.14}$$

Note that for each diagram, we can obtain a surface $S(D)$ (without boundary) by gluing a disk along each boundary component as shown in Figure 7.26.

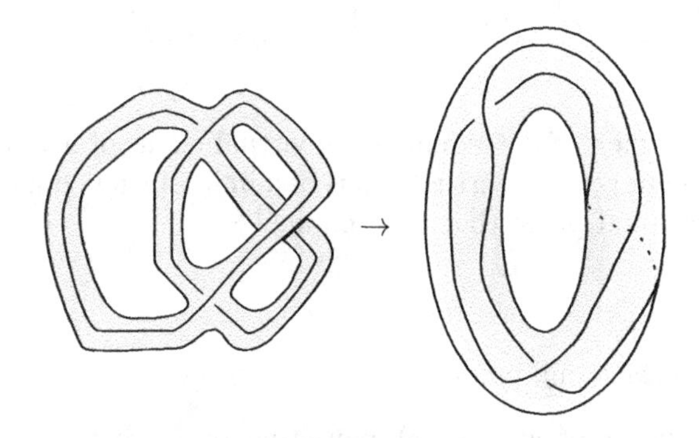

FIGURE 7.26: Surface containing a virtual trefoil

Using Equation 7.14, if we know the genus of S_D and the number of crossing n, we compute the number of boundary components. If $g(S_D) = 0$ then

$$b = n + 2. \tag{7.15}$$

Corollary 7.14. *$G(K)$ is an invariant of virtual links.*

This theorem does not answer the question of how to calculate the virtual genus of a virtual link diagram. The diagram shown in Figure 7.27 demonstrates that a diagram of a fixed link can be manipulated to give an arbitrarily high genus. However, individual knots can have a high genus.

Exercises

1. Construct the abstract link diagram for the virtual knot K in Figure 7.27. Calculate the genus of the surface link pair, $(S(K), \hat{K})$, obtained from the ALD. Remember that boundary components of the ALD are counted as 2-cells.
2. Compute how the analogs of the Reidemeister moves affect the genus of the surface.
3. Draw a virtual knot diagram with genus five.
4. Prove that a virtual link diagram D with no virtual crossings has $g(S_D) = 0$.
5. Prove Theorem 7.13.
6. Prove Theorem 7.14.

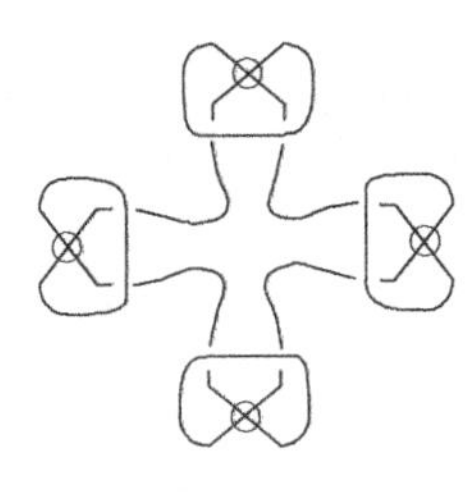

(a) A knot with genus 4

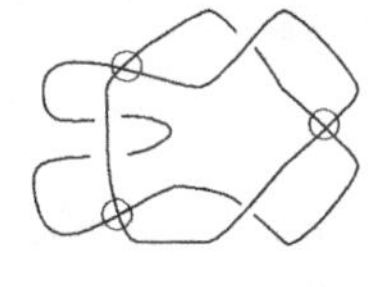

(b) A diagram with genus 2

FIGURE 7.27: Genus: knot versus diagram

7.4 OPEN PROBLEMS AND PROJECTS

Open problems

1. Find a simple method of determining the virtual genus of a virtual link. It is easy to find an upper bound on the virtual genus of a link, but much more challenging to find the virtual genus and prove that it is minimal.

Projects

Practice visualizing surface link pairs.

1. Construct a sequence of surface link pairs that prove that virtual crossing number is not bounded by the virtual genus.
2. Construct a sequence of virtual links that have virtual crossing number n and virtual genus n. (Do not try to prove that the virtual genus is n.)
3. Consider families of virtual pretzel knots (in tangle notation: $w_1, w_2, \dots w_n$). Give a conjecture about their virtual genus with supporting examples.
4. Construct a family of rational knots and give a conjecture about their virtual genus. Support your conjecture with examples.

Further Reading

1. J. Scott Carter, Seiichi Kamada, and Masahico Saito. Stable equivalence of knots on surfaces and virtual knot cobordisms. *J. Knot Theory Ramifications*, 11(3):1–12, 2008
2. Louis H. Kauffman. Virtual knot theory. *European J. Combin.*, 20(7):663–690, 1999
3. Greg Kuperberg. What is a virtual link? *Algebr. Geom. Topol.*, 3:587–591 (electronic), 2003
4. Robert Messer and Philip Straffin. *Topology now!* Classroom Resource Materials Series. Mathematical Association of America, Washington, DC, 2006
5. Naoko Kamada and Seiichi Kamada. Abstract link diagrams and virtual knots. *J. Knot Theory Ramifications*, 9(1):93–106, 2000
6. C. E. Burgess. Classification of surfaces. *The American Mathematical Monthly*, 92(5):349–354, 1985

7. Peter Andrews. The classification of surfaces. *The American Mathematical Monthly*, 95(9):861–867, 1988

CHAPTER 8

Bracket polynomial II

The goal in this chapter is to improve our weak version of the Kauffman-Murasugi-Thistlethwaite theorem. For a virtual link diagram D, with $D \sim K$ that satisfies certain conditions, the span of $f_K(A)$ is bounded in terms of the number of classical crossings and the genus of the diagram. A bound on the span of the f-polynomial for virtual link diagrams would: 1) determine bounds on the crossing number and genus of a virtual knot and 2) distinguish classical versus non-classical knots.

The key to this extension is counting the number of components in a state s of a virtual knot diagram D and its dual $\overline{s}$. We use the fact that virtual links are in one-to-one correspondence with knots in surfaces (modulo the moves discussed in the last chapter) and our results also apply to classical link diagrams in surfaces.

Recall the following theorem.

Theorem 8.1 (Kauffman-Murasugi-Thistlethwaite). *For all connected, classical link diagrams D with n classical crossings then*

1. $\mathrm{Span}(f_D(A)) \leq 4n$.
2. *If D is an alternating and reduced diagram then* $\mathrm{Span}(f_D)(A) = 4n$.
3. *If D is non-alternating and prime then* $\mathrm{Span}(f_D(A)) < 4n$.

For a virtual knot diagram with n classical crossings, we proved a weak bound on the number of components of a state and its dual, $|s| + |\overline{s}| \leq 2n$.

8.1 STATES AND THE BOUNDARY PROPERTY

For a virtual link diagram D, a state s of the diagram is determined by selecting a smoothing type for each classical crossing of the diagram. The dual state, $\overline{s}$, is determined by making the opposite choice for each smoothing. In a **marked state**, we include trace marks at each smoothed classical crossing, as shown in Figure 8.1. From a marked state, we can construct any other state of the diagram by making local changes to the diagram at each trace as shown in Figure 8.2. A state s that has been marked with traces also encodes its dual state $\overline{s}$.

Given a connected link diagram D, let (S_D, D) denote the abstract link diagram obtained from D. The surface $S(D)$ is obtained by attaching a disk to each boundary component of S_D. Near each classical crossing, the boundary components of S_D correspond to the two possible smoothings of the classical crossing. Further, we can construct a cell decomposition C_D of $S(D)$ directly from the diagram (S_D, D) as shown in Figure 8.3.

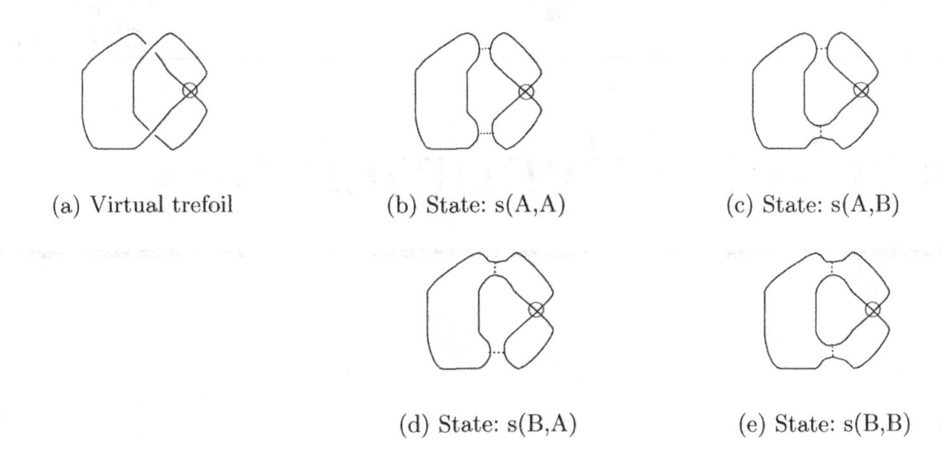

(a) Virtual trefoil (b) State: s(A,A) (c) State: s(A,B)

(d) State: s(B,A) (e) State: s(B,B)

FIGURE 8.1: Marked states of the trefoil

FIGURE 8.2: Trace exchange move

For some virtual link diagrams, the local correspondence between the boundary of S_D at crossings and the two possible smoothings of the crossings extends to the entire diagram, resulting in a one-to-one correspondence between the components of ∂S_D and the components of a state s_∂ and its dual state $\overline{s_\partial}$. If $\partial S_D = s_\partial \cup \overline{s_\partial}$ then the diagram D has the **boundary property**. In Figure 8.4, we see two diagrams where the boundary components of S_D correspond to the components of a state s_∂ and its dual $\overline{s_\partial}$. Not all virtual link diagrams have the boundary property, as shown in Figure 8.5.

If D is a connected virtual link diagram with the boundary property then there is a correspondence between the 2-cells in a cellular decomposition of $S(D)$ and the components of s_∂ and $\overline{s_\partial}$.

Lemma 8.2. *For all connected virtual link diagrams D with the boundary property, $S(D)$ has a cell decomposition such that the 2-cells are in one-to-one correspondence with the components of the state s_∂ and its dual $\overline{s_\partial}$.*

Proof. Let D be a connected virtual link diagram with the boundary property. Let (S_D, D)

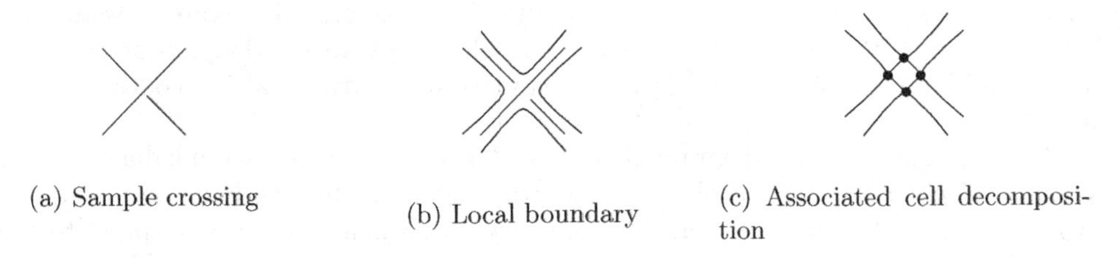

(a) Sample crossing (b) Local boundary (c) Associated cell decomposition

FIGURE 8.3: Crossing with boundary and cell decomposition

(a) Trefoil

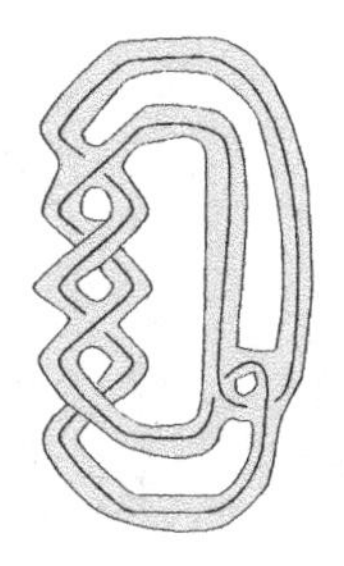

(b) Virtual knot

FIGURE 8.4: Correspondence between ∂S_D and states $s_\partial \cup \overline{s_\partial}$

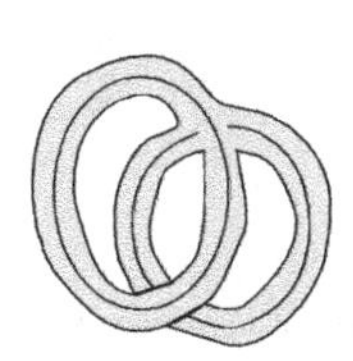

(a) Virtual Hopf link

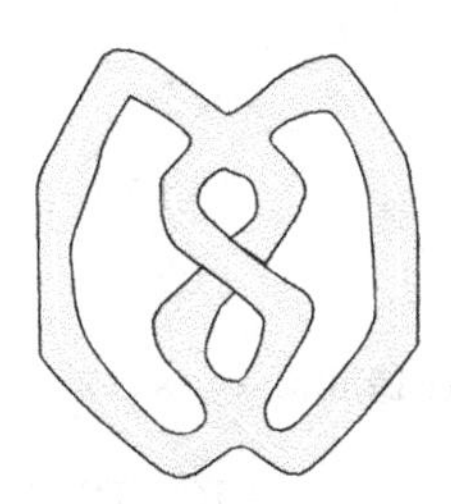

(b) Virtual trefoil

FIGURE 8.5: Links without the boundary property

denote the abstract link diagram obtained by applying the thickening map to D. The surface $S(D)$ is formed by attaching disks to the boundary of S_D. We obtain a cell decomposition of $S(D)$ from D as shown in Figure 8.4. Reduce the cell decomposition as shown in Figure 8.7. The remaining 0-cells are in one-to-one correspondence with the classical crossings in D. Similarly, the 1-cells are in one-to-one correspondence with the edges of D and the 2-cells are in one-to-one correspondence with the components of s and $\overline{s_\partial}$. □

Theorem 8.3. *For all connected, virtual link diagrams D with n classical crossings, the boundary property, and $g(S(D)) = g$,*

$$|s_\partial| + |\overline{s_\partial}| = n + 2 - 2g. \tag{8.1}$$

Proof. The diagram D has the boundary property and $g(S(D)) = g$. There is a cell decomposition of $S(D)$ with n 0-cells, $2n$ 1-cells, and $|s_\partial| + |\overline{s_\partial}|$ 2-cells. Recall that

$$\chi(S(D)) = 2 - 2g \tag{8.2}$$

and

$$\chi(S(D)) = |s_\partial| + |\overline{s_\partial}| - 2n + n. \tag{8.3}$$

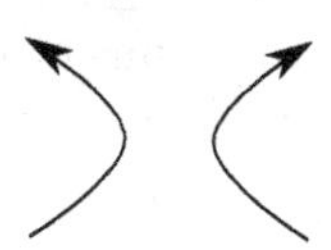

FIGURE 8.6: Smooth crossing vertically

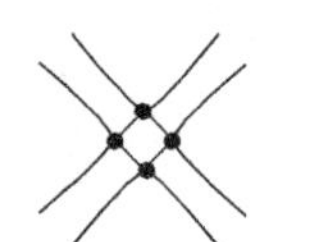

(a) From a crossing

(b) Reduced cell decomposition

FIGURE 8.7: Cell decompositions at a crossing

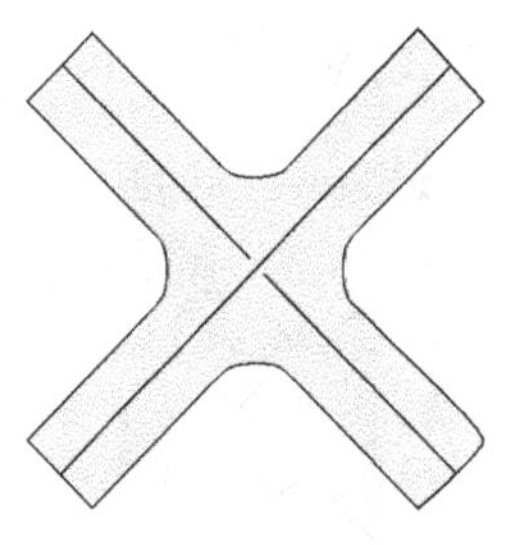

(a) Boundary: $s \cup \overline{s}$

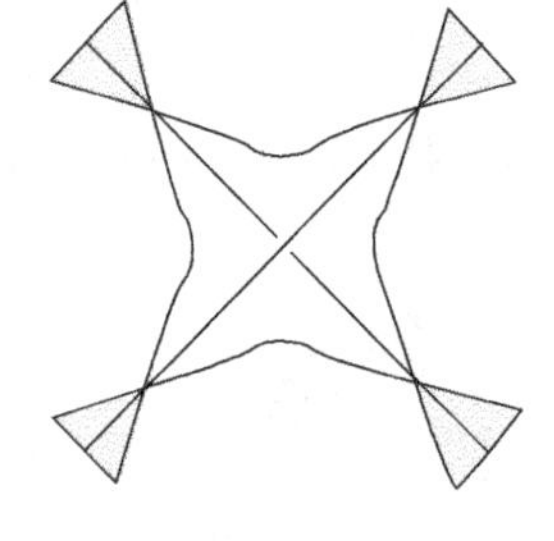

(b) Boundary after trace exchange: $s' \cup \overline{s}'$

FIGURE 8.8: Boundaries and trace exchange

Combining Equations 8.2 and 8.3,

$$|s_\partial| + |\overline{s_\partial}| = n + 2 - 2g. \tag{8.4}$$

□

For a connected virtual link diagram D, each state s and dual state $\overline{s}$ defines a surface containing the diagram D. For a diagram D with the boundary property, one of these surfaces is the ALD, (S_D, D). The surface S_D is bounded by s_∂ and $\overline{s_\partial}$. Let the state s' be related to s by a single resmoothing at classical crossing c. To form the surface associated with the state s' and $\overline{s'}$, we remove the "X" shape from the abstract link diagram that contains crossing c and then we give a half twist to the legs of the "X" and re-attach the "X". The boundaries of the surface have been modified and now correspond to s' and $\overline{s'}$ as shown in Figure 8.8. We have the following theorem.

Theorem 8.4. *For all connected virtual link diagrams D with the boundary property, n classical crossings, and $g(S(D)) = g$, if s' is state is obtained from s_∂ by k trace exchange moves then*

$$|s'| + |\overline{s'}| \leq \min\{n + 2 - 2g + 2k, n + 2\}. \tag{8.5}$$

Proof. The virtual link diagram D has the boundary property and $|s_\partial|+|\overline{s_\partial}| = n+2-2g$. If s' is obtained from s_∂ by a single trace exchange move then $|s'| \leq |s_\partial| + 1$. Consequently, if s' is obtained from s_∂ by k trace exchange moves then $|s'| \leq |s_\partial|+k$. Similarly, $\overline{s'}$ is obtained from $\overline{s_\partial}$ by k trace exchange moves so that $|\overline{s'}| \leq |\overline{s_\partial}|+k$. Then $|s'|+|\overline{s'}| \leq 2n-2g+2k$. Each trace exchange move possibly reduces the genus of the surface. As a result, the maximum number of possible boundary components is $n + 2$. □

Corollary 8.5. *For all connected virtual link diagrams D with the boundary property and n classical crossings, for any state s,*

$$2 \leq |s| + |\overline{s}| \leq n + 2. \tag{8.6}$$

Proof. The number of the components of a state and its dual determine the genus of the surface associated to the state and its dual. The sphere has the maximum number of components, $n+2$. In the worst case, the state s and its dual $\overline{s}$ each have at most one component. □

The crossing number of a virtual link forms a lower bound on the genus of the virtual link.

Corollary 8.6. *For all connected virtual link diagrams D with the boundary property, and n classical crossings,*

$$\frac{n}{2} \geq g. \tag{8.7}$$

Proof. Let D be a virtual link diagram with $c_d(D) = n$. Let $g = g(S(D))$ then

$$n + 2 - 2g = |s_\partial| + |\overline{s_\partial}|. \tag{8.8}$$

Each state contains at least one component, so that $n + 2 - 2g \geq 2$ or

$$\frac{n}{2} \geq g. \tag{8.9}$$

□

The number of classical crossings limits the genus. We calculate the maximum and minimum degree of terms in $\langle D \rangle$. The maximum and minimum degrees are denoted d_{max} and d_{min}, respectively.

Lemma 8.7. *For all connected virtual link diagrams D with the boundary property, if D has n classical crossings then*

$$d_{max} \leq n + 2(|s_\alpha| - 1) \tag{8.10}$$

and

$$d_{min} \geq -n - 2(|s_\beta| - 1). \tag{8.11}$$

Proof. We calculate the contribution of the s_α state to $f_D(A)$. Recall that

$$\langle D \rangle = \sum_{s \in S} A^{\alpha(s) - \beta(s)} d^{|s|-1}. \tag{8.12}$$

The state s_α contributes a monomial with degree $n + 2(|s_\alpha| - 1)$. Suppose that s' is obtained from s_α by making i trace exchanges. Then $|s'|$ can't be greater than $|s_\alpha| + i$ because a single trace exchange increases the number of components in a state by at most one. If the state s' is related to s_α by i trace exchanges then $\alpha(s') - \beta(s') = n - 2i$. Now, s' contributes a monomial with degree less than or equal to $n - 2i + 2(|s_\alpha| + (i - 1))$, which reduces to

$$n + 2(|s_\alpha| - 1). \tag{8.13}$$

The proof of the bound on the minimum degree is similar. □

Theorem 8.8. *For all connected virtual link diagrams D with the boundary property, n classical crossings, and $g(S(D)) = g$,*

$$\text{Span}(f_D(A)) \leq 2n + 2(|s_\alpha| + |s_\beta| - 2). \tag{8.14}$$

Proof. Let D be a virtual link diagram with n classical crossings and virtual genus g. By Lemma 8.7, maximum possible degree of $f_D(A)$ is

$$(-3w(K)) + n + 2(|s_\alpha| - 1). \tag{8.15}$$

The minimum possible degree is

$$(-3w(K)) - n - 2(|s_\beta| - 1). \tag{8.16}$$

The span is bounded by the difference between the maximum and minimum. We conclude that

$$\mathrm{Span}(f_D(A)) \leq 2n + 2(|s_\alpha| + |s_\beta| - 2). \tag{8.17}$$

.

□

We have the following corollary.

Corollary 8.9. *For all connected virtual link diagrams D with the boundary property, n classical crossings, and $g(S(D)) = g$*

1. *If $s_\partial = s_\alpha$ or $s_\partial = s_\beta$ then* $\mathrm{Span}(f_D(A)) \leq 4n - 4g$.
2. *If s_α can be obtained from s_∂ or $\overline{s_\partial}$ with k trace exchanges then* $\mathrm{Span}(f_D(A)) \leq 4n - 4g + 2k$.

Proof. We apply Theorem 8.4 to the result in Theorem 8.8. □

Exercises

1. Construct all the marked states of a trefoil.
2. Compute $|s| + |\overline{s}|$ for each state of the trefoil. Use marked states so that it is easy to determine $\overline{s}$.
3. For the link diagrams shown in Figure 8.9, which diagrams D have the boundary property?
4. For the diagrams in Figure 8.9, compute the genus of the diagram. Use Theorem 8.3 to compute the number of components in s_∂ and $\overline{s}_\partial$. If the knot does not have the boundary property, compute the number of components in each state and dual state.

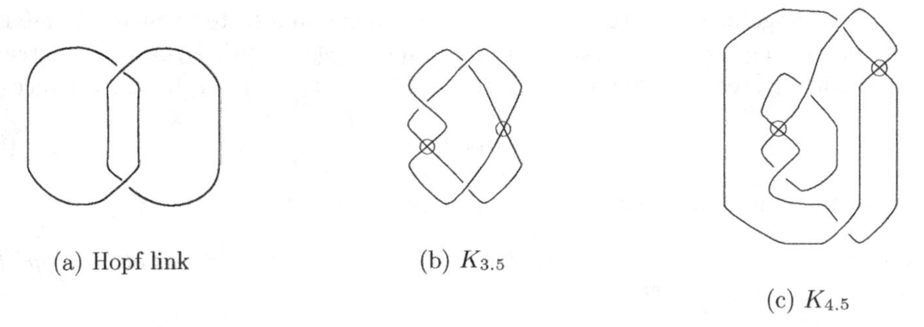

(a) Hopf link (b) $K_{3.5}$ (c) $K_{4.5}$

FIGURE 8.9: Determine if the boundary corresponds to s and $\overline{s}$

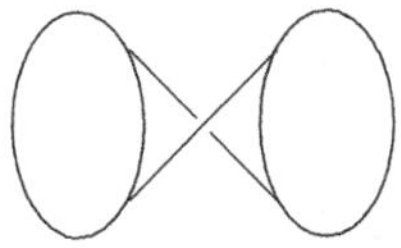

FIGURE 8.10: Nugatory crossing

8.2 PROPER STATES

In this section, we introduce proper states. These states play a key role in sharpening the bound in Theorem 8.1. Proper states are a generalization of adequate states from classical knot theory. A state s is a **proper state** if four different components of s and $\overline{s}$ meet at every classical crossing in the diagram. A diagram D has the **proper boundary property** if s_∂ is a proper state.

In a proper state s, if the state s' is obtained from s by a single resmoothing then $|s| = |s'| + 1$.

A **nugatory crossing** is shown in Figure 8.10. A connected, classical link diagram without nugatory crossings is a reduced classical link diagram. A reduced, alternating classical diagram has the proper boundary property. Unfortunately, for virtual link diagrams is possible to have a reduced, alternating diagram that is not proper. For this reason, we use the more general condition that a virtual link diagram has the proper boundary property.

Theorem 8.10. *For all connected virtual link diagrams D with the proper boundary property, $g(S(D)) = g$, and n classical crossings, if $s_\partial = s_\alpha$ or $s_\partial = s_\beta$ then*

$$\text{Span}(f_k(A)) = 4(n-g). \tag{8.18}$$

Proof. Suppose that D is a connected, virtual link diagram with the proper boundary property. By hypothesis, the state s_∂ is either s_α or s_β.

The contribution from s_α to $\langle D \rangle$ has a maximum degree of

$$n + 2(|s_\alpha| - 1). \tag{8.19}$$

Suppose that s' is a state obtained from s_α by changing exactly one smoothing to a B-type smoothing. The degree of the contribution from s' to $\langle D \rangle$ is:

$$(n-2) + 2(|s'| - 1). \tag{8.20}$$

Since s_α is proper, $|s'| = |s_\alpha| - 1$ so that the maximum degree of the contribution from s' is

$$n - 2 + 2(|s_\alpha| - 2) = n + 2(|s_\alpha| - 3). \tag{8.21}$$

As a result, the maximum contribution from the states related by one trace exchange does not cancel out the maximum contribution from s_α. The maximum degree of $\langle D \rangle$ is

$$n + 2(|s_\alpha| - 1). \tag{8.22}$$

A similar argument applies to s_β, which contributes the minimum degree to $\langle D \rangle$, denoted d_{min}:

$$d_{min} = -(n + 2(|s_\beta| - 1)). \tag{8.23}$$

The span of $f_D(A)$ is

$$d_{max} - d_{min} = 2n + 2(|s_\alpha| + |s_\beta| - 2). \tag{8.24}$$

Then since $|s_\alpha| + |s_\beta| = n + 2 - 2g$, we see that

$$d_{max} - d_{min} = 2n + 2((n + 2 - 2g) - 2) \tag{8.25}$$
$$= 4(n - g). \tag{8.26}$$

□

The question that remains is "when do the states s_α and s_β correspond with s_∂ and $\overline{s_\partial}$?" Recall that a link is **alternating** if, following the orientation of each component, the crossings alternate between over- and underpasses. The states correspond when the knot diagram has the proper boundary property and is alternating.

Lemma 8.11. *For all alternating, connected virtual link diagrams D with the proper boundary property, $s_\partial = s_\alpha$ or $s_\partial = s_\beta$.*

Proof. Consider the diagram D in $S(D)$. Suppose that $s_\partial \neq s_\alpha$ and $s_\partial \neq s_\beta$. There are two classical crossings in D that are connected by an edge such that in s_∂, one crossing has a type A smoothing and one crossing has a type B smoothing. We sketch the possible configuration and observe that D is not alternating. This is a contradiction, so $s_\partial = s_\alpha$ or $s_\partial = s_\beta$. □

Corollary 8.12. *For all alternating, connected virtual link diagrams D with the proper boundary property, n classical crossings, and $g(S(D)) = g$, $span(D) = 4(n - g)$.*

Proof. If D has the proper boundary property then $s_\partial(D)$ is a proper state. If the diagram is alternating then the states s_α and s_β correspond to the components of ∂S_D. □

We have now proved a version of the first two parts of the KMT Theorem for virtual link diagrams with the boundary property. In the next section, we consider a class of diagrams that do not have the boundary property.

Exercises

1. Construct an infinite family of link diagrams with the boundary property.
2. Determine if the diagrams in Figure 8.11 have 1) the boundary property, 2) the proper boundary property, and 3) are alternating.
3. Construct two link diagrams with the proper boundary property.
4. Compute a bound on the span of $f_K(A)$ for the link diagram in Figure 8.12.

8.3 DIAGRAMS WITH ONE VIRTUAL CROSSING

In this section, we consider a class of diagrams that do not have the boundary property. We begin with a virtual link diagram that has the proper boundary property, n classical crossings, and $g(S(D)) = g$. We then modify the diagram by changing a single classical crossing into a virtual crossing. This theorem and proof was originally given by Naoko Kamada in her 2003 paper.

Theorem 8.13. *Let D be an alternating, connected virtual link diagram with the proper boundary property, n classical crossings, and $g(S(D)) = g$. Let c be a classical crossing in D and construct a non-trivial diagram D_c from D by changing c into a virtual crossing. Then* $\text{span} f_{D_c}(A) \leq 4(n - g) - 6$.

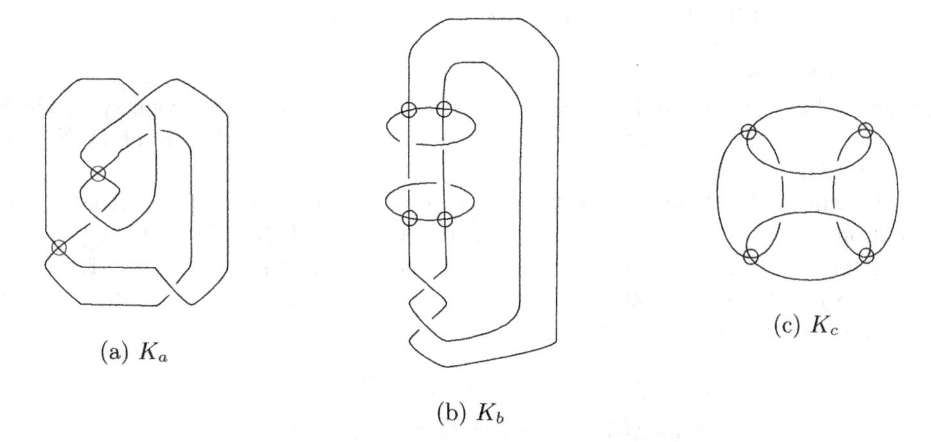

FIGURE 8.11: Determine if the links have the boundary property

FIGURE 8.12: Compute bounds on $f_K(A)$

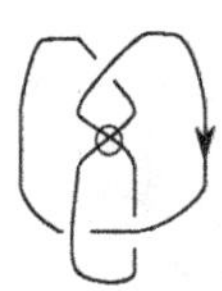

FIGURE 8.13: Compute a bound on Span($f_K(A)$)

Proof. Let D be a virtual link diagram with the proper boundary property, n classical crossings, and $g(S(D)) = g$. Let $s_\partial(D)$ denote the Seifert smoothed state of D. Note that $|s_\partial(D)| + |\overline{s_\partial}(D)| = n + 2 - 2g$.

Let D_c denote the diagram obtained by converting crossing c to a virtual crossing. The notation $s_\partial(D_c)$ denotes the Seifert smoothed state of D_c. Since D is proper, we know that the trace at c on $s_\partial(D)$ does not self-touch and $|s_\partial(D_c)| = |s_\partial| - 1$. Similarly, $|\overline{s_\partial}(D_c)| = |\overline{s_\partial}| - 1$. We compute that $|s(D_c)| + |\overline{s}(D_c)| = n - 2g$. The total number of crossings has been reduced by 1, and

$$d_{max}(\langle D_c \rangle) = (n-1) + 2(|s(D_c)| - 1)$$
$$d_{min}(\langle D_c \rangle) = -((n-1) + 2(|\overline{s}(D_c)| - 1).$$

Hence,

$$\begin{aligned} \text{span}(f_K(D_c)) &= 2(n-1) + 2(|s_\partial(D_c)| + |\overline{s_\partial}(D_c)| - 2) \\ &= 2n - 6 + 2(|s_\partial(D_c)| + |\overline{s_\partial}(D_c)|) \qquad = 2n - 6 + 2(n-2g) \\ &= 4(n-g) - 6. \end{aligned}$$

□

Exercises

1. Let T denote a virtual trefoil. Prove that Span($f_T(A)$) ≤ 6 without computing $f_T(A)$.
2. Construct a bound Span($f_K(A)$) for the diagram in Figure 8.13.
3. Let D be a link diagram with m disjoint link diagrams. Suppose each component has the boundary property. Prove that Span($f_D(A)$) $= 4(n - (g_1 + g_2 \ldots g_m)) - m + 1$ where g_i is the genus of the ith link diagram.

8.4 OPEN PROBLEMS AND PROJECTS

Open problems

1. Clarify the relationship between proper and reduced. Notice that a proper diagram is reduced (contains no nugatory crossings). Is there a move that introduces/removes areas of the diagram that are not proper?

Projects

1. Construct a family of diagrams that have genus n and have the boundary property.

2. Construct a family of diagrams that have genus n and have the proper boundary property.

3. Calculate bounds on the span for a family of knots.

Further Reading

1. Colin C. Adams. *The knot book.* American Mathematical Society, Providence, RI, 2004

2. Peter R. Cromwell. *Knots and links.* Cambridge University Press, Cambridge, 2004

3. Naoko Kamada. Span of the Jones polynomial of an alternating virtual link. *Algebraic & Geometric Topology,* 4(November):1083–1101, 2004

4. Naoko Kamada. On the Jones polynomials of checkerboard colorable virtual links. *Osaka J. Math.*, 39(2):325–333, 2002

5. Louis H. Kauffman. State models and the Jones polynomial. *Topology,* 26(3):395–407, 1987

6. Kunio Murasugi. Jones polynomials and classical conjectures in knot theory. *Topology,* 26(1):187–194, 1987

7. Morwen B. Thistlethwaite. A spanning tree expansion of the Jones polynomial. *Topology,* 26(3):297–309, 1987

CHAPTER 9

The checkerboard framing

At the end of the last chapter, we determined an upper bound on the span of the bracket polynomial for diagrams with the boundary property. We prove that a virtual link diagram D has the boundary property if and only if D can be checkerboard colored. We introduce an abstraction of the checkerboard coloration for oriented virtual links and use this to calculate the number of components in a state and its dual. We use a variety of techniques, including algorithms, proof by induction, and contradiction.

9.1 CHECKERBOARD FRAMINGS

In the last section, we considered virtual link diagrams D with the boundary property. From a virtual link diagram D, construct the abstract link diagram (S_D, D) by applying the thickening map. To form the closed, compact two-dimensional surface $S(D)$, we attach disks to the boundary components of S_D. If D has the boundary property, the boundary components of S_D are in one-to-one correspondence with the components of s_{∂} and $\overline{s_{\partial}}$. Designate disks bounded by curves of s_{∂} as black regions of $S(D)$ and disks bounded by curves of $\overline{s_{\partial}}$ as white regions. No two regions of the same color can meet at an edge of the underlying diagram of D. To do so would imply that two components of a state share the same edge.

Examine the underlying oriented diagram of a trefoil and a checkerboard coloration in Figure 9.1. The checkerboard coloration of the oriented diagram in the plane transfers to a checkerboard coloration of an abstract link diagram (ALD) as shown in Figure 9.1c.

Following the orientation of the knot diagram in the ALD, we can replace the coloration with a labeling. Place a label of "1" on edges with black regions to the right (in the direction of orientation) and a "0" on edges with a black region to the left. For convenience, the edges are also marked with thick dashes to indicate an edge with a "0" label (see Figure 9.2). (In your own work, use red to mark the "0" labeled edges instead.)

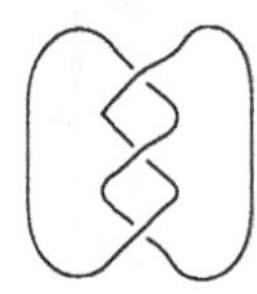

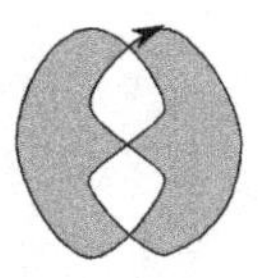

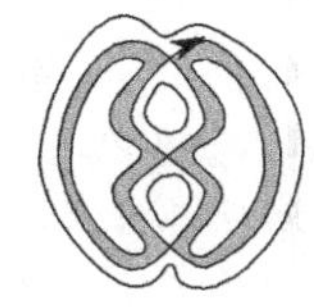

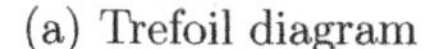

(a) Trefoil diagram (b) Checkerboard coloring (c) Checkerboard colored ALD

FIGURE 9.1: Trefoil and checkerboard colorations

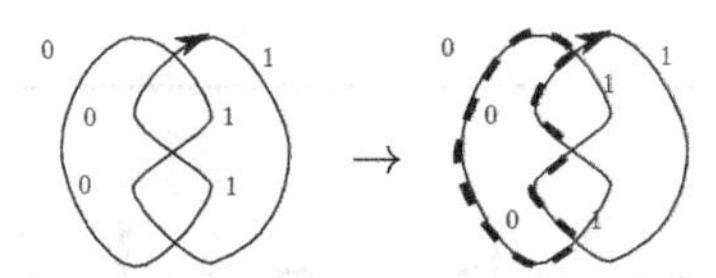

FIGURE 9.2: Labeled diagram to marked diagram

(a) Crossing possibility

(b) Crossing possibility

FIGURE 9.3: Labelings of crossings

A **cut point** or **cut loci** is a point on an edge of a virtual link diagram. The cut point subdivides the edge of the virtual link diagram into two edges. In a virtual link diagram with cut points, **edges** are bounded by 1) classical crossings, 2) a classical crossing and a cut point, or 3) two cut points. A **checkerboard framing** of a virtual link diagram consists of adding cut points to the diagram and labeling the resulting edges subject to the following conditions.

1. Each edge is labeled with either 0 or 1.
2. On each component, the edge labels alternate.
3. At a classical crossing, the edge labels satisfy one of the two possibilities shown in Figure 9.3.

In a checkerboard framing, we label the edges with 0's and 1's but for ease of notation, we mark edges labeled 0 with thick dashes as shown in Figure 9.4. (Some virtual link diagrams with cut points may not have checkerboard framings.) Three checkerboard framings of a virtual trefoil are shown in Figure 9.5.

A checkerboard framing F of the diagram D is denoted as the ordered pair (D, F).

Three different checkerboard framings of a virtual link diagram are shown in Figure 9.6. In Figure 9.6a, the cut points are placed in pairs underneath classical crossings. In Figure 9.6b, the framing has the minimum possible number of cut points; a pair of cut points is

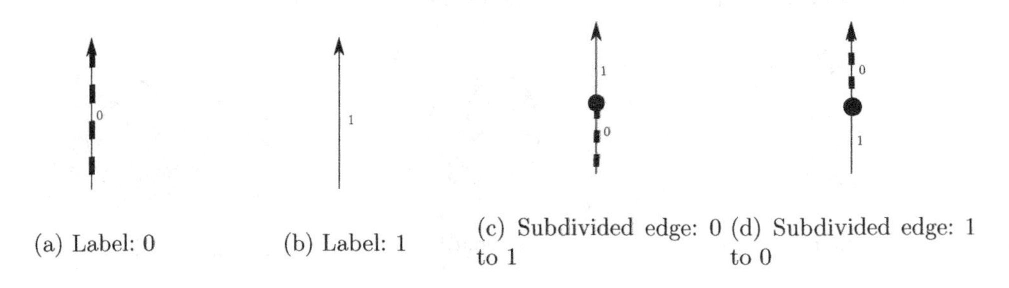

(a) Label: 0

(b) Label: 1

(c) Subdivided edge: 0 to 1

(d) Subdivided edge: 1 to 0

FIGURE 9.4: Edge labels

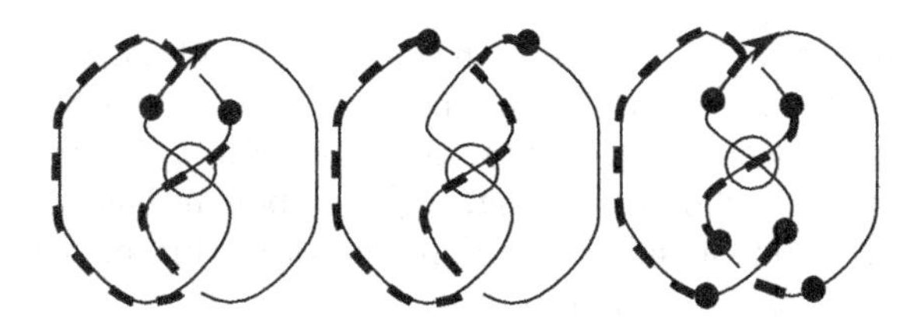

FIGURE 9.5: Three framings of a virtual trefoil

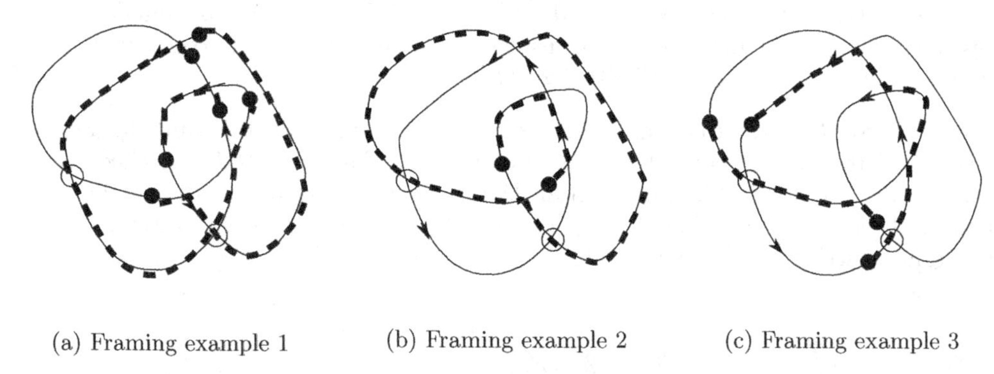

(a) Framing example 1 (b) Framing example 2 (c) Framing example 3

FIGURE 9.6: Framing examples

placed to one side of a classical crossing. In Figure 9.6c, pairs of cut points are placed near each virtual crossing. Placing cut points on the diagram near virtual crossings simulates the labeling of classical crossings. Every classical link diagram is checkerboard colorable, so this results in a checkerboard framing of the virtual link diagram.

Use $\mathcal{P}((D,F))$ to denote the number of cut points in (D,F). We define $\mathcal{P}_d(D)$ to be the minimum number of cut points of any checkerboard framing of the diagram D.

$$\mathcal{P}_d(D) = \min\{\mathcal{P}((D,F)) | F \text{ is a framing of } D\}. \tag{9.1}$$

We have the following theorem about the number of cut points in a framing.

Theorem 9.1. *For all virtual link diagrams D.*

1. *For all checkerboard framings F of D, $\mathcal{P}((D,F))$ is even.*

2. *For all checkerboard framings F of a diagram D with n classical crossings, $0 \leq \mathcal{P}((D,F)) \leq 2n$.*

3. *There is a checkerboard framing F of D such that $\mathcal{P}((D,F)) \leq 2\mathrm{v}_d(D)$.*

Proof. We prove that a virtual link diagram D must have an even number of cut points. Let D be an oriented virtual knot diagram. If the diagram has n classical crossings; then the diagram has $2n$ edges. Choose a start point and follow the orientation of the knot, the edge labels alternate as we pass through crossings and cut points. Returning to the start point, the edge has the same label. The checkerboard framing has an even number of edges (possibly subdivided from the edges of the knot diagram). Consequently, there were an even number of cut points in the framing.

In a multi-component virtual link diagram with n classical crossings, two components interlinked by a virtual crossing may have an odd number of edges, as in the virtual Hopf link. The checkerboard framing of the component must have an odd number of cut points to create an even number of edges and satisfy the condition that 0 and 1 labels alternate on the component. In the checkerboard framing, each component has an even number of edges and components with an odd number of cut points occurring in pairs. The checkerboard framing has an even number of cut points.

We prove that $0 \leq \mathcal{P}((D,F)) \leq 2c_d(D)$. Let D be a diagram with n crossings. At worst, we would have to add a cut point to every edge of D to satisfy the classical crossing condition.

We prove that for some checkerboard framing F of D, $\mathcal{P}((D,F))$ is bounded by two times the number of virtual crossings. Construct a framing of D by inserting cut points at each virtual crossing. Then, $\mathcal{P}((D,F)) \leq 2v_d(D)$. □

We want to understand how different checkerboard framings of the same diagram and checkerboard framings of equivalent diagrams may be related to (D,F). A checkerboard framing can be constructed for any diagram D, but we want control over the number of cut points. A checkerboard framing of a diagram can be modified as shown in Figure 9.7 and still obtain a checkerboard framing.

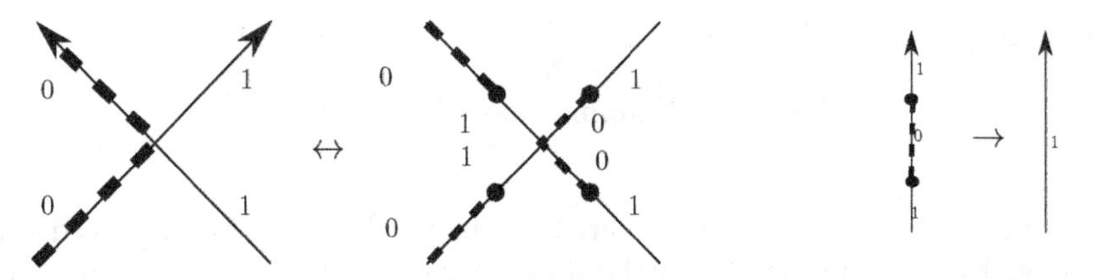

(a) Modification at crossing

(b) Reducing cut points on edge

FIGURE 9.7: Modifying the cut points

We determine the effect of the diagrammatic moves on the checkerboard framing. A checkerboard framing F is not altered by the virtual Reidemeister moves. These moves involve only virtual crossings and do not affect the classical crossings. The Reidemeister moves I–III do have the potential to interact with the checkerboard framing. We have the following result about the Reidemeister II move.

Theorem 9.2. *For all virtual link diagrams D with a framing F, if $D' \sim D$ then there is a framing F' of D' induced by the Reidemeister moves. Further, only the Reidemeister II move changes the number of cut points.*

Proof. For each classical Reidemeister move, consider the left hand side of the move. We construct all possible checkerboard framings and examine how the move affects the framing.

We examine a Reidemeister II move (with one of four possible framings) in Figure 9.8. The number of cut points in the induced framing increases as a result of the Reidemeister II move. The remaining moves and possible checkerboard framings are left as an exercise. □

For a virtual link diagram D with checkerboard framing F, the states of D inherit a framing from (D,F). In fact, if virtual link diagram D' is obtained by smoothing a set of classical crossings in a diagram D with checkerboard framing F, the diagram D' inherits a checkerboard framing from F'. The diagram D' is called a **partially expanded state**.

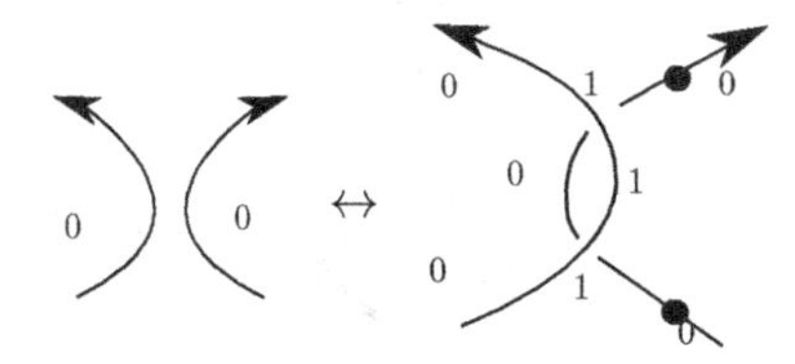

FIGURE 9.8: Framing under a Reidemeister II move

Theorem 9.3. *For all virtual link diagrams with a checkerboard framing, (D, F), a partially expanded state D' inherits a checkerboard framing. A state of the diagram, $s(D)$, also inherits a checkerboard framing and each component of the state contains an even number of cut points.*

Proof. Let D be a virtual link diagram with n classical crossings and a framing F.

In a link diagram with 0 classical crossings, each component has an even number of cut points.

If we vertically smooth a classical crossing in D, we obtain the virtual link diagram D_V with an induced checkerboard framing as shown in Figure 9.9.

If we horizontally smooth a classical crossing in D, we obtain the diagram D_H. To obtain an oriented virtual link diagram, we reverse the orientation inherited from D on half of the diagram. We then reverse the framing (inherited from D) on the same half of the diagram where the orientation was reversed. There are two possibilities for this reversal as shown in Figure 9.10b. Two examples of this process are shown, the virtual Hopf link in Figure 9.11 and the trefoil in Figure 9.12.

At classical crossings that contain edges with the original orientation, the original framing from D remains. At crossings that contain an edge with the original orientation and the reverse orientation, the modification of the framing given above satisfies the checkerboard framing conditions. At classical crossings where both edges have the reverse orientation, the modification of checkerboard framing satisfies the checkerboard framing conditions. We say that this framing is inherited from D.

We can continue this process until all classical crossings have been eliminated. The resulting state $s(D)$ has the same number of cut points as the original diagram. □

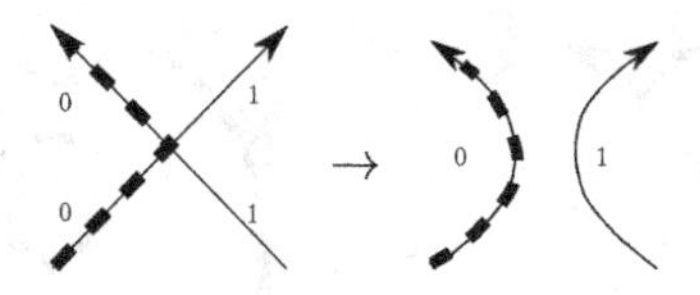

FIGURE 9.9: D_V with induced framing

Exercises

1. Find a checkerboard framing of the diagrams in Figure 9.13.
2. Compute $\mathcal{P}_d(D)$ for the diagrams in Figure 9.13.

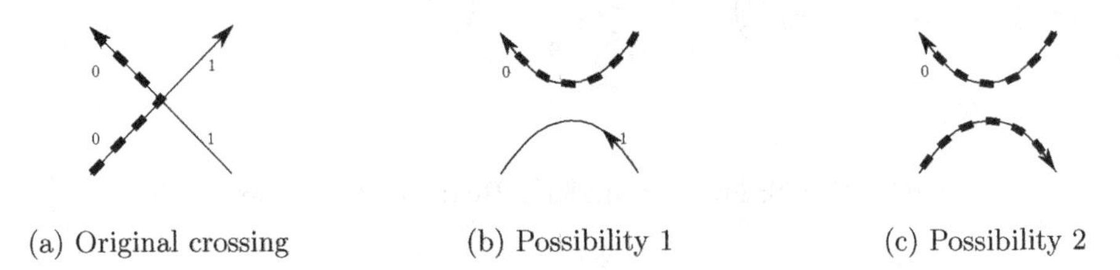

(a) Original crossing (b) Possibility 1 (c) Possibility 2

FIGURE 9.10: D_H with possible induced framings

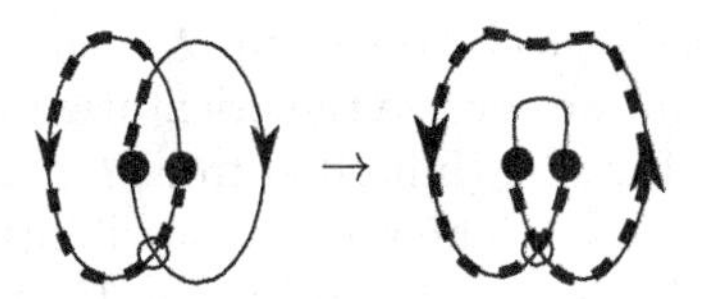

FIGURE 9.11: Horizontal smoothing of virtual Hopf link

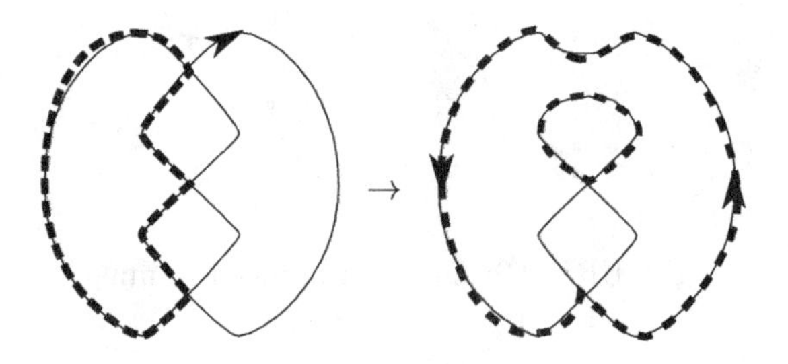

FIGURE 9.12: Horizontal smoothing of a trefoil

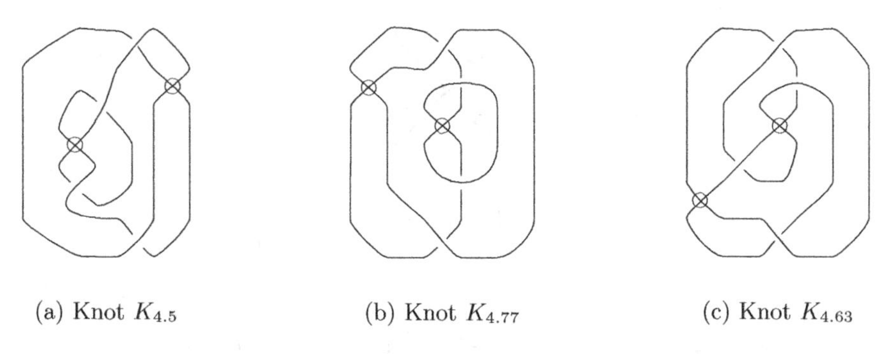

(a) Knot $K_{4.5}$ (b) Knot $K_{4.77}$ (c) Knot $K_{4.63}$

FIGURE 9.13: Construct checkerboard framings and compute $\mathcal{P}(D, B)$

3. Consider the left hand side of the Reidemeister moves I and III. Prove that a framing on the left hand side induces a framing on the right hand side of the move, and vice versa.
4. For a virtual Hopf diagram with a checkerboard framing, recursively expand the crossings and verify that any component in a state has an even number of cut points.
5. For a virtual trefoil diagram with a checkerboard framing, recursively expand the crossings and verify that any component in a state has an even number of cut points.

9.2 CUT POINTS

A checkerboard framing F of a diagram D is **minimal** if

$$\mathcal{P}(D, F) = \mathcal{P}_d(D). \tag{9.2}$$

A diagram D is **checkerboard colorable** if there is a checkerboard framing F of D such that $\mathcal{P}(D, F) = 0$.

The **cut point number** of K, $\mathcal{P}(K)$, to be the minimal number of cut points in any framing of any diagram equivalent to K. $\mathcal{P}(K)$ can be defined as

$$\mathcal{P}(K) = \min\{\mathcal{P}_d(D) | D \sim K\}. \tag{9.3}$$

This is an invariant of virtual links.

Theorem 9.4. *$\mathcal{P}(K)$ is an invariant of the virtual links.*

Proof. The proof is left as an exercise. □

Each virtual link has a diagram with a minimal number of cut points, but we don't know exactly what that diagram is—and there may be multiple diagrams with that number of cut points.

Corollary 9.5. *For all classical links K, $\mathcal{P}(K) = 0$.*

Proof. Let K be a classical link. Then there is a link diagram D of K with no virtual crossings. The classical diagram D has checkerboard framing with no cut points. □

Remark 9.1. *Advanced students may be interested in reading Vassily O. Manturov's work on atoms and virtual links. His work provides a graduate level view of this material.*

FIGURE 9.14: Cut point placement

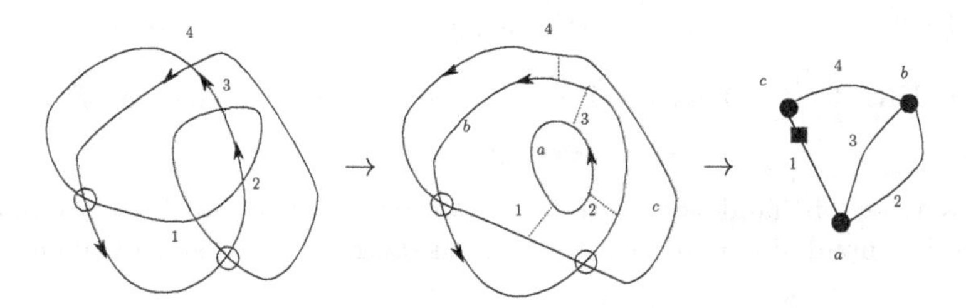

FIGURE 9.15: Cut point placement

Our next goal is to construct a framing F of the virtual link diagram D so that the cut points are positioned in a specific manner. The state s_B is determined by the checkerboard framing F. A different framing results in a different s_B. Given (D, F), we bound $|s_B| + |\overline{s_B}|$.

The **Seifert state** of a diagram D, $s_{seif}(D)$, is obtained by smoothing all crossings in the diagram vertically. The **Seifert graph** of a diagram is constructed by taking the components of the Seifert state as vertices and the traces of the Seifert state as edges. A **cycle** is an ordered list of edges where the vertices do not repeat. The length of the cycle is determined by the number of edges in the list. In a classical knot, the Seifert graph has only **even cycles**. In the Seifert graph, identify odd cycles and insert an extra vertex on specific edges to ensure that all cycles have even length. On the link diagram, place pairs of cut points as shown at crossings that correspond to edges (traces) with an extra vertex as shown in Figure 9.14. This process is illustrated using the diagram from Figure 9.6b. In Figure 9.15, we see a flat version of the diagram with numbered crossings. Next, we see the Seifert smoothed state. The traces are numbered and the components are labeled with the letters a, b, and c. Then, the Seifert graph has vertices a, b, and c with edges numbered $1, 2, 3$, and 4. An extra vertex (marked as a square) has been inserted on the edge number 1 so that the graph does not contain an odd cycle. In the virtual link diagram, we insert a pair of cut points on one side of the corresponding classical crossing. The resulting diagram and framing is given in Figure 9.16.

FIGURE 9.16: Framed diagram

Note that the state s_B can be manipulated by moving the cut points to the other side of the crossing.

Exercises

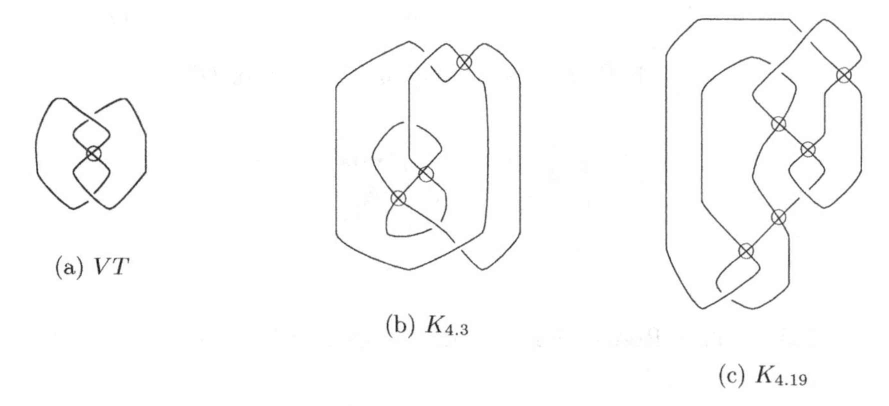

(a) VT

(b) $K_{4.3}$

(c) $K_{4.19}$

FIGURE 9.17: Bound $\mathcal{P}(K)$

1. Find a checkerboard framing using the strategy given in this section for the diagrams in Figure 9.17.
2. Calculate a bound on $\mathcal{P}(K)$ for each diagram in Figure 9.17.
3. Construct the state s_s and $\overline{s}_s$ for the diagrams in Figure 9.17 and compare with ∂S_D. Calculate $|s_s| + |\overline{s}_s|$ and $|\partial S_D|$.
4. Prove Theorem 9.4.

9.3 EXTENDING THE THEOREM

We use cut points to extend the Kauffman-Murasugi-Thistlethwaite Theorem to virtual link diagrams without the boundary property.

Theorem 9.6. *For all connected virtual link diagrams D with n classical crossings, $g(S(D)) = g$, and a checkerboard framing F with $\mathcal{P}((D,F)) = 2c$ obtained from the Seifert graph*

$$|s_B(D)| + |\overline{s_B}(D)| \leq (n+2) - 2g + c. \tag{9.4}$$

Proof. We prove this theorem by induction. In the base case, we have a virtual link diagram with n classical crossings, genus g, and a framing with 0 cut points. Hence $|s_B| + |\overline{s_B}| \leq (n+2) - 2g$.

Let D be a connected virtual link diagram D with n classical crossings, genus g and a checkerboard framing F where there are $2c$ cut points ($c > 0$) and that cut points are placed in pairs, adjacent to a crossing as shown in Figure 9.18a.

Select a crossing with a pair of cut points and construct a new diagram D_V by vertically smoothing the crossing. The diagram D_V has $n-1$ crossings, genus g or $g-1$, and an inherited framing with $2(c-1)$ cut points as shown in Figure 9.18.

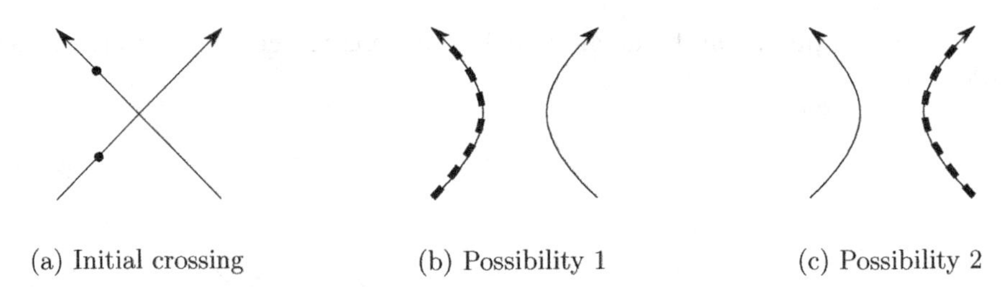

(a) Initial crossing (b) Possibility 1 (c) Possibility 2

FIGURE 9.18: Modifying D to obtain D_V

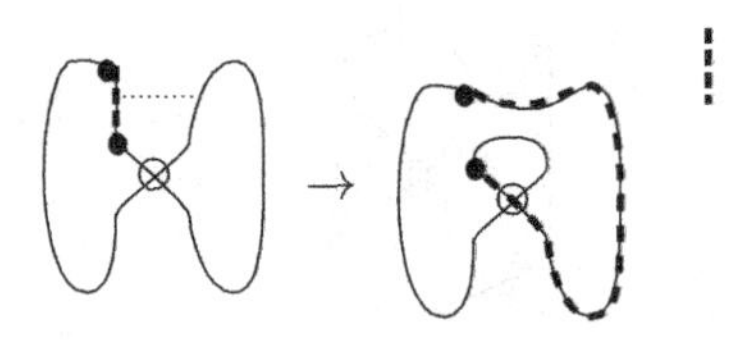

FIGURE 9.19: Resmoothing a single component into a single component

Now, the diagram D_V has $n-1$ crossings and $2(c-1)$ cut points. If we assume that D_V has genus $g-1$, then by the induction hypothesis,

$$|s_B(D_V)| + |\overline{s_B}(D_V)| \leq (n-1) + 2 - 2(g-1) + (c-1) \tag{9.5}$$
$$= n + 2 - 2g + c. \tag{9.6}$$

If we assume that D_V has genus g then

$$|s_B(D_V)| + |\overline{s_B}(D_V)| \leq (n-1) + 2 - 2(g) + (c-1) \tag{9.7}$$
$$= n - 2g + c. \tag{9.8}$$

We conclude from Equations 9.5 and 9.7 that

$$|s_B(D_V)| + |\overline{s_B}(D_V)| \leq n + 2 - 2g + c. \tag{9.9}$$

Now, $|s_B(D)| = |s_B(D_V)|$. We need to calculate $|\overline{s_B}(D)|$ in terms of $|\overline{s_B}(D_V)|$.

The state $\overline{s_B}(D_v)$ is related to $\overline{s_B}(D)$ by resmoothing at a single trace. If the trace in $\overline{s_B}(D_V)$ connects two components,

$$|\overline{s_B}(D)| + 1 = |\overline{s_B}(D_V)|. \tag{9.10}$$

If the trace in $\overline{s_B}(D_V)$ connects a single component, then resmoothing results in one component or two components. We recall the inherited framing described in Theorem 9.3, leaving all cut points in position as expansions occur. We first consider a special case. Suppose that the component of $\overline{s_B}(D_v)$ has no cut points except at the position of the smoothed crossing. Then, the corresponding state of D has a single pair of cut points. Resmoothing into two components would place one cut point into each component which is not possible. If the component of $\overline{s_B}(D_v)$ does have cut points, the position of the cut points required for resmoothing into two components indicates that D did not have a checkerboard framing.

This is a contradiction. We conclude that the resmoothing must result in a single component. Hence, $|\overline{s_B}(D_V)| = |\overline{s_B}(D)|$. We conclude that

$$|s_B(D)| + |\overline{s_B}(D)| \leq |s_B(D_V)| + |\overline{s_B}(D_V)|. \tag{9.11}$$

From Equations 9.9 and 9.11, we calculate that

$$|s_B(D)| + |\overline{s_B}(D)| \leq n + 2 - 2g + c$$

□

We can now state and prove our extension of the Kauffman-Murasugi-Thistlethwaithe theorem.

Theorem 9.7. *For all virtual link diagrams D with n classical crossings, $g(S(D)) = g$, with a framing F from the Seifert graph such that $\mathcal{P}_d(D) = 2c$ and $s_B = s_\alpha$ or $s_B = s_\beta$.*

$$\text{Span}(f_K(A)) \leq 4(n - g) + 2c. \tag{9.12}$$

Proof. Let D be a connected virtual link diagram with n classical crossings, genus g, and a framing from the Seifert graph with $\mathcal{P}((D, F)) = 2c$. The cut point can be adjusted so that $s_B = s_\alpha$ or $s_B = s_\beta$. We assume that $s_B = s_\alpha$. The maximum degree is from the state s_{alpha}: $d_{max} = n + 2(|s_B| - 1)$. The lowest possible degree is from the state $\overline{s_B}$: $d_{min} = -n - 2(|\overline{s_B}| - 1)$. Subtracting the minimum possible degree from the maximum:

$$\begin{aligned} d_{max} - d_{min} &= n + 2(|s_B| - 1) - (-n - 2(|\overline{s_B}| - 1)) \\ &= 2n + 2(|s_B| + |\overline{s_B}|) - 4. \end{aligned} \tag{9.13}$$

Applying Theorem 9.6 to Equation 9.13:

$$\begin{aligned} M - m &= 2n + 2(n + 2 - 2g + c) - 4 \\ &= 4(n - g) + 2c. \end{aligned} \tag{9.14}$$

□

Corollary 9.8. *For all virtual link diagrams where D Span$(f_D(A))$ is not divisible by 4 then D is not equivalent to a classical diagram.*

Proof. Suppose that D is a classical link diagram with no virtual crossings. Let s denote the all A state. The contribution from the all A state can be expressed as: $A^n d^{(|s|-1)}$. Letting $|s| = k$, the maximum and minimum degrees contributed by this all A state are $n + 2k - 2$ and $n - 2k + 2$. (Note that the difference is a multiple of 4.) Let the state s' be obtained by resmoothing a single crossing. Since D is a diagram with no virtual crossings: $|s'| = |s| \pm 1$. Hence the maximum degree of a contribution from s' is:

$$n + 2k - 2 \text{ or } n + 2k - 6. \tag{9.15}$$

Both possibilities differ by a factor of 4. Since D is a classical diagram, each change in smoothing type either divides a loop into two or fusions two components. □

Exercises

1. For the virtual trefoil, VT, compute the bound on Span$(f_{VT}(A))$.
2. Compute a bound on Span$(f_D(A))$ for the diagrams in Figure 9.20.

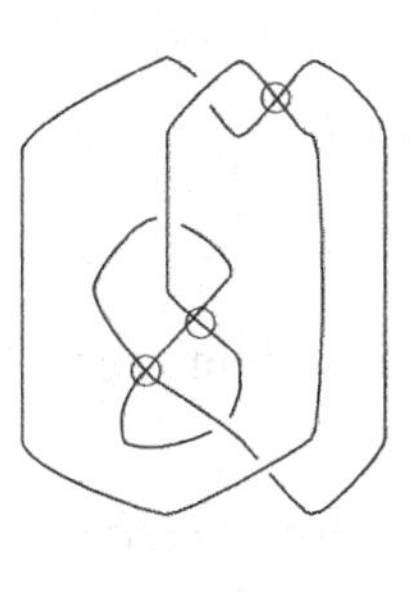

(a) Knot $K_{4.3}$

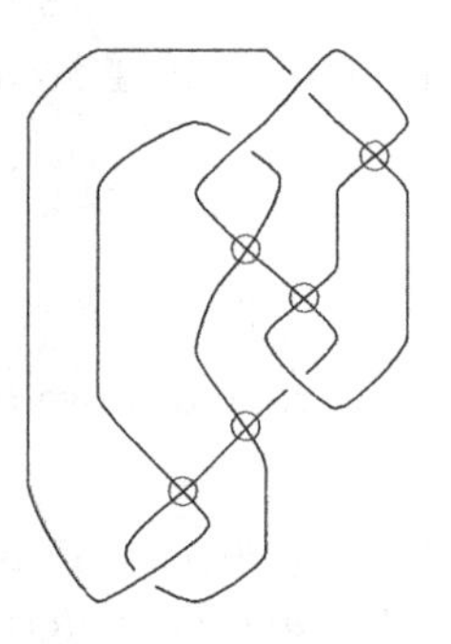

(b) Knot $K_{4.19}$

FIGURE 9.20: Compute the bounds

9.4 OPEN PROBLEMS AND PROJECTS

Open problems

1. Is $\mathcal{P}(K)$ always obtained on the diagram D where $\mathsf{c}_d(D) = \mathsf{c}(D)$?
2. Is there a relationship between $\mathcal{P}(K)$ and the genus of K?
3. Can we formulate a relationship between: $\mathsf{c}(K_1) + \mathsf{c}(K_2)$ and $\mathsf{c}(K_1 \sharp K_2)$ using $\mathcal{P}(K_1 \sharp K_2)$?

Projects

1. Rewrite the theorem to include virtual links that are not connected.
2. Select a family of virtual links (such as $(1, i)$, $(1, 1, i)$, ...) and compute the bounds on this family.
3. Find a virtual link diagram D such that $\text{Span}(f_D(A)) = 4(n-g)+2c$ where $n = \mathsf{c}(D)$, $g = g(S(D))$, and $2c = \mathcal{P}(D)$.

Further Reading

1. Naoko Kamada. On the Jones polynomials of checkerboard colorable virtual links. *Osaka J. Math.*, 39(2):325–333, 2002
2. Naoko Kamada. Span of the Jones polynomial of an alternating virtual link. *Algebraic & Geometric Topology*, 4(November):1083–1101, 2004
3. Colin C. Adams. *The knot book.* American Mathematical Society, Providence, RI, 2004
4. Peter R. Cromwell. *Knots and links.* Cambridge University Press, Cambridge, 2004
5. J. Scott Carter. *How surfaces intersect in space*, volume 2 of *Series on Knots and Everything.* World Scientific, May 1995

CHAPTER 10

Modifications of the bracket polynomial

In this section, we modify $f_K(A)$. This leads to several different versions of the f-polynomial.

10.1 THE FLAT BRACKET

The flat f-polynomial is obtained by evaluating the f-polynomial at $A = 1$.

$$F(K) = f_K(1). \tag{10.1}$$

We can also rewrite the axioms of the bracket polynomial to compute the flat bracket directly. This computation is simpler than the bracket polynomial since $A = A^{-1}$. Let U denote the unknot and let $U \cup K$ denote the disjoint union of the link K and the unknot. Then the skein relation is rewritten as:

$$\langle \times \rangle_f = \langle \times \rangle_f = \langle)(\rangle_f + \langle \asymp \rangle_f .$$

The other bracket polynomial axioms simplify as

$$\langle U \rangle_f = 1 \tag{10.2}$$

and

$$\langle U \cup K \rangle_f = -2 \langle K \rangle_f. \tag{10.3}$$

Recall that $w(K)$ denotes the writhe of K. The flat bracket polynomial is not invariant under the Reidemeister I move; each Reidemeister I move multiplies the polynomial by a factor of -1. In this context, the flat f-polynomial is

$$F(K) = (-1)^{w(K)} \langle K \rangle_f. \tag{10.4}$$

Based on the axioms, we immediately obtain the following theorem.

Theorem 10.1. *For all classical links K with n components then $F(K) = (-2)^{n-1}$.*

Proof. Let K be a classical link. Then there exists a diagram D of K with zero virtual crossings. Through a sequence of crossing changes, we can transform K into the n unknotted, unlinked components. Then

$$F(K) = F(\bigcup_{i=1}^{n} U). \tag{10.5}$$

Now, $F(\bigcup_{i=1}^{n} U) = (-2)^{n-1}$. □

Corollary 10.2. *For all classical knots K, $F(K) = 1$.*

We compute the flat bracket for the virtual links in Figure 10.1:

$$F(VH) = -2, \qquad F(K_8) = -1, \qquad F(K_{2.1}) = 1. \tag{10.6}$$

Notice that since $F(K_8) = -1$, this fact proves that K_8 is not a classical link and not equivalent to the unknot.

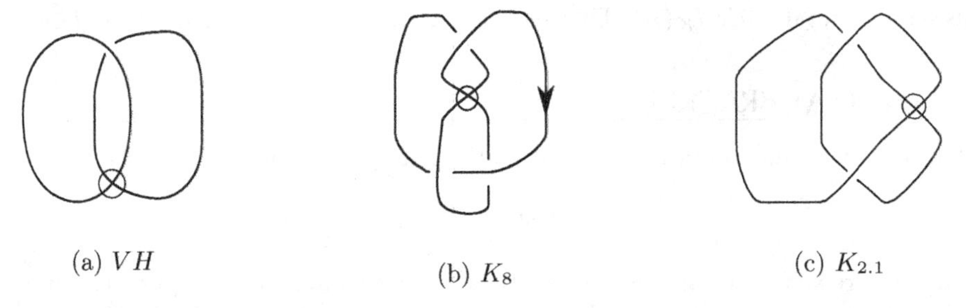

(a) VH (b) K_8 (c) $K_{2.1}$

FIGURE 10.1: Flat bracket examples

We have the following theorems about $F(K)$.

Theorem 10.3. *The flat f-polynomial $F(K)$ is an invariant of flat virtual links*

Theorem 10.4. *The flat bracket polynomial is invariant under the virtualization move. Further, the flat f-polynomial is an invariant of free links.*

In some cases, the flat f-polynomial provides just enough information to differentiate a link diagram from the unknot and is substantially easier to compute than the f-polynomial.

Exercises

1. Apply the flat skein relation to a Reidemeister I move and compute that the correct normalization factor is -1.

2. Prove that the flat bracket is invariant under the diagrammatic moves.

3. Compute the flat bracket for the diagrams in Figure 10.2.

4. Prove Theorem 10.3.

5. Prove Theorem 10.4.

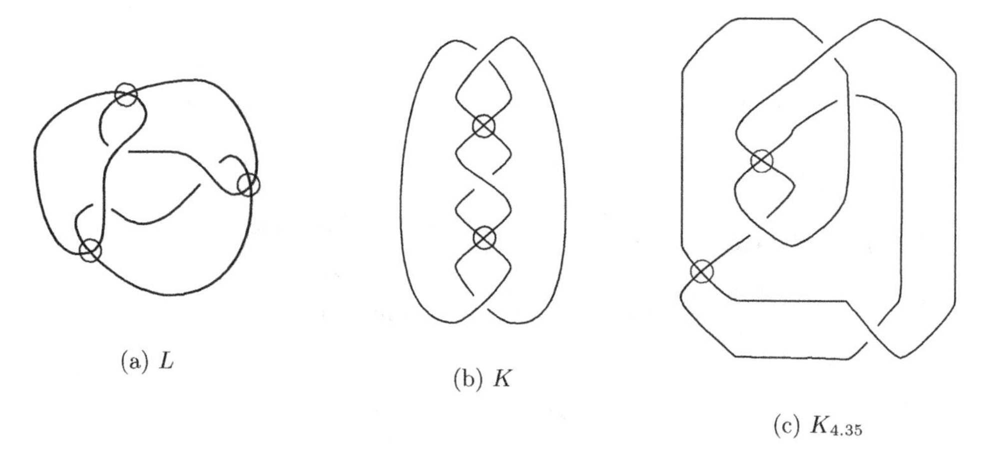

(a) L

(b) K

(c) $K_{4.35}$

FIGURE 10.2: Compute the flat bracket

10.2 THE ARROW POLYNOMIAL

The arrow polynomial is a decorated version of the f-polynomial introduced by Dye and Kauffman. The invariant incorporates extra information at each crossing and captures some information about the virtual crossings. The arrow polynomial is equivalent to the Miyazawa polynomial. The Miyazawa polynomial was introduced by Yasuyuki Miyazawa; the Miyazawa polynomial's definition begins with magnetic graph diagrams instead of virtual link diagrams. These two different approaches lead to equivalent invariants of virtual links. References to the Miyazawa polynomial and the arrow polynomial are included in the further reading list. The arrow polynomial is a Laurent polynomial in the variables A, A^{-1} and $K_1, K_2, \ldots$. The variable K_i captures the information about the virtual crossings. To construct the arrow polynomial, we modify the skein relation as shown in Figure 10.3.

(a) Positive crossing

(b) Negative crossing

FIGURE 10.3: Skein relation for the arrow polynomial

For an oriented virtual link diagram, the **states of the arrow polynomial** are determined by choosing a smoothing type (see Figure 10.4) for each classical crossing in the diagram. The states consist of closed curves (possibly containing virtual crossings) that are marked by an even number of arrows. We evaluate the states in a two step process. First the number of arrows in each closed curve is reduced by recursively canceling adjacent arrows with the same orientation (see Figure 10.5). After reduction, if a curve C has $2n$ arrows, the **arrow number** of the component is n (denoted $a(C) = n$). We evaluate a single component

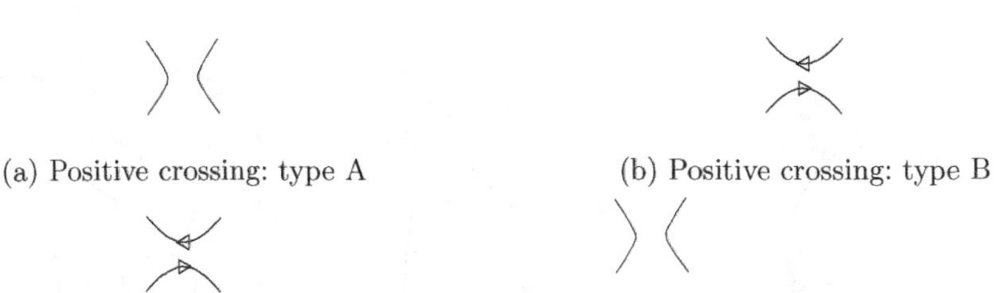

(a) Positive crossing: type A

(b) Positive crossing: type B

(c) Negative crossing: type A

(d) Negative crossing: type B

FIGURE 10.4: Smoothing type

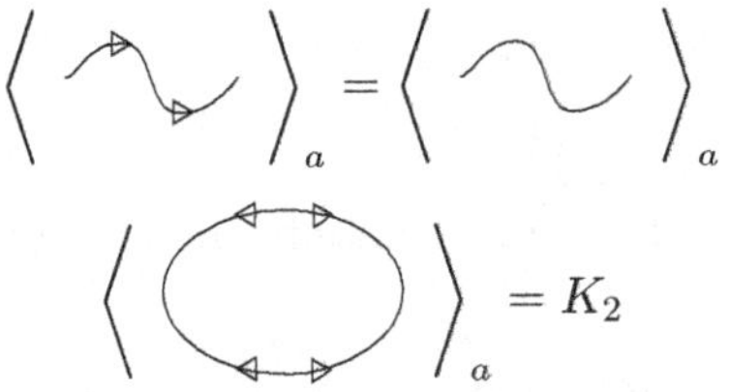

$= K_2$

FIGURE 10.5: Reducing arrows and evaluating loops

state as follows:

$$\langle C \rangle_a = \begin{cases} 1 \text{ if } a(C) = 0 \\ K_{a(C)} \text{ if } a(C) > 0. \end{cases} \tag{10.7}$$

We show a sample reduction and evaluation in Figure 10.5. In a state s with multiple components $C_1, C_2, \ldots, C_n$, we evaluate the state as

$$\langle s \rangle_a = d^{n-1} \prod_{i=1, a(C_i) \neq 0}^{n} K_{a(C_i)}. \tag{10.8}$$

For a virtual link diagram D, let s denote a state of D, let $\alpha(s)$ (respectively $\beta(s)$) denote the number of type A (respectively type B) smoothings in state s, and $|s|$ denotes the number of loops in s. Now,

$$\langle D \rangle_a = \sum_{s \in S} A^{\alpha(s) - \beta(s)} d^{|s|-1} \langle s \rangle_a. \tag{10.9}$$

Let D be a virtual link diagram with writhe $w(D)$. We define the normalized arrow polynomial of D as

$$Ar_D(A, K_i) = (-A^{-3})^{w(D)} \langle D \rangle_a. \tag{10.10}$$

We compute the arrow polynomial of the virtual trefoil, VT, shown in Figure 10.6

$$Ar_{VT}(A, K_i) = A^4 + A^6(2K_1 - 1) + A^{10} K_1. \tag{10.11}$$

Theorem 10.5. *For all virtual link diagrams D, if $D \sim D'$ then $Ar_D(A, K_i) = Ar_{D'}(A, K_i)$.*

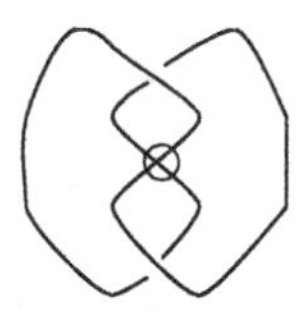

FIGURE 10.6: The virtual trefoil

Proof. We prove invariance under the diagrammatic moves. We begin with the Reidemeister I move. Apply the arrow skein relation to the right hand side of a Reidemeister I move in Equation 10.12.

$$\langle \;\; \rangle_a = A\langle \;\; \rangle_a + A^{-1}\langle \;\; \rangle_a = (-A)^{-3}\langle \;\; \rangle_a . \tag{10.12}$$

There are two possible orientations of the Reidemeister II moves. We compute both sides of Reidemeister II move on two co-oriented strands in Equation 10.13.

$$\langle \;\; \rangle_a = \langle \;\; \rangle_a + A^{-1}\langle \;\; \rangle_a + A\langle \;\; \rangle_a + \langle \;\; \rangle_a = \langle \;\; \rangle_a . \tag{10.13}$$

Next, consider the Reidemeister II move with two oppositely oriented strands (Equation 10.14). The remainder of this computation is left as an exercise.

$$\langle \;\; \rangle_a = \langle \;\; \rangle_a . \tag{10.14}$$

Examine the Reidemeister III move by expanding the left hand side (Equation 10.15) and reducing the resulting expression. We then compute the right hand side (Equation 10.17).

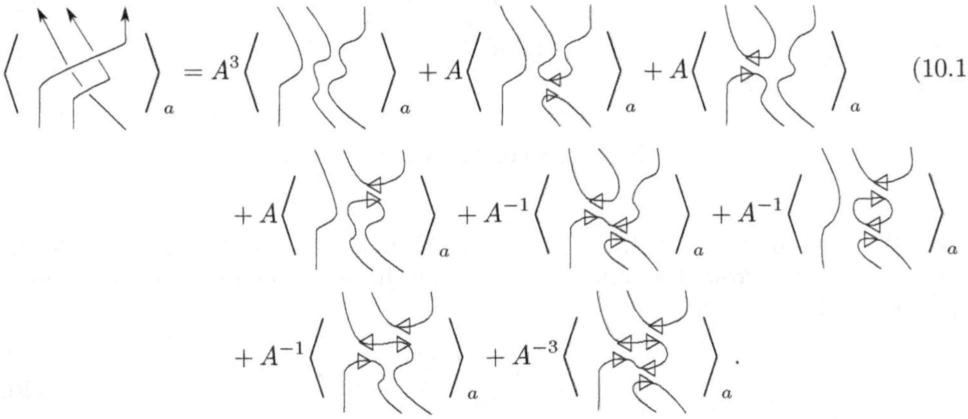

$$\langle \; \rangle_a = A^3\langle \; \rangle_a + A\langle \; \rangle_a + A\langle \; \rangle_a + A\langle \; \rangle_a + A^{-1}\langle \; \rangle_a + A^{-1}\langle \; \rangle_a + A^{-1}\langle \; \rangle_a + A^{-3}\langle \; \rangle_a. \quad (10.15)$$

After reduction, this becomes

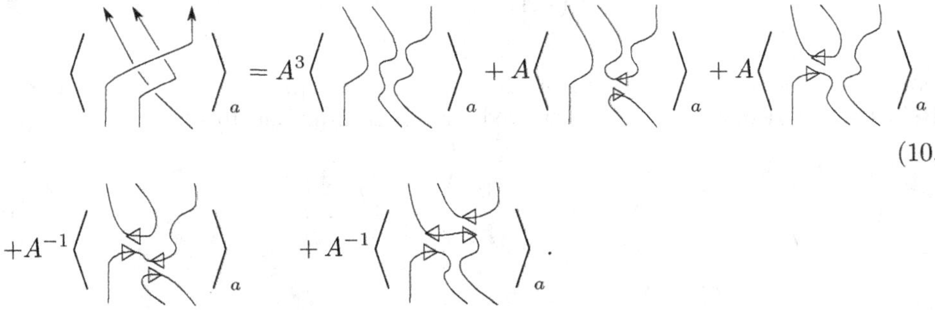

$$\langle \; \rangle_a = A^3\langle \; \rangle_a + A\langle \; \rangle_a + A\langle \; \rangle_a + A^{-1}\langle \; \rangle_a + A^{-1}\langle \; \rangle_a. \quad (10.16)$$

The expansion of the right hand side of the Reidemeister III move:

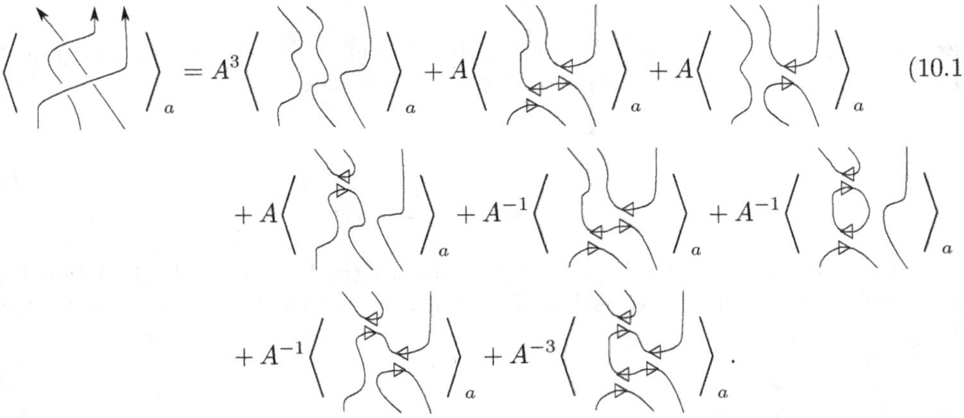

$$\langle \; \rangle_a = A^3\langle \; \rangle_a + A\langle \; \rangle_a + A\langle \; \rangle_a + A\langle \; \rangle_a + A^{-1}\langle \; \rangle_a + A^{-1}\langle \; \rangle_a + A^{-1}\langle \; \rangle_a + A^{-3}\langle \; \rangle_a. \quad (10.17)$$

After further reduction of the right hand side, we see that the two expansions are equivalent when reduced. The virtual Reidemeister moves I–III do not involve classical crossings and both sides of the moves are equivalent when expanded. We only need to examine the virtual Reidemeister move IV (Equation 10.18).

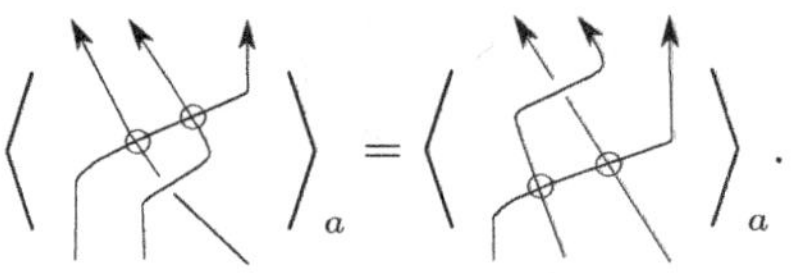

$$\langle \; \rangle_a = \langle \; \rangle_a. \quad (10.18)$$

□

A way virtualization is detected by the arrow polynomial in Equation 10.19:

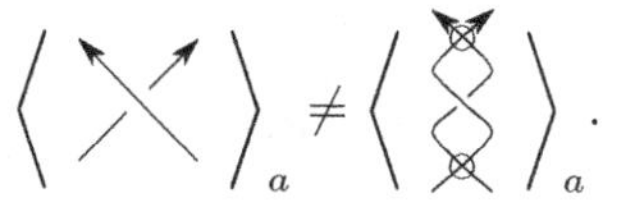

$$\left\langle \quad \right\rangle_a \neq \left\langle \quad \right\rangle_a . \tag{10.19}$$

But, two applications of virtualization are not detected.

Theorem 10.6. *The arrow polynomial is invariant under double virtualization:*

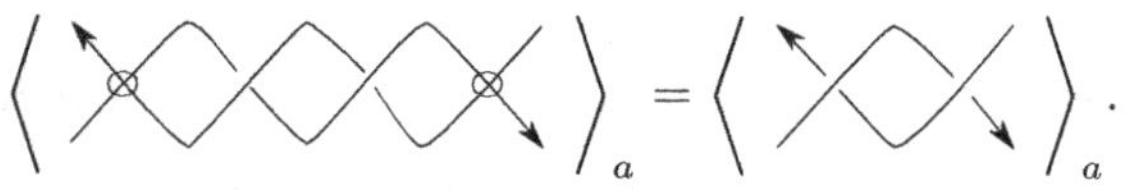

$$\left\langle \quad \right\rangle_a = \left\langle \quad \right\rangle_a . \tag{10.20}$$

We prove that the arrow polynomial is sensitive to the planarity of the diagram. The arrow polynomial of a virtual link is a sum of monomials of the form:

$$A^n (K_{i_1}^{j_1} K_{i_2}^{j_2} \dots K_{i_m}^{j_m}). \tag{10.21}$$

The **k-degree** of the monomial in Equation 10.21 is

$$d = i_1 \times j_1 + i_2 \times j_2 + \dots i_m \times j_m. \tag{10.22}$$

If the monomial has no K_i terms, the k-degree is zero.

We define the **arrow set** of a virtual link K (denoted as $AS(K)$). The polynomial $Ar_K(A, K_i)$ has the following form:

$$Ar_K(A, K_i) = \sum_{i_1, i_2, \dots i_n \in \mathbb{N}} \sum_{j \in \mathbb{Z}} c_j A^j K_{i_1} K_{i_2} \dots K_{i_n}. \tag{10.23}$$

The number of non-zero c_j is finite. We call these terms the **surviving summands** and for each surviving summand of the form, $c_j A^j K_{i_1} K_{i_2} \dots K_{i_n}$, we obtain an associated k-degree. The arrow set, $AS(K)$, is the set of k-degrees obtained from the arrow polynomial. The arrow set has a finite number of elements and it is an invariant of virtual links since the k-degrees are obtained from the arrow polynomial.

Corollary 10.7. *For all virtual link diagrams D, $AS(D)$ is invariant under the diagrammatic moves.*

We prove several theorems about the k-degree and the arrow set. Let D be a diagram with a checkerboard framing F. A state s of the arrow polynomial inherits a labeling from the checkerboard framing and the arrows in skein relation. The edges of s are determined by the cut points and arrows. The edges are alternately labeled with 0's and 1's. We eliminate cut points on the labeled state as shown in Figure 10.7 by reversing the 0 and 1 labels between any pairs of cut points and then removing the cut points. (Note that there may be an arrow between pairs of cut points.)

After the reduction, edges are defined by arrows and labeled with either 0 or 1. Each arrow has an associated weight of ± 1 as shown in Figure 10.8. The weight of an arrow a is denoted as $\mathrm{wt}(a)$.

Theorem 10.8. *For all checkerboard colorable link diagrams D, $AS(D) = 0$.*

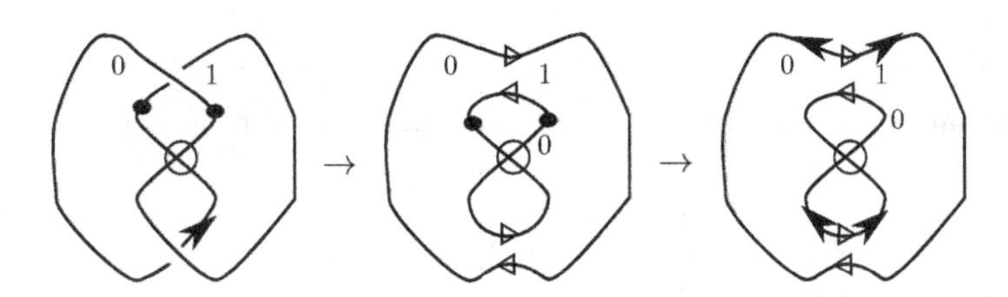

FIGURE 10.7: Constructing a labeled state

(a) Arrow with weight: 1

(b) Arrow with weight: −1

FIGURE 10.8: Weighted arrows

Proof. The strategy in this proof is due to N. Kamada and Y. Miyazawa. Let D be a checkerboard colorable virtual link diagram in the surface $S(D)$. The diagram D has a checkerboard framing with zero cut points. We mark the edges of the diagram with the framing. Without loss of generality, we assume crossing in the diagram is marked as shown in Figure 10.9. The state with all vertical smoothings has k-degree zero because there are zero arrows.

Consider a state with exactly one horizontal smoothing. Using the checkerboard framing, we see that a component of the state has either two arrows or zero arrows. If a component has two arrows, then the arrows' weight will sum to zero. (We see this by realizing the closed curve has only two edges and determine the direction of the second arrow based on the checkerboard coloration.) The arrows cancel and the state has k-degree zero. By the inductive hypothesis, we assume that a state with $k - 1$ horizontal smoothings also has k-degree zero. This indicates the arrows' weight on a component sums to zero; every component has zero arrows after reduction.

We resmooth a single crossing to obtain a state with k horizontal smoothings. The trace of the resmoothing either 1) connects two components or 2) connects two points on the same component. In the two component case, the resmoothing connects two loops with an arrow number of zero. Using the checkerboard coloring, we observe that the resmoothing introduces two oppositely signed arrows.

In the one component case, resmoothing must produce two components. (Resmoothing and obtaining one component is not possible if we have a checkerboard coloring.) Sketching out a possible checkerboard coloring of a component, we observe that the upper part of a component has a net of 1 positive arrow and the lower part has a net of 1 negative arrow. Then each new loop will have 0 arrows. □

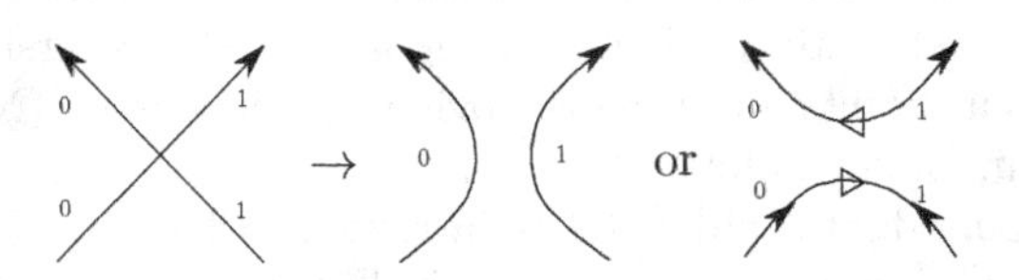

FIGURE 10.9: Checkerboard and arrows

Corollary 10.9. *For all virtual link diagrams D, if $AS(D) \neq 0$ then D is a non-classical virtual link diagram.*

Theorem 10.10. *For all virtual link diagrams D,* $\max(AS(D)) \leq \mathsf{v}(D)$. *The arrow set, $AS(D)$, provides a lower bound on the number of virtual crossings.*

Proof. Let D be a virtual link diagram with a checkerboard framing F. Let s be a state of the diagram marked with traces and the checkerboard framing.

From each state, we construct a classical link diagram $\lambda(s, F)$. The state s is a collection of curves containing arrows and cut points. The arrows and cut points divide the curves into a collection of segments labeled with either 1 or 0. Reduce the labeling by reversing the labeling between any pair of cut points and then eliminating the cut points. Continue until all cut points have been eliminated.

We construct a classical diagram, $\lambda(s)$, from the state s. Examine each pair of arrows that is connected by a trace. Insert a horizontal smoothing or a classical crossing as indicated in the Figure 10.10. Two connected arrows with positive (respectively negative) weight result in a positive (respectively negative) crossing. Two connected arrows with opposite weight result in a vertical smoothing. Next, construct a classical crossing at each virtual crossing by changing the edge labeled "1" to an over passing edge.

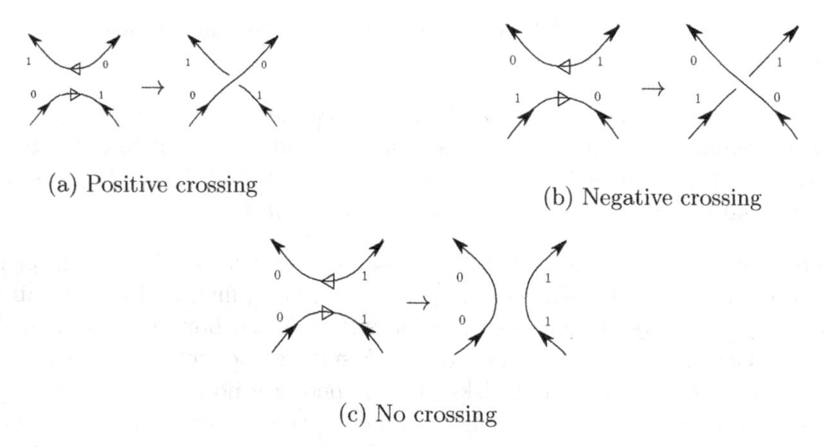

(a) Positive crossing

(b) Negative crossing

(c) No crossing

FIGURE 10.10: Constructing a classical diagram

Now, $\lambda(s)$ is a classical link diagram. The only crossings are between different components and each component is either labeled "0 " or "1". The components have an orientation obtained from the original virtual link.

Let $\lambda(s)_v$ denote the set of crossings obtained from the virtual crossings, where a component labeled 1 overpasses. Respectively, let $\lambda(s)_c$ denote the set of crossings obtained from the smoothed classical crossings, where a component labeled 0 overpasses. We compute that

$$L(\lambda(s)) = \sum_{c \in \lambda(s)_v} \operatorname{sgn}(c). \tag{10.24}$$

By construction, $L(\lambda(s)) \leq \mathsf{v}(K)$. We also see that if a_1 and a_2 are arrows corresponding to the constructed crossing c in the set $\lambda(s)_c$ then $\operatorname{sgn}(c) = wt(a_1) = wt(a_2)$. Note that if we switch the over passing strand at the virtual crossings, then we have an unlink. As a result,

$$\sum_{c \in \lambda(s)_v} \operatorname{sgn}(c) = \sum_{c \in \lambda(s)_c} \operatorname{sgn}(c). \tag{10.25}$$

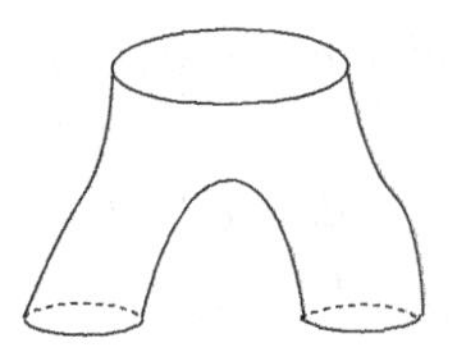

FIGURE 10.11: Pair of pants surface

We need to prove that for some framing F, $L(\lambda(s)) = d$, where d is the k-degree of s. For each component C_i of the state s, label the edges so that the weight from each component is non-negative. We construct the framing F by adding cut points as needed at individual crossings. We use a to indicate an arrow from the set of arrows in the state s. Then

$$d = \frac{1}{2}\sum_i \sum_a \mathrm{wt}(a) = \sum_{c \in \lambda(s)_v} \mathrm{sgn}(c). \tag{10.26}$$

□

We need the following lemma about cutting a two-dimensional surface into surfaces with boundary.

Lemma 10.11. *For all oriented, closed, two dimensional surfaces F with genus $g \geq 1$ if $g = 1$ then F contains at most 1 non-intersecting curve that does not bound a disk. If $g > 1$ then F contains at most $3g - 3$ non-intersecting curves that 1) do not bound a disk in the surface and 2) no pair of curves that bound an annulus in the surface.*

Proof. Note that these curves are allowed to separate the surface. If we cut a torus along a curve that does not bound a disk, then we produce a twice punctured sphere (an annulus). If there are two non-intersecting curves in a torus that do not bound disks, then the curves form the boundary of an annulus in the surface. A surface F of genus g with $g > 1$ contains $3g - 3$ curves that 1) do not bound disks and 2) there are no pairs of curves that bound an annulus. We can cut the surface along such a collection of curves to form a collection of surfaces with boundary that are called "pairs of pants". If the surface contains a curve C that does not intersect this set of curves, then C co-bounds an annulus with an element of the set. □

We use Lemma 10.11 in the proof of Theorem 10.12.

Theorem 10.12. *Let K be a virtual link. Now, let m be the maximum number of different K_i in a surviving summand of $AS(K)$. Then $1/3m + 1 \geq g$ where g is a lower bound on the virtual genus of K.*

Proof. Let K be a virtual link. Let D be a diagram equivalent to K with surface $S(D)$ such that $\}(S(D))$ is minimal. Suppose that $g = \}(S(D))$. Suppose that a summand of $AS(K)$ contains m factors: $K_{i_1}, K_{i_2}, \ldots K_{i_m}$ where $i_j \neq i_k$. This summand corresponds to a state $s(D)$ which lies in the surface $S(D)$. Each K_i term represents a curve in $S(D)$ that does not bound a disk. Further, since $s(D)$ is a state of D, the curves do not intersect. Then $S(D)$ contains m distinct curves that do not bound disks and can not cobound an annulus. We note that $m \geq 3g - 3$ for $m > 1$. □

We define the normalized flat arrow polynomial by evaluating $Ar_K(1, K_i)$. We can also compute a flat arrow bracket polynomial as $\langle K \rangle_{fa}$ directly from a skein relation. The normalized flat arrow polynomial can distinguish between two diagrams that are not distinguished by the f-polynomial. Recall that $\langle K_1 \sharp K_2 \rangle = \langle K_1 \rangle \langle K_2 \rangle$. We define a modification of connected sum by inserting a single classical crossing as shown in Figure 10.12. We denote

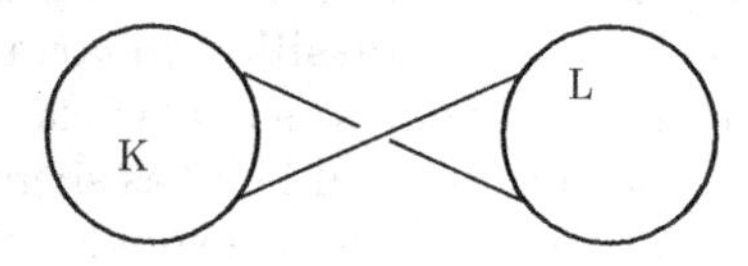

FIGURE 10.12: Modification of connect sum: $K \sharp_t L$

this modification as $K \sharp_t L$.

If either K or L is a classical diagram, then $K \sharp_t L \sim K \sharp L$ and $\langle K \sharp_t L \rangle_a = (-A^3) \langle K \sharp L \rangle$. If $K \sharp_t L \sim K \sharp L$ for all virtual knot diagrams K and L then $\langle K \sharp_t L \rangle_{fa} = (-1) \langle K \sharp L \rangle_{fa}$. We easily prove that these $K \sharp L$ and $K \sharp_t L$ are not equivalent for all virtual link diagrams K and L by considering Kishino's knot and a twisted version of Kishino's knot (see Figure 10.13). We compute that

$$\langle K \rangle_{fa} = -3 + K_2 + K_1^2$$
$$\langle K_t \rangle_{fa} = 1 + K_2 - K_1^2.$$

The flat arrow polynomial gives a lower bound on the genus (the minimal genus surface is at least a torus) and a bound on the number of virtual crossings ($\mathsf{v}(K) \geq 2$ and $\mathsf{v}(K_t) \geq 2$).

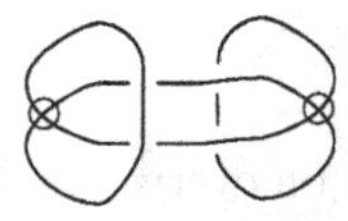

(a) Kishino's knot, K

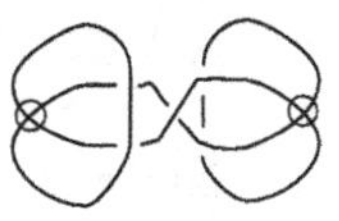

(b) Twisted Kishino's knot, K_t

FIGURE 10.13: Applying the flat arrow polynomial

Exercises

1. Compute the arrow polynomial of Kishino's knot.
2. Compute the arrow polynomial of the link diagrams shown in Figure 10.2.
3. Calculate $AS(D)$ for the diagrams in Figure 10.2.
4. Use the arrow polynomial of the link diagrams shown in Figure 10.2 to determine a bound on the genus of the virtual link and the virtual crossing number.
5. Prove that Equation 10.14 is true.
6. Prove that Equation 10.18 is true.

7. Prove that Equation 10.19 is true.
8. Prove that Equation 10.20 is true.

10.3 VASSILIEV INVARIANTS

We consider another evaluation of $f_K(A)$, the Taylor expansion of $f_K(e^x)$. From the coefficients of this expansion, we obtain a set of **Vassiliev invariants** of K. Vassiliev invariants have been defined on singular virtual links and singular classical links. In addition to the classical and virtual crossings, a singular virtual link has singular points and a set of singular moves (see Figure 10.14). The f-polynomial is extended to singular virtual links using

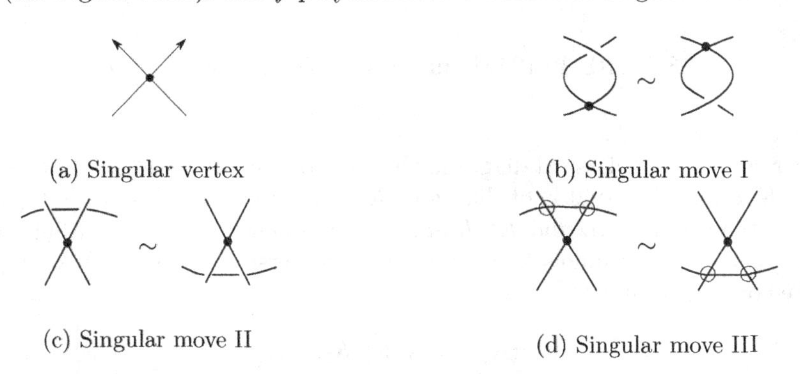

(a) Singular vertex (b) Singular move I

(c) Singular move II (d) Singular move III

FIGURE 10.14: Singular vertex and moves

the following skein relation:

$$\left\langle \times_{\bullet} \right\rangle = \left\langle \times_{+} \right\rangle - \left\langle \times_{-} \right\rangle. \tag{10.27}$$

A Vassiliev invariant v of **degree** $\leq k$ vanishes (is zero) on every singular link diagram with $k+1$ or more singular points. A Vassiliev invariant of degree k vanishes on every singular link with $k+1$ or more singular points, but is non-zero on some singular link diagram with exactly k singular points. For example, a Vassiliev invariant of degree 0 has the following property:

$$v\left(\times_{+} \right) - v\left(\times_{-} \right) = 0. \tag{10.28}$$

A Vassiliev invariant of degree $\leq n$ satisfies the following condition. Let $K_{\bullet^{n+1}}$ denote a virtual link diagram with $n+1$ singular points: $p_1, p_2, \ldots, p_{n+1}$. Let $\overline{x}$ denote a vector of length $n+1$ such that $x_i = \pm 1$. (There are 2^{n+1} such vectors.) Use $K_{\overline{x}}$ to denote a virtual link diagram where the singular point p_i is resolved as a positive crossing with sign x_i. Let $|\overline{x}|$ denote the number of $x_i = -1$. Then if v is a Vassiliev invariant of degree $\leq n$:

$$v(K_{\bullet^{n+1}}) = \sum_{\overline{x}} (-1)^{|\overline{x}|} v(K_{\overline{x}}) \tag{10.29}$$

$$= 0. \tag{10.30}$$

Define

$$V_K(x) = f_K(e^x). \tag{10.31}$$

Recall the expansion of e^x at $x = 0$ from calculus:

$$e^x = 1 + x + \frac{x^2}{2!} + \frac{x^3}{3!} \cdots \tag{10.32}$$

$$= \sum_{n=0}^{\infty} \frac{1}{n!} x^n. \tag{10.33}$$

Rewrite $V_K(x)$ using Equation 10.32 as:

$$V_K(x) = \sum_{n=0}^{\infty} v_n(K) x^n. \tag{10.34}$$

Extend the f-polynomial to singular virtual links using the following definition:

$$\langle \times_{\bullet} \rangle = \langle \times_{+} \rangle - \langle \times_{-} \rangle. \tag{10.35}$$

A sample computation involving the virtual trefoil knot, T:

$$f_T(A) = A^{-4} + A^{-6} - A^{-2}. \tag{10.36}$$

Let $A = e^x$ and

$$f_T(e^x) = e^{-4x} + e^{-6x} - e^{-2x}. \tag{10.37}$$

From Equation 10.37, calculate the Laurent series expansion:

$$V_T(x) = 1 - 8x + 24x^2 - \frac{136x^3}{3} + 64x^4 - \frac{1096x^5}{15} + \frac{352x^6}{5} + O[x^7] \tag{10.38}$$

where $O[x^7]$ is a polynomial containing all terms $v_i(T)x^i$ where $i \geq 7$.

Theorem 10.13. *For all singular virtual links K, $v_n(K)$ (the coefficient of x^n in $V_K(x)$) is a Vassiliev invariant of degree $\leq n-1$.*

Proof. This fact is proved inductively using a method first used in J. S. Birman's and X.-S. Lin's "Knot polynomials and Vassiliev's invariants" published in *Inventiones Mathematicae* in 1993. Let $K_\bullet$ be a knot with one singular crossing. Let K_{+1} (respectively K_{-1}) denote the resolution of the singular crossing as a positive (respectively negative) crossing. Let K_0 denote the vertical smoothing of the crossing and let K_∞ denote the horizontal. Now, we compute that

$$f_{K+}(A) = (-A^{-3})^{k+1}(A\langle K_0 \rangle + A^{-1}\langle K_\infty \rangle), \tag{10.39}$$

$$f_{K-}(A) = (-A^{-3})^{k-1}(A^{-1}\langle K_0 \rangle + A^{1}\langle K_\infty \rangle). \tag{10.40}$$

Rewriting Equations 10.39 and 10.40

$$f_{K+}(A) = (-A^{-3})^{+1}(A f_{K_0}(A) + A^{-1} f_{K_\infty}(A)), \tag{10.41}$$

$$f_{K-}(A) = (-A^{-3})^{-1}(A^{-1} f_{K_0}(A) + A^{1} f_{K_\infty}(A)). \tag{10.42}$$

Then, simplify Equations 10.41 and 10.42:

$$f_{K+}(A) = -A^{-2} f_{K_0}(A) - A^{-4} f_{K_\infty}(A), \tag{10.43}$$

$$f_{K-}(A) = -A^{2} f_{K_0}(A) - A^{4} f_{K_\infty}(A). \tag{10.44}$$

Combine the two equations to see that

$$f_{K_\bullet}(A) = (-A^2 + A^{-2}) f_{K_0}(A) + (-A^{-4} + A^4) f_{K_\infty}(A). \tag{10.45}$$

Let $A = e^x$ and compute the Laurent expansion of $A^m - A^{-m}$:

$$\left(\sum_{n=0}^{\infty} \frac{1}{n!}(mx)^n \right) - \left(\sum_{n=0}^{\infty} \frac{1}{n!}(-mx)^n \right) = \sum_{n=0}^{\infty} \frac{2}{(2n+1)!}(mx)^{2n+1}. \tag{10.46}$$

This indicates that $F_{K_\bullet}(x)$ is divisible by x or that $v_0(K_\bullet) = 0$.

By the induction hypothesis, we assume that $F_{K_{\bullet^n}}(x)$ is divisible by x^n. Now,

$$f_{K_{\bullet^{n+1}}}(A) = (-A^2 + A^{-2}) f_{K_{\bullet^n 0}}(A) + (-A^{-4} + A^4) f_{K_{\bullet^n \infty}}(A). \tag{10.47}$$

Then $F_{K_{\bullet^{n+1}}}(x)$ is divisible by x^{n+1} and $v_n(K_{\bullet^{n+1}})$ vanishes. □

Exercises

1. Compute $V_K(x)$ (up to degree 6) for the knot diagrams shown in Figure 10.15.
2. Prove that the f-polynomial is invariant under the singular moves with the expansion

   (10.48)

   and the usual bracket axioms.
3. Prove that a Laurent expansion of the arrow polynomial for $A = e^x$ produces Vassiliev invariants.

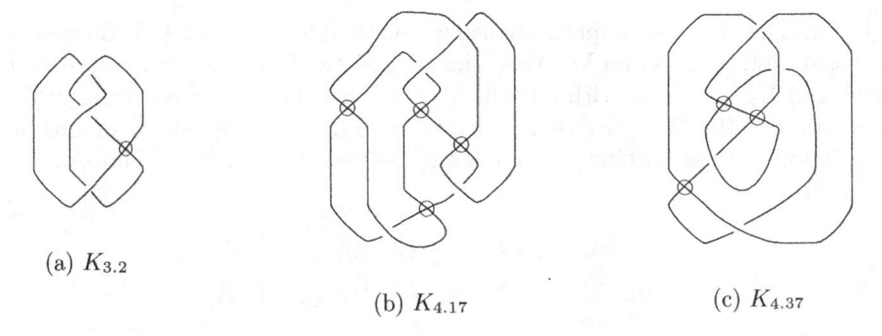

(a) $K_{3.2}$

(b) $K_{4.17}$

(c) $K_{4.37}$

FIGURE 10.15: Compute Vassiliev invariants

10.4 OPEN PROBLEMS AND PROJECTS

Open problems

1. Is there a relationship between the crossing weights and arrows?
2. What computational methods can be used to compute the arrow polynomial of a family of virtual knot diagrams?
3. Is there a relationship between $g(K)$, $\mathcal{P}(K)$, and the arrow numbers?

Projects

1. Apply the arrow polynomial to a family of knots obtained through symmetry. Determine a formula for the arrow polynomial.
2. Construct a non-trivial family of virtual knots and compute the flat arrow bracket.
3. Construct a family of virtual links (K_n) such that $AS(K_n) = \{0\}$.
4. Compute the span of the arrow polynomial for a family of knot diagrams.

Further Reading

1. H. A. Dye and Louis H. Kauffman. Virtual crossing number and the arrow polynomial. *J. Knot Theory Ramifications*, 18(10):1335–1357, 2009
2. Kumud Bhandari, H. A. Dye, and Louis H. Kauffman. Lower bounds on virtual crossing number and minimal surface genus. In *The mathematics of knots*, volume 1 of *Contrib. Math. Comput. Sci.*, pages 31–43. Springer, Heidelberg, 2011
3. Yasuyuki Miyazawa. A virtual link polynomial and the virtual crossing number. *J. Knot Theory Ramifications*, 18(5):605–623, 2009
4. Yasuyuki Miyazawa. A multi-variable polynomial invariant for virtual knots and links. *J. Knot Theory Ramifications*, 17(11):1311–1326, 2008
5. Naoko Kamada. An index of an enhanced state of a virtual link diagram and Miyazawa polynomials. *Hiroshima Math. J.*, 37(3):409–429, 2007
6. Louis H. Kauffman. An extended bracket polynomial for virtual knots and links. *J. Knot Theory Ramifications*, 18(10):1369–1422, 2009
7. S. Chmutov, S. Duzhin, and J. Mostovoy. *Introduction to Vassiliev knot invariants.* Cambridge University Press, Cambridge, 2012
8. Micah W. Chrisman. Twist lattices and the Jones-Kauffman polynomial for long virtual knots. *J. Knot Theory Ramifications*, 19(5):655–675, 2010

III

Algebraic structures

CHAPTER 11

Quandles

In this chapter, we introduce an algebraic structure named the quandle. A quandle consists of a set X and an operation $\triangleright$ that satisfies algebraic versions of the Reidemeister moves. A quandle can be associated with an oriented virtual knot diagram. Equivalent oriented virtual knot diagrams are associated with the same quandle structure; the quandle is an invariant of oriented virtual knots. Invariants based on the quandle are defined using information from the classical crossings; a quandle invariant can distinguish between Kishino's knot and its flip. This result negatively answers the question that we asked in Chapter 5: for all virtual knot diagrams K, is $K \sim K^F$?

11.1 TRICOLORING

Before introducing quandles, we study tricoloring of virtual knot diagrams. Tricoloring a virtual knot diagram allows us to distinguish the trefoil (and other knots) from the unknot. (Tricoloring is actually an example of labeling a knot diagram with the elements of a quandle.) To tricolor a virtual knot diagram, we label (or color) the arcs of a knot diagram with elements of the set $\{r, g, b\}$. Recall that an arc of the diagram begins and terminates at the underpasses of classical crossings. In most of the literature on quandles, an arc is referred to as an edge. In this chapter, an **edge** begins and terminates at the underpasses.

To color a virtual knot diagram, we select an element from the set $\{r, g, b\}$ for each edge in the diagram. Since we select edge labels or colors from a three element set, this is referred to as a tricoloring. The edge colors are selected using the following rules.

1. Each edge is labeled with an element of the set: $\{r, g, b\}$.
2. Each classical crossing is labeled so that either all three colors **or** exactly one color appears at the crossing.

In Figure 11.1, three virtual knot diagrams have been colored following these rules. The trefoil knot in Figure 11.1a is labeled with three colors, but it can also be labeled with a single color. The unknot (which consists of a single edge) is labeled with a single color. The edges of the figure 8 knot are colored with one color in Figure 11.1c. Attempting to label the figure 8 knot with three colors will convince you that it is not possible to label the figure 8 knot with all three colors.

A coloring of a diagram with only one element is a **trivial coloring**. A coloring using all three colors is a **non-trivial coloring**. Tricolorability is a knot invariant.

Theorem 11.1. *For all virtual knot diagrams K with a non-trivial tricoloring, if L is a*

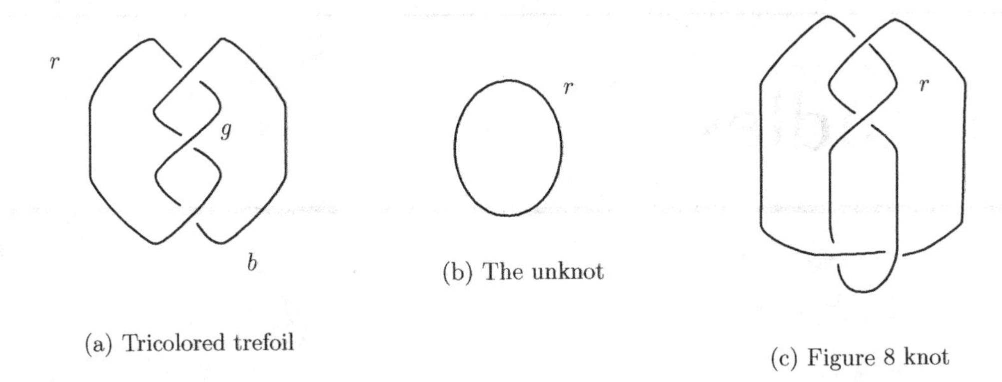

(a) Tricolored trefoil (b) The unknot (c) Figure 8 knot

FIGURE 11.1: Labeling diagrams with a 3 element set

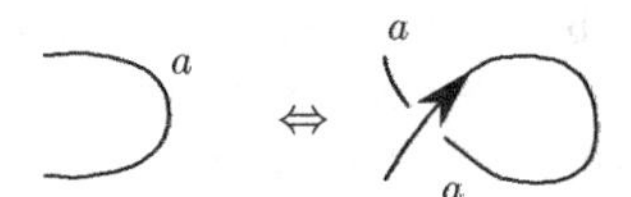

FIGURE 11.2: Labeling and Reidemeister I

virtual knot diagram related to K by a single diagrammatic move then L has a non-trivial tricoloring.

Proof. Let K be a virtual knot diagram with a non-trivial coloring. Let L be a diagram related to K by a single diagrammatic move. Consider a labeled Reidemeister I move, as shown in Figure 11.2. The edge in K is labeled with an a. We see that both edges in L must be labeled with a. The rest of the coloring is unchanged.

Consider the Reidemeister II move in Figure 11.3. Suppose that two edges in K have different colors. There are a total of four edges in the diagram of L. From the initial colors in K and the axiom that each crossing has three colors, the coloring of L is determined. The colors on the remainder of the diagrams K and L correspond.

Next, consider a Reidemeister II move where the two edges of the diagram in K have the same color. Then the edges in L are determined by the axioms and all edges have the same color.

The labeling of the Reidemeister III move has been left as an exercise. The virtual Reidemeister moves do not change the coloring of the edges. □

This theorem leads to the following corollaries.

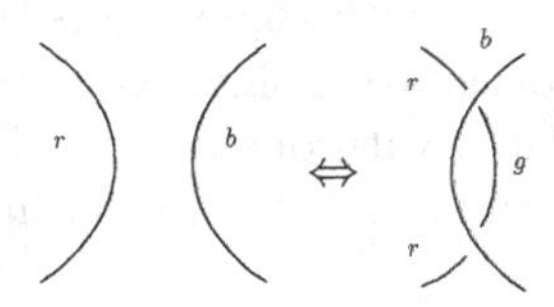

FIGURE 11.3: Labeling and Reidemeister II

Corollary 11.2. *For all equivalent virtual knot diagrams K and K', K has a non-trivial tricoloring if and only if K' has a non-trivial tricoloring.*

Corollary 11.3. *For all virtual link diagrams K with a non-trivial tricoloring, K is not equivalent to the unknot.*

Proof. Suppose that K can be non-trivially tricolored and that K is equivalent to U, the unknot. Then there are a sequence of diagrammatic moves that transform K into the unknot. But the Reidemeister moves preserve tri-coloring, meaning that two diagrams related by a single move must both be colored with three colors. This is a contradiction. □

We now answer a question about the symmetries of the knot. We demonstrate that there is virtual knot diagram K where K is not equivalent to K^F. Kishino's knot, K_{Kish}, and its flip, K^F_{Kish} are not distinguished from each other or the unknot by the crossing weight or the f-polynomial. (The arrow polynomial distinguishes Kishino's knot and its flip from the unknot.) In Figure 11.4a, we see K_{Kish} and K^F_{Kish}. Kishino's knot cannot be colored with three colors. However, the flip of Kishino's knot can be colored with three colors. Tricoloring proves that the flip of Kishino's knot is not equivalent to the unknot and Kishino's knot.

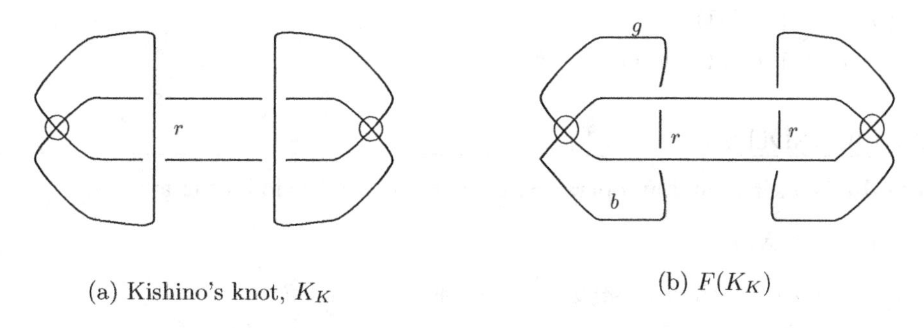

(a) Kishino's knot, K_K

(b) $F(K_K)$

FIGURE 11.4: Kishino's knot and its flip

Exercises

1. Determine if it is possible to non-trivially tricolor the diagrams in Figure 11.5 with three colors.

2. Prove that if the tangle

can be tricolored then the tangle

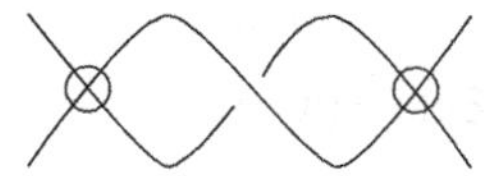

can be tricolored.

3. Prove that the Reidemeister III move preserves tricolorability.

4. Prove that numerator closures of the following tangles can be tricolored.

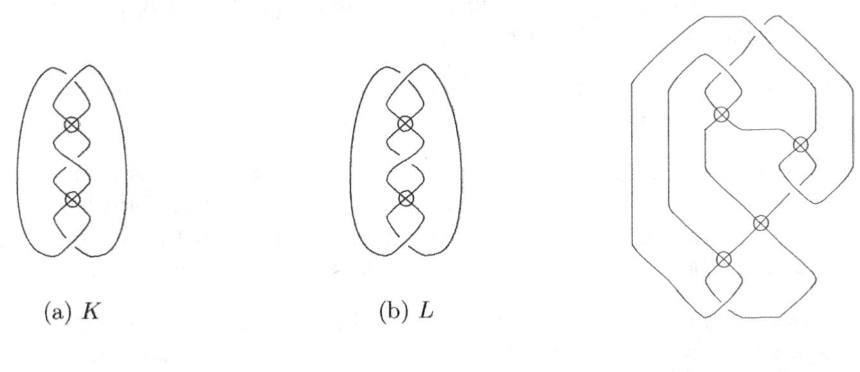

(a) K

(b) L

(c) Knot $K_{4.58}$

FIGURE 11.5: Can these diagrams be colored?

(a) $(1, 1, 1)$

(b) $(1, 1, 1, 1, 1, 1)$

(c) $(1, 1, 1, 1, 1, 1, 1, 1, 1)$

11.2 QUANDLES

A **quandle** is a set X and an operation $\triangleright$ that satisfies the following axioms:

1. For all $a \in X$, $a \triangleright a = a$.
2. For all $b, c \in X$, there exists a unique a such that $a \triangleright b = c$.
3. For all a, b, and $c \in X$, $(a \triangleright b) \triangleright c = (a \triangleright c) \triangleright (b \triangleright c)$.

These axioms do not include associativity or commutativity ($a \triangleright b \neq b \triangleright a$!). Many different sets and operations satisfy the quandle axioms.

The tricoloring structure defined in the last section has the structure of a quandle. Instead of using the set $\{r, g, b\}$, we let $X = \{0, 1, 2\}$. We use a Cayley table to define the operation $\triangleright$. In a Cayley table, read off the value of $a \triangleright b$ by finding a in the left most column and then finding b in the top row. The position of these two elements identifies the entry in the table containing $a \triangleright b$.

$\triangleright$	0	1	2
0	0	2	1
1	2	1	0
2	1	0	2

TABLE 11.1: Cayley table: $X = \{0, 1, 2\}$

From Table 11.1, we see that $1 \triangleright 2 = 0$. Inspecting the table, we observe that for $a \neq b$, $a \triangleright b$ always evaluates to the remaining element of the set, just as the two colors meeting at a crossing in the tricoloring always produced the third color. We also observe that $a \triangleright a = a$ for all a in the set $\{0, 1, 2\}$. We verify the second axiom by inspection—for every ordered pair of elements b, c (which are possibly equal) there is a unique element a such that $a \triangleright b = c$.

The third axiom is verified by checking all nine cases. We verify one case. Let $a = 0, b = 1$, and $c = 2$. Begin with the left hand side of the third axiom:

$$(a \triangleright b) \triangleright c = (0 \triangleright 1) \triangleright 2 \quad (11.1)$$
$$= 2 \triangleright 2 \quad (11.2)$$
$$= 2.$$

Now evaluate the right hand side of the third axiom:

$$(a \triangleright c) \triangleright (b \triangleright c) = (0 \triangleright 2) \triangleright (1 \triangleright 2) \quad (11.3)$$
$$= 1 \triangleright 0 \quad (11.4)$$
$$= 2.$$

After checking all nine cases, we can conclude that $(a \triangleright b) \triangleright c = (a \triangleright c) \triangleright (b \triangleright c)$ is true. Then we can conclude that $X = \{0, 1, 2\}$ and $\triangleright$ form a quandle. We can also define the operation $\triangleright$ computationally as $a \triangleright b = 2b - a \mod 3$. With this definition, we obtain the same Cayley table as in Table 11.1.

Two integers a and b are said to be equivalent "modulus n" (or "mod n") (denoted $a \equiv b \mod n$) if there exists an integer x such that $a = nx + b$. This means that every integer is equivalent mod n to one of the elements of the set $\{0, 1, \ldots n - 1\}$. We define the **dihedral quandle** on the set $\mathbb{Z}_n$ with $a \triangleright b$ defined as $2b - a \mod n$. The dihedral quandle on $\mathbb{Z}_n$ is denoted as R_n.

The tricoloring quandle has the same structure as the dihedral quandle R_3. We can verify that each entry in Table 11.1 satisfies $a \triangleright b \equiv 2b - a \mod 3$.

Next, we study the quandle R_5. In R_5, $X = \mathbb{Z}_5$ and $a \triangleright b$ is defined as $2b - a \mod 5$. A Cayley table of R_5 is illustrated in Table 11.2.

$\triangleright$	0	1	2	3	4
0	0	2	4	1	3
1	4	1	3	0	2
2	3	0	2	4	1
3	2	4	1	3	0
4	1	3	0	2	4

TABLE 11.2: Cayley table: R_5

We prove that all three quandle axioms hold for R_5. The simplest approach is to make a list of all possible cases and check that each case is true. We easily check (using Table 11.2) that $a \triangleright a = a$ for all five elements of $\mathbb{Z}_5$. Next, we verify that for each pair b and c that there is a unique a such that $a \triangleright b = c$. Examining the Table 11.2, we see that each element of $\mathbb{Z}_5$ occurs only once in each column. This proves that there is a unique a such that $a \triangleright b = c$. We also need to verify the third axiom, $(a \triangleright b) \triangleright c = (a \triangleright c) \triangleright (b \triangleright c)$, but there are 125 cases to verify. This suggests that we should try a different approach.

We use algebra to verify the third axiom in Equations 11.5 and 11.7. Rewrite the left hand side of the third axiom using the definition of $\triangleright$.

$$(a \triangleright b) \triangleright c = ((2b - a) \mod 5) \triangleright c \quad (11.5)$$
$$= (2c - (2b - a) \mod 5) \mod 5 \quad (11.6)$$
$$= (2c - 2b + a) \mod 5.$$

Next, we evaluate the right hand side of the third axiom using the definition of $\triangleright$.

$$(a \triangleright c) \triangleright (b \triangleright c) = ((2c - a) \mod 5) \triangleright ((2c - b) \mod 5) \tag{11.7}$$
$$= (2((2c - b) \mod 5) - (2c - a) \mod 5) \mod 5 \tag{11.8}$$
$$= (2c - 2b + a) \mod 5.$$

We conclude that $(a \triangleright b) \triangleright c = (a \triangleright c) \triangleright (b \triangleright c)$ for all a, b, and c in R_n.

Each of the three axioms can be verified algebraically. This allows us to prove that the axioms hold for all dihedral quandles R_n, regardless of the value of n. Quandles do not contain some of the elements that we are familiar with from algebra. For example, quandles do not contain an identity element. Suppose that the quandle R_5 contains an identity element. If so, there exists an element I_r in $\mathbb{Z}_5$ such that $a \triangleright I_r = a$, for all a in R_n. Suppose that $a \triangleright I_r = a$. Then $2I_r - a \equiv a \mod 5$. Then $2I_r \equiv 2a \mod 5$ or $I_r \equiv a \mod 5$. We conclude that $I_r = a$, but an identity element applies to each element of the set.

We define other examples of quandles. The first example is the **trivial quandle** with the set $X = \{x_1, x_2 \ldots, x_n\}$ and $x_i \triangleright x_j = x_i$ for all i, j. If $X = \{a\}$ then all three axioms are easily shown to be satisfied since $a \triangleright a = a$.

The **Alexander quandle**, X, is the set of Laurent polynomials with variable t and coefficients in $\mathbb{Z}$. Elements of X have the form

$$\sum_{i=-n}^{m} c_i t^i. \tag{11.9}$$

There are only a finite number of non-zero c_i terms in the sum. This set of Laurent polynomials is denoted as $\mathbb{Z}[t, t^{-1}]$. Elements of the set include $1 - 2t$ and $-t^{-2} + t + 7$. The operation $\triangleright$ is defined as $a \triangleright b = ta + (1 - t)b$. We only need to prove that all three axioms hold for X and this definition of $\triangleright$. This is left as an exercise.

The **free quandle** is defined using a finite set of elements called generators. Let F_n denote the free quandle with generators $\{x_1, x_2, \ldots, x_n\}$. The set X consists of all expressions involving $\{x_1, x_2, \ldots, x_n\}$ generated by the quandle operation $\triangleright$ and the three axioms. Consider F_2. This free quandle has two generators. We use the labels a and b to denote the two generators, since it is easy to lose track of subscripts. The elements of length one are a and b. The elements of length two are $a \triangleright b$ and $b \triangleright a$. The expression $a \triangleright a$ can be reduced to the element a using the quandle axioms, so it is not a new element of length two. Construct an element of length n by applying $\triangleright$ $n - 1$ times. For example, begin with the two elements of length two in X:

$$a \triangleright b, \qquad b \triangleright a.$$

Then, we apply $\triangleright$ to construct eight expressions of length three:

$$\begin{array}{ll} (a \triangleright b) \triangleright a, & (a \triangleright b) \triangleright b, \\ (b \triangleright a) \triangleright a, & (b \triangleright a) \triangleright b, \\ a \triangleright (b \triangleright a), & a \triangleright (a \triangleright b), \\ b \triangleright (a \triangleright b), & b \triangleright (b \triangleright a). \end{array}$$

But, $(a \triangleright b) \triangleright a = a \triangleright (b \triangleright a)$ and $(b \triangleright a) \triangleright b = b \triangleright (a \triangleright b)$. So there are actually only six elements of length three in F_2. The axioms result in equivalences between the expressions, even though we can not explicitly evaluate $a \triangleright b$ or $b \triangleright a$.

We can also define a quandle by imposing additional relations on the free quandle F_n. This is a **quandle presentation**. The additional relations are equations that can be used to determine equality between expressions. A quandle presentation is defined using set builder notation: a set of generating elements $x_1, x_2, \ldots, x_n$ and a list of equations (usually called relations) r_1 through r_m, and

$$Q = \{x_1, x_2, \ldots x_n | r_1, r_2, \ldots, r_m\}. \tag{11.10}$$

A specific example is

$$Q = \{a, b \mid (a \rhd b) \rhd b = a\}. \tag{11.11}$$

Recall from the previous example that there are six elements of length three in F_2:

$$\begin{aligned} &(a \rhd b) \rhd a, && (a \rhd b) \rhd b, \\ &(b \rhd a) \rhd a, && (b \rhd a) \rhd b, \\ &a \rhd (a \rhd b), && b \rhd (b \rhd a). \end{aligned} \tag{11.12}$$

The relation $(a \rhd b) \rhd b = a$ reduces the number of elements of length three to a list of five elements.

$$\begin{aligned} &(a \rhd b) \rhd b, && (b \rhd a) \rhd a, \\ &(b \rhd a) \rhd b, && a \rhd (a \rhd b), \\ &b \rhd (b \rhd a). \end{aligned} \tag{11.13}$$

Given an oriented knot diagram, we can determine a quandle associated with the diagram using a presentation of the quandle. This process is introduced in the next section.

Exercises

1. Construct a Cayley table for R_6.
2. Construct a Cayley table for R_7.
3. Prove that $a \rhd a = a$ for all a in R_n.
4. Prove that for all b, c, there is a unique element a such that $a \rhd b = c$ for all a, b, c in R_n.
5. Show that an Alexander quandle satisfies all three axioms.
6. Determine all elements of length four in the free quandle F_2.

11.3 KNOT QUANDLES

We associate a quandle to an oriented virtual knot diagram. Let K be a virtual knot diagram. An *edge* of the diagram begins and terminates at a classical under passing. In a diagram with n classical crossings, we label the edges of the diagram $a_1, a_2, \ldots a_{2n}$. We enumerate the crossings as $c_1, c_2, \ldots c_n$. The classical crossing c_i in the diagram determines a relation r_i on the elements as in Figure 11.6. The **knot quandle** associated with a virtual knot K (denoted $Q(K)$) is the free quandle on elements $a_1, a_2, \ldots, a_{2n}$ modulo the relations $r_1, r_2, \ldots r_n$ as determined by the crossings. Hence, the associated quandle $Q(K)$ is described by:

$$Q(K) = \{a_1, a_2, \ldots a_{2n} | r_1, r_2, \ldots r_n\}. \tag{11.14}$$

FIGURE 11.6: Quandle relation: $x \triangleright y = z$

The set $Q(K)$ consists of expressions of finite length (frequently called words) modulo the relationships given by the crossings.

The unknot diagram, U, has one edge and no crossings. As a result, $Q(U)$ is the trivial quandle.

We compute two examples of a knot quandle using the oriented, labeled knot diagrams in Figure 11.7. In Figure 11.7a, the three edges in the knot diagram are labeled a, b, and c.

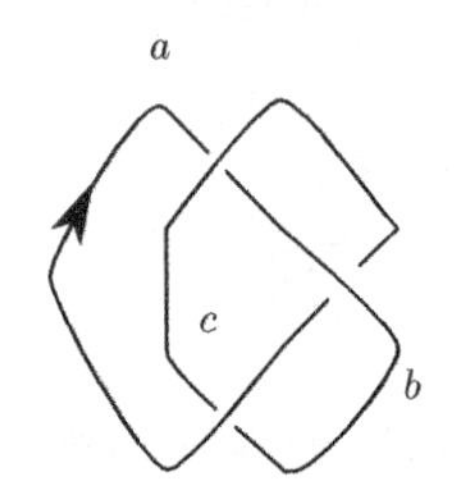

(a) Quandle labeling of VT

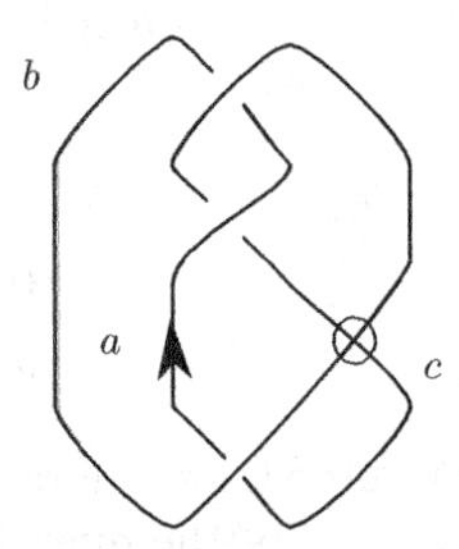

(b) Quandle labeling of $K_{3.2}$

FIGURE 11.7: Quandle labelings

The three relations generated by the crossings are:

$$a \triangleright b = c, \qquad b \triangleright c = a, \qquad c \triangleright a = b. \tag{11.15}$$

We conclude that $Q(VT) = \{a, b, c \mid a \triangleright b = c, b \triangleright c = a, c \triangleright a = b\}$. We eliminate the generator $c = a \triangleright b$ and obtain a reduced presentation, $Q(VT) = \{a, b | b \triangleright (a \triangleright b) = a, (a \triangleright b) \triangleright a = b\}$.

Examine Figure 11.7b. This knot diagram has three edges (labeled a, b, and c) and three crossings. The three crossings determine the relations:

$$b \triangleright b = a \qquad c \triangleright a = b \qquad b \triangleright a = c. \tag{11.16}$$

However, since $b \triangleright b = b$, we see that $a = b$ and then $a = c$ by the quandle axioms. We compute that $Q(K) = \{a | a \triangleright a = a\}$, the trivial quandle.

Let $(Q_1, \triangleright_1)$ and $(Q_2, \triangleright_2)$ denote two quandles. A map $f : Q_1 \rightarrow Q_2$ is a **quandle homomorphism** if it satisfies:

$$f(a \triangleright_1 b) = f(a) \triangleright_2 f(b). \tag{11.17}$$

Two quandles are isomorphic (equivalent) if we can find a bijective quandle homomorphism between the two quandles.

Two different diagrams of an oriented virtual knot K, say D and D', determine isomorphic quandles even though we obtain two different presentations from the diagrams. To prove that the quandles $Q(D)$ and $Q(D')$ are isomorphic, we need to construct a bijective (one-to-one and onto) quandle homomorphism f from $Q(D)$ to $Q(D')$. We show that such an isomorphism exists between diagrams that differ by exactly one diagrammatic move. Then, $Q(D)$ and $Q(D')$ are isomorphic if the diagrams are related by finite sequence of diagrammatic moves. However, in general, determining if two presentations $Q(D)$ and $Q(D')$ are equivalent can be challenging. Our focus in the next chapter is on proving that two quandles $Q(D)$ and $Q(D')$ are not equivalent and then concluding that $D \nsim D'$.

Theorem 11.4. *Let D and D' be equivalent oriented virtual knot diagrams. Then $Q(D)$ and $Q(D')$ are isomorphic quandles.*

Proof. We let D represent the left hand side of a diagrammatic move and let D' represent the right hand side. We first consider the Reidemeister I move, which introduces or removes a single classical crossing. We see that the Reidemeister I move (Figure 11.8) introduces a

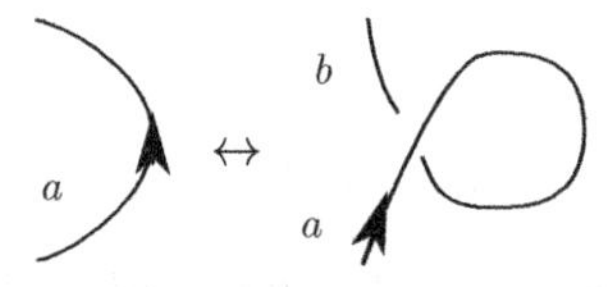

FIGURE 11.8: Reidemeister I with labeling

new edge label on the right hand side. However, the crossing introduces the relation $a \triangleright a = b$ so that $a = b$. In this case, $f : Q(D) \to Q(D')$ is the identity map.

Next, we consider the Reidemeister II move in Figure 11.9. The right hand side of

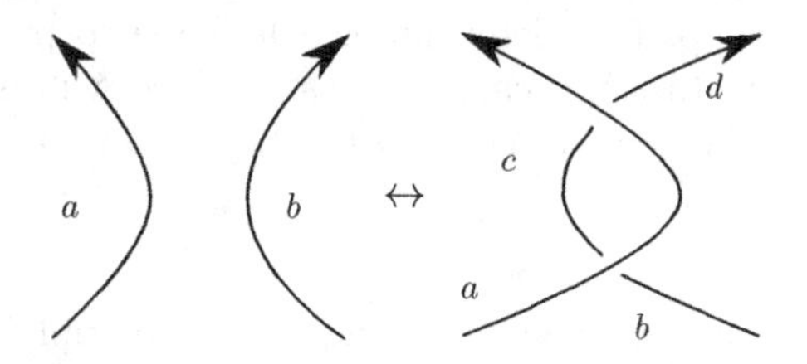

FIGURE 11.9: Reidemeister II (co-oriented) with labeling

the Reidemeister II move introduces two new edge labels: c and d. However, the two new crossings introduce two additional relationships:

$$b \triangleright a = c \text{ and } d \triangleright a = c. \tag{11.18}$$

By the quandle axioms, there is a unique element x such that $x \triangleright a = c$. Then, $b = d$ and $c = b \triangleright a$ which were already elements of both quandles. As a result, we can map these elements of $Q(D)$ to the two additional generators in $Q(D')$, c and d. The Reidemeister II move with oppositely oriented strands is left as an exercise.

Consider the Reidemeister III move in Figure 11.10. We consider the generators of $Q(D)$. In Equation 11.19, we list the three relations from the Reidemeister III move.

$$a_{2l} = a \triangleright b \qquad a_{3l} = a_2 \triangleright c \qquad b_{2l} = b \triangleright c. \tag{11.19}$$

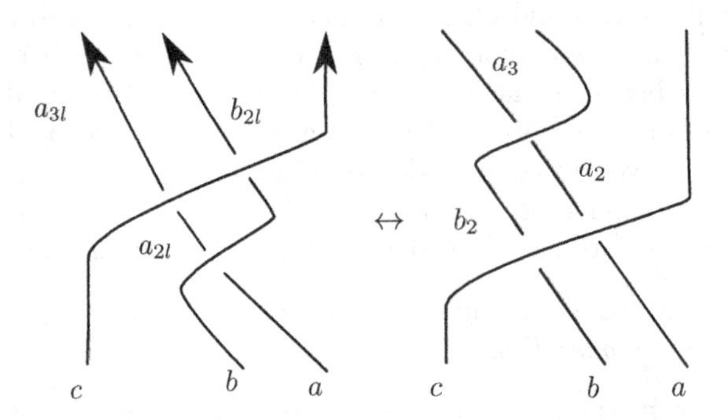

FIGURE 11.10: Reidemeister III with labeling

We can reduce the equation for a_{3l},

$$a_{3l} = (a \rhd b) \rhd c. \tag{11.20}$$

On the right hand side, $Q(D')$ is defined by the relations list in Equation 11.21.

$$a_{2r} = a \rhd c \qquad a_{3r} = (a \rhd c) \rhd (b \rhd c) \qquad b_{2r} = b \rhd c. \tag{11.21}$$

As a result, the map f is defined by $f(a_{3l}) = a_{3r}$. Using the quandle axioms, we see that f defines a quandle isomorphism. The virtual Reidemeiester moves I–IV do not change the relations determined by the classical crossings. □

We've shown that individual diagrammatic moves do not change $Q(K)$; they only change the presentation of the group. But determining whether two presentations result in isomorphic quandles is tricky. In some cases, we can reduce a presentation by eliminating generators. Let's consider the generators: $\{a, b, c\}$ and relationships

$$b \rhd a = c \qquad b \rhd c = a \qquad c \rhd a = b. \tag{11.22}$$

We can eliminate the generator c by substituting $b \rhd a$ and applying quandle axiom 3:

$$b \rhd (b \rhd a) = a \qquad a(b \rhd a) = b. \tag{11.23}$$

Just as in the case of the trefoil, we are able to reduce this quandle presentation to two generators. We would like to know if we can reduce to one relation. Then, if the relations are equivalent the quandle associated to the trefoil is equivalent to the quandle defined by the relations in Equation 11.23. Our casual inspection leads us to believe that these quandles are two different mathematical objects, but this is not proof.

In our calculational example, the knot $K_{3.2}$ (see Figure 11.7b), $Q(K_{3.2})$ is the trivial quandle. This demonstrates that some non-trivial virtual knots are associated with the trivial quandle. However, the quandle is an excellent invariant of classical knots. In fact, we have the following theorem due to Joyce from his 1982 article, "A classifying invariant of knots, the knot quandle," published in the *Journal of Pure and Applied Algebra.*

Theorem 11.5. *Let K and L be classical knot diagrams. If $Q(K)$ is isomorphic to $Q(L)$ then $K \sim L$, $K \sim L^I$, or $K \sim (L^M)^I$.*

The quandle distinguishes classical knots up to mirror and inverse.

Determining whether or not a quandle can be reduced to the trivial quandle or if two quandles are isomorphic is a challenge that we return to in the next chapter. Given an oriented virtual knot diagram D and the knot quandle $Q(D)$, we find a solution set to a system of equations obtained from the relations in $Q(D)$. If a quandle contains a solution, then each edge of the diagram D can be labeled with an element of the quandle and the labels satisfy the system of equations.

Exercises

1. Compute the knot quandle of the oriented virtual knot diagrams shown in Figure 11.11.
2. Prove that the Reidemeister II move with oppositely oriented strands results in isomorphic $Q(D)$ and $Q(D')$.

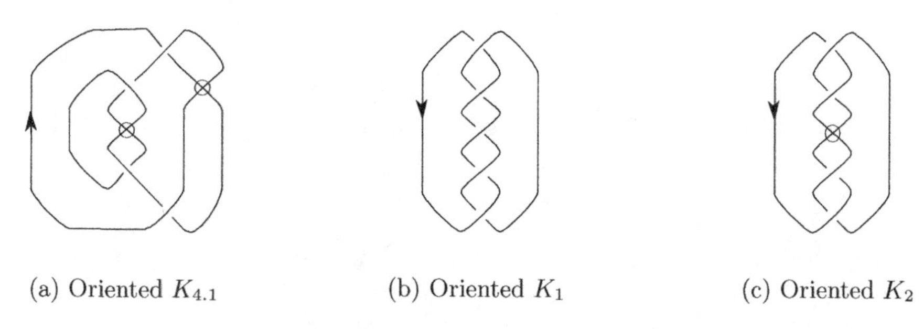

(a) Oriented $K_{4.1}$ (b) Oriented K_1 (c) Oriented K_2

FIGURE 11.11: Compute the quandles

11.4 OPEN PROBLEMS AND PROJECTS

Open problems

1. Characterize all virtual knots K such that $Q(K)$ is the trivial quandle.
2. K is a virtual knot such that $Q(K)$ is the trivial quandle. Let K_n denote the n-cabling of this knot. Is $Q(K_n)$ also trivial?
3. Is there a quandle structure for pseudo knots?

Projects

1. Find an infinite family of virtual knot diagrams that admit non-trivial 3-colorings.
2. Find an infinte family of virtual knot diagrams that only admit trivial 3-colorings.
3. Using the knot table in the book, determine which of the knots with trivial f-polynomial have a non-trivial coloring.

Further Reading

1. Peter Andersson. The color invariant for knots and links. *Amer. Math. Monthly*, 102(5):442–448, 1995

2. David Joyce. A classifying invariant of knots, the knot quandle. *J. Pure Appl. Algebra*, 23(1):37–65, 1982

3. Seiichi Kamada. Knot invariants derived from quandles and racks. *Invariants of knots and 3-manifolds (Kyoto, 2001)*, 4(July 2002):103–117, 2002

4. W. Edwin Clark, Mohamed Elhamdadi, Masahico Saito, and Timothy Yeatman. Quandle colorings of knots and applications. *J. Knot Theory Ramifications*, 23(6):1–29, 2014

CHAPTER 12

Knots and quandles

Our goal in this chapter is to determine if the edges of an oriented knot diagram can be labeled with the elements of a quandle. The labeling must satisfy the relations determined by the crossings in the diagram. We begin by studying labelings of a virtual knot diagram by elements of the quandle R_n. This type of labeling is sometimes called a Fox n-coloring, named after the topologist R. H. Fox. We use linear algebra to find a solution to the system of equations determined by the knot quandle. Finding these solutions is challenging because not all systems of equations have a non-trivial integer solution. Determinants and Cramer's rule are used to calculate non-trivial solution sets in R_n. The Alexander polynomial of an oriented virtual knot diagram is computed using the same strategy.

12.1 A LITTLE LINEAR ALGEBRA AND THE TREFOIL

We begin with a review of linear algebra. Given a system of equations with n equations and n variables, the system of equations can be rewritten as a matrix equation of the form $A\overline{x} = \overline{b}$. The matrix A is an $n \times n$ matrix.

We use the following notation for vectors of length n:

$$\overline{x} = \begin{bmatrix} x_1 \\ x_2 \\ \vdots \\ x_n \end{bmatrix} \qquad \overline{0} = \begin{bmatrix} 0 \\ 0 \\ \vdots \\ 0 \end{bmatrix} \qquad \overline{1} = \begin{bmatrix} 1 \\ 1 \\ \vdots \\ 1 \end{bmatrix}. \tag{12.1}$$

A homogeneous system of equation has the form $A\overline{x} = \overline{0}$. The **null set** of A (denoted $Nul(A)$) is the set $\{\overline{x} | A\overline{x} = \overline{0}\}$.

Two matrices A and A' are row equivalent if they are related by a sequence of **row operations**. If A is row equivalent to A', then $Nul(A) = Nul(A')$. There are three row operations.

1. Multiplication: A can be obtained from A' by scalar multiplying a row of A'.
2. Row addition: A can be obtained from A' by adding a row of A' to an another row of A'.
3. Row exchange: A can be obtained from A' by exchanging two rows of A'.

If the matrix A can be obtained from $n \times n$ matrix A' by adding a column of zeros and a row representing the equation

$$x_{n+1} = \sum_{i=1}^{n} c_i x_i \tag{12.2}$$

then the solution sets of A and A' are equivalent but not equal. The matrix A has dimension $(n+1) \times (n+1)$ and

$$A = \begin{bmatrix} & & A' & & & 0 \\ & & & & & 0 \\ & & & & & \vdots \\ & & & & & 0 \\ -c_1 & -c_2 & \dots & -c_{n-1} & -c_n & 1 \end{bmatrix}. \tag{12.3}$$

This process is called matrix augmentation. In a solution to the equation $A\overline{x} = \overline{0}$, the variable x_{n+1} is determined by the value of the variables $x_1, x_2, \dots, x_n$.

We consider a special case. Suppose that A represents the same system of equations as the matrix A' and the additional equation $-x_n + x_{n+1} = 0$. Then, although $Nul(A) \neq Nul(A')$, there is a one-to-one correspondence between $Nul(A)$ and $Nul(A')$. For each solution of the equation $A\overline{x} = \overline{0}$, we obtain a solution to $A'\overline{x} = \overline{0}$. In this case, we write $Nul(A) \cong Nul(A')$ to indicate that there is a one-to-one correspondence between the two sets.

Row operation and augmentation also have a well-defined effect on the determinant of a matrix.

1. Multiplication: Change the determinant by a scalar.
2. Row addition: Does not change the determinant.
3. Row exchange: Change the determinant by a factor of -1.
4. Augmentation: Changes the determinant by a factor of ± 1.

If A is an $n \times n$ matrix, there are two types of solution sets for a homogeneous system of equations.

1. $A\overline{x} = \overline{0}$ has a unique solution, namely $\overline{x} = \overline{0}$. Then $\det(A) \neq 0$.
2. $A\overline{x} = \overline{0}$ has infinitely many solutions and $\det(A) = 0$. In this case, we express each solution as a linear combination of m linearly independent vectors of length n:

$$\overline{x} = \sum_{i=1}^{m} c_i \overline{v}_i \tag{12.4}$$

where c_i is a constant and $m \leq n$.

For reasonable values of n, a computational algebra program can solve $A\overline{x} = \overline{0}$. The dimension of the null space of A is m which is the number of linearly independent vectors in $Nul(A)$. Further, the $\text{Rank}(A) = n - m$.

Remark 12.1. *If we were solving a non-homogeneous equation, $A\overline{x} = \overline{b}$, there is also the possibility that the system of equations has no solution.*

We can find a solution to $A\overline{x} = \overline{b}$ using Cramer's rule. For $1 \leq i \leq n$, the entry x_i in the vector $\overline{x}$ is determined by the formula:

$$x_i = \frac{\det(A(b)_i)}{\det(A)} \tag{12.5}$$

where $A(b)_i$ is the matrix obtained from A by replacing the i-th column with $\overline{b}$. If the entries in $\overline{b}$ are multiples of $\det(A)$ then $\det(A(b)_i)$ is divisible by $\det(A)$ and x_i is an integer. Let C_i denote the columns of A. The solution to $A\overline{x} = \det(A)\overline{b}$ can be written as a linear combination of the column vectors:

$$\sum_{i=1}^{n} x_i C_i = \det(A)\overline{b}. \tag{12.6}$$

In Equation 12.6, the coefficients x_i are integers and if $\det(A) = n$ then

$$\sum_{i=1}^{n} x_i C_i \equiv \overline{0} \mod n. \tag{12.7}$$

We begin our exploration of the solution set obtained from the trefoil. We introduce some notation for our convenience. The matrix obtained by deleting the ith row and j^{th} column is denoted as $A_{i,j}$. The determinant of the matrix $A_{i,j}$ is called a minor and is denoted as $M_{i,j}$. Let T denote the trefoil and recall that $Q(T) = \{a_1, a_2, a_3 | r_1, r_2, r_3\}$ where

$$r_1 : a_3 = a_1 \triangleright a_2, \tag{12.8}$$

$$r_2 : a_1 = a_2 \triangleright a_3, \tag{12.9}$$

$$r_3 : a_2 = a_3 \triangleright a_1. \tag{12.10}$$

We can rewrite these relations in the context of $\mathbb{Z}_n$:

$$a_3 = 2a_2 - a_1 \mod n, \tag{12.11}$$

$$a_1 = 2a_3 - a_2 \mod n, \tag{12.12}$$

$$a_2 = 2a_1 - a_3 \mod n. \tag{12.13}$$

From Equations 12.11–12.13, we obtain the homogeneous matrix equation $A(T)\overline{x} = \overline{0}$ where

$$A(T) = \begin{bmatrix} -1 & 2 & -1 \\ -1 & -1 & 2 \\ 2 & -1 & -1 \end{bmatrix}. \tag{12.14}$$

A quick inspection shows that the rows and columns of $A(T)$ sum to the zero vector. Note that $\det(A(T)) = 0$ and the elements of $Nul(A(T))$ are all scalar multiples of $\overline{1}$. This solution indicates that we can label all edges in the diagram with the same element of R_n. However, we want a non-trivial labeling of the diagram with elements of R_n for some n. A non-trivial labeling would distinguish the trefoil from the unknot.

We delete the last row and column of $A(T)$,

$$A(T)_{3,3} = \begin{bmatrix} -1 & 2 \\ -1 & -1 \end{bmatrix}. \tag{12.15}$$

The determinant of $A(T)_{3,3}$ is 3. Applying Cramer's rule, we find a solution to the equation

$$A(T)_{3,3}\overline{y}_1 = 3 \begin{bmatrix} 1 \\ 0 \end{bmatrix}. \tag{12.16}$$

Then

$$\overline{y}_1 = \begin{bmatrix} 1 \\ -1 \end{bmatrix}. \tag{12.17}$$

Next, apply Cramer's rule and find a solution to the equation

$$A(T)_{3,3}\overline{y}_2 = 3 \begin{bmatrix} 0 \\ 1 \end{bmatrix} \tag{12.18}$$

and determine that

$$\overline{y}_2 = \begin{bmatrix} 2 \\ 1 \end{bmatrix}. \tag{12.19}$$

Form the vector

$$\overline{y}_2' = \begin{bmatrix} 2 \\ 1 \\ 0 \end{bmatrix}. \tag{12.20}$$

Next,

$$A(T)\overline{y}_2' = \begin{bmatrix} -1 & 2 & -1 \\ -1 & -1 & 2 \\ 2 & -1 & -1 \end{bmatrix} \begin{bmatrix} 2 \\ 1 \\ 0 \end{bmatrix} = \begin{bmatrix} 0 \\ -3 \\ 3 \end{bmatrix}. \tag{12.21}$$

The last entry in the product vector is a 3 and $\overline{y}_2'$ is a solution to $A\overline{x} = \overline{0} \mod 3$. This means that the edges of the trefoil diagram can be labeled with elements of R_3 as follows: $a_1 = 2, a_2 = 1$, and $a_3 = 0$. Any scalar multiple of this vector also represents an edge labeling of the trefoil. Any linear combination of the vectors $\overline{1}$ and $\overline{y_2}'$ is a solution to the equation $A(T)\overline{x} = \overline{0} \mod 3$. The solution $\overline{y}_1$ is eliminated since $\overline{y}_1 \equiv \overline{y}_2 \mod 3$. All possible labelings of the oriented trefoil diagram can be expressed as a linear combination of the form

$$\overline{x} = c_1 \begin{bmatrix} 2 \\ 1 \\ 0 \end{bmatrix} + c_2 \begin{bmatrix} 1 \\ 1 \\ 1 \end{bmatrix} \tag{12.22}$$

where c_i is an element of R_3. Counting the possible combinations of coefficients, there are 9 possible colorings of the trefoil, including the 3 trivial colorings in R_3.

12.2 THE DETERMINANT OF A KNOT

We use the sequence of calculations that we just completed to define an oriented knot invariant. Given an oriented knot diagram K with n classical crossings, let $A(K)$ denote the $n \times n$ matrix obtained from relations determined by the classical crossings in the diagram. Let $M(K)_{i,j}$ denote the minors of $A(K)$. We define $det(K)$, the **determinant of** K, to be the greatest common divisor of the minors of $A(K)$,

$$det(K) = GCD\{M(K)_{i,j} | 1 \leq i \leq n, 1 \leq j \leq n\}. \tag{12.23}$$

The edges of the virtual knot diagram K can be labeled with elements of $R_{det(K)}$.

We first examine the relationship between the minors, $M(K)_{i,j}$, of a matrix $A(K)$. Next, for two virtual knot diagrams, D and D', which are related by a single diagrammatic move, we prove that the minors of matrices $A(D)$ and $A(D')$ have the same greatest common divisor up to a factor of ± 1. Then we conclude that $det(K)$ is a knot invariant.

Let C_i denote the columns of the matrix $A(K)$ and let R_i denote the rows of the matrix. Let $\overline{e}_i$ denote the standard basis vectors of $\mathbb{R}^n$. We will prove that $M(K)_{i,1} = \pm M(K)_{i,k}$ for all $1 \leq k \leq n$ since the columns of A are linearly dependent.

Theorem 12.1. *For all virtual knot diagrams K, $M(K)_{i,1}$ divides $M(K)_{i,j}$ for $1 \leq i \leq n, 1 \leq j \leq n$.*

Proof. We prove that $M(K)_{n,1} = \pm M(K)_{n,2}$. The proof of the rest of the cases is analogous to this case. Let M denote $M(K)_{n,1}$. We use C'_i to denote the columns of $A(K)$ with the last row deleted. Since the columns of $A(K)$ sum to the zero vector,

$$\sum_{i=1}^{n} C'_i = \overline{0}. \tag{12.24}$$

Rewriting equation 12.24,

$$-C'_2 = \sum_{i=1, i \neq 2}^{n} C'_i. \tag{12.25}$$

By Cramer's rule, there exists integers k_i

$$\sum_{i=2}^{n} k_i C'_i = M\overline{e_l}. \tag{12.26}$$

The vector

$$\begin{bmatrix} k_2 \\ k_3 \\ \vdots \\ k_n \end{bmatrix} \tag{12.27}$$

is a solution to the equation $A(K)_{n,1}\overline{x} = M\overline{e}_l$.

Rewrite Equation 12.26,

$$k_2 C'_2 + \sum_{i=3}^{n} k_i C'_i = M\overline{e}_l. \tag{12.28}$$

Applying Equation 12.25 to Equation 12.28,

$$-k_2 \sum_{i=1, i \neq 2}^{n} C'_i + \sum_{i=3}^{n} k_i C'_i = M\overline{e}_l, -k_2 C'_1 + \sum_{i=3} (k_i - k_2) C'_i = M\overline{e}_1. \tag{12.29}$$

This gives a vector solution with integer coefficients of the form

$$\begin{bmatrix} -k_2 \\ k_3 - k_2 \\ \vdots \\ k_n - k2 \end{bmatrix} \tag{12.30}$$

to the equation $A(K)_{n,2}\overline{x} = M\overline{e}_l$.

Using Cramer's rule, we replace the jth row of $A(K)_{1,1}$ with $M\overline{e}_l$ and calculate that

$$k_{j+1} = \frac{det(A(K)_{1,1}(Me_l)_j)}{M} \tag{12.31}$$

for $1 \leq j \leq n-1$. The indices are off by one term because the jth row of $A(K)_{n,1}$ corresponds to the $j+1$th row of $A(K)$.

We apply Cramer's rule to $A(K)_{1,2}$:

$$-k_2 = \frac{det(A(K)_{n,2}(Me_l)_1)}{M_{1,2}(K)}. \tag{12.32}$$

We notice that $A(K)_{n,2}(Me_l)_1 = A_{n,1}(K)(Me_l)_1$. From Equations 12.31 and 12.32, we compute that $M_{1,2}(K) = -M$. □

We use the transpose to prove that if the rows of $A(K)$ are linearly dependent then $M(K)_{1,1} = M(K)_{1,j}$ for $1 \leq j \leq n$.

Theorem 12.2. *For all virtual knot diagrams K, if the rows (R_i) of $A(K)$ have integer coefficients y_i such that*

$$\sum_{i=1}^{n} y_i R_i = \overline{0} \tag{12.33}$$

then $det(A_{i,j}) = \pm k$ for some integer k.

Proof. Let $\overline{y}$ denote an integer valued vector of length n. such that

$$A^T \overline{y} = \overline{0}. \tag{12.34}$$

Note that $det(A^T_{j,i}) = det(A_{i,j})$. Following the same argument as in Theorem 12.3, we see that $det(A_{i,j}) = \pm det(A_{1,j})$. □

Theorem 12.3. *For all classical knot diagrams K, there exist coefficients $c_i = \pm 1$, such that the rows R_i of $A(K)$ satisfy*

$$\sum_{i=1}^{n} c_i R_i^T = \overline{0}. \tag{12.35}$$

Proof. This proof is due to Charles Livingston. Select a point exterior to the knot diagram. Label each overcrossing with a point. Draw a path from the overcrossing point to the exterior point. If the path intersects the diagram an even number of times, the row corresponding to the crossing has coefficient 1. If the path intersects the diagram an odd number of times the row corresponding to the crossing has coefficient -1. □

Theorem 12.4. *Let A and A' be square matrices related by one of the following.*

1. *Augmentation*
2. *Row exchange*
3. *Row addition*

Then the greatest common divisor of the minors of A and A' is equivalent up to a factor of ± 1.

Proof. Let $M'_{i,j}$ denote $det(A'_{i,j})$.

Suppose that A' is a $(n+1) \times (n+1)$ matrix obtained from the $n \times n$ matrix A by augmentation. For $1 \leq i \leq n$ and $1 \leq j \leq n$, $M_{i,j} = M'_{i,j}$. The minor $M'_{i,n+1}$ is a linear combination of $M_{i,n}$ for $i \leq n$. Next, $M'_{n+1,j} = 0$ for $1 \leq j \leq n+1$. Finally, $M'_{i,j} = M_{i,j}$ for $i \leq n$ and $j \leq n$. The minors of A' include 0, the minors of A, and linear combinations of the minors of A. There is no change in the GCD of the minors since $GCD\{a, a+b\} = GCD\{a, b\}$.

Suppose that A' is obtained from A by exchanging row l and row k. Then $M'_{i,j} = -M_{i,j}$

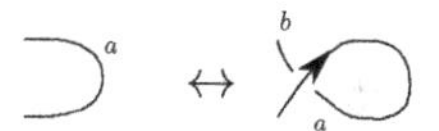

FIGURE 12.1: Reidemeister I move

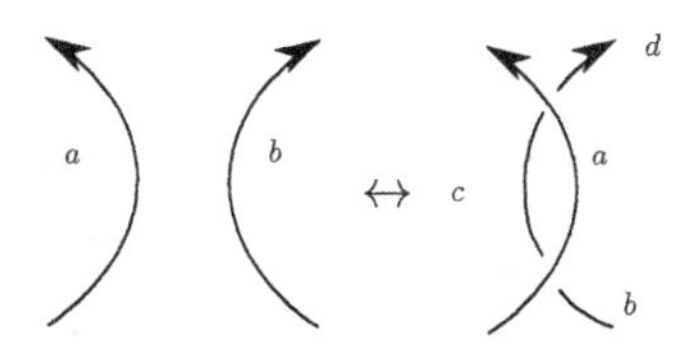

FIGURE 12.2: Reidemeister II move

for $i \neq l, k$. Next, we observe that $M'_{l,j} = \pm M_{k,j}$. We conclude row exchange does not affect the GCD.

Suppose that A' is obtained from A by adding a copy of row k to row 1. Then $M'_{1,j} = M_{1,j}$ and $M'_{k,j} = M_{k,j} \pm M_{1,j}$. For $i \neq 1, k$, $M'_{i,j} = M_{i,j}$. Again, $GCD\{a, a+b\} = GCD\{a, b\}$ and there is no effect on the GCD of the minors. □

Theorem 12.5. *For all virtual knot diagrams K and K', if $K \sim K'$ then $A(K)$ and $A(K')$ are related by a sequence of augmentations, row exchanges, and row addition.*

Proof. Let K and K' be related by a single Reidemeister move, and let K' represent the left hand side of the Reidemeister move. The order of the rows of $A(K)$ is arbitrary. Any two matrices obtained from the same diagram are related by a sequence of row exchanges. We assume without loss of generality that the rows related to the Reidemeister moves are placed at the bottom of the matrices $A(K)$ and $A(K')$.

In a Reidemeister I move (shown in Figure 12.1), the diagram K' has an additional edge and crossing. This introduces a new variable b and the new equation $a - b = 0$. The matrices $A(K)$ and $A(K')$ are related by augmentation and row addition.

The co-oriented Reidemeister II move (shown in Figure 12.2) introduces the new variables, c and d, and the equations $d + c = 2a$ and $b + c = 2a$. The matrices $A(K)$ and $A(K')$ are related by augmentation and row addition. The contra-oriented Reidemeister II move has been left as an exercise.

In a Reidemeister III move, we can assume that the matrices $A(K)$ and $A(K')$ differ only in the last three rows. From the left hand side of the Reidemeister III move, we obtain the equations $a + e = 2c, b + f = 2c$, and $e + g = 2f$. We construct the last three rows of the matrix $A(K')$ in Equation 12.36. The columns are labeled with the variables for

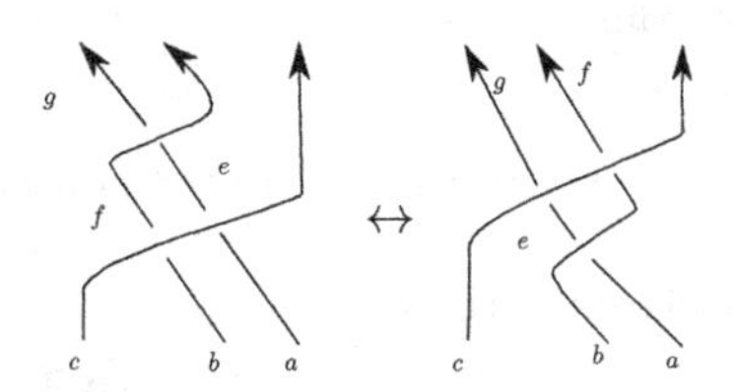

FIGURE 12.3: Reidemeister III move

convenience.

$$\begin{bmatrix} a & b & c & f & g & e \\ 0 & -1 & 2 & -1 & 0 & 0 \\ 0 & 0 & 0 & 2 & -1 & -1 \\ -1 & 0 & 2 & 0 & 0 & -1 \end{bmatrix}. \tag{12.36}$$

After row addition, we obtain

$$\begin{bmatrix} a & b & c & f & g & e \\ 0 & -1 & 2 & -1 & 0 & 0 \\ -1 & -2 & 2 & 0 & -1 & 0 \\ -1 & 0 & 2 & 0 & 0 & -1 \end{bmatrix}. \tag{12.37}$$

From the right hand side of the Reidemeister III move, we obtain the equations $a+e = 2b, b+f = 2c$, and $e+g = 2c$. This results in the matrix:

$$\begin{bmatrix} a & b & c & f & g & e \\ 0 & -1 & 2 & -1 & 0 & 0 \\ 0 & 0 & 2 & 0 & -1 & -1 \\ -1 & 2 & 0 & 0 & 0 & -1 \end{bmatrix}. \tag{12.38}$$

After row addition, we obtain the matrix:

$$\begin{bmatrix} a & b & c & f & g & e \\ 0 & -1 & 2 & -1 & 0 & 0 \\ -1 & -2 & 2 & 0 & -1 & 0 \\ -1 & 2 & 0 & 0 & 0 & -1 \end{bmatrix}. \tag{12.39}$$

The matrices in Equations 12.37 and 12.39 differ only in the last row, which is an augmentation row. The GCD of the minors is the same. □

The following corollary immediately follows.

Corollary 12.6. $\det(K)$ *is an oriented knot invariant.*

The number $\det(K)$ does not necessarily represent the smallest number of colors needed to label a diagram of K. The value of $det(K)$ tells us that K can be labeled (or colored) using the elements of R_n. This number can distinguish a diagram K from the unknot if $n \neq 0$. To further distinguish knot diagrams with the same value for $det(K)$, we can count the number of colorings of the diagram with R_n. Let $Col_X(K)$ denote the number of ways that K can be colored with the quandle X. The following theorem is immediate.

Theorem 12.7. *For all virtual knot diagrams K, with n classical crossings and $det(K) = p$, $p \leq Col_{R_p}(K) \leq p^n$.*

Proof. We see that we can obtain p trivial colorings of K, so that $p \leq Col_{R_p}(K)$. At best, we utilize every possible color combination (choose one of p colors for each on n edges) so that $Col_{R_p}(K) \leq p^n$. □

We compute two examples: the figure 8 knot and the non-classical knot $K_{4.62}$ (see Figure 12.4).

Determine the matrix associated with K_8:

$$A(K_8) = \begin{bmatrix} -1 & -1 & 2 & 0 \\ 2 & 0 & -1 & -1 \\ 0 & -1 & -1 & 2 \\ -1 & 2 & 0 & -1 \end{bmatrix}. \tag{12.40}$$

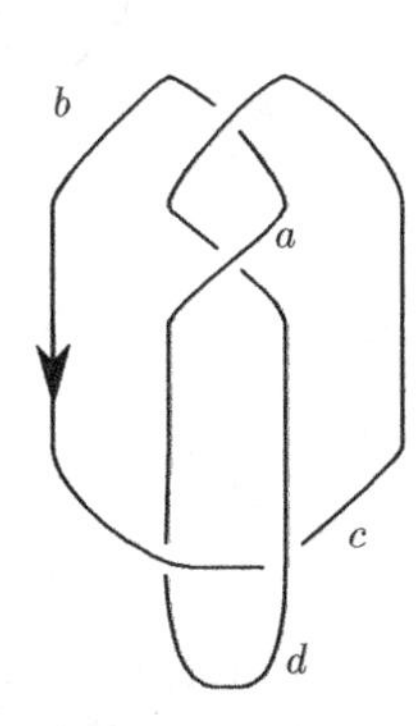

(a) Labeled figure eight knot: K_8

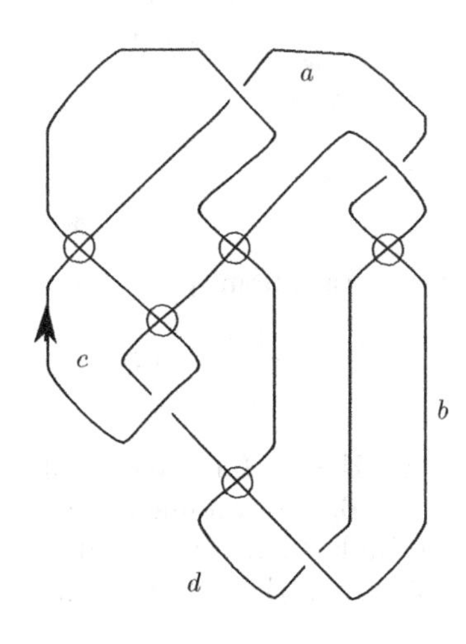

(b) Knot: $K_{4.62}$

FIGURE 12.4: Knots with labels

From $A(K_8)$, we compute that $\det(K_8) = 5$. For $1 \leq i \leq 3$, the solutions to the equation $A_{4,4}(K_8)\overline{y}_i = \pm 5\overline{e}_i$ are

$$\overline{y}_1 = \begin{bmatrix} 1 \\ -2 \\ 2 \end{bmatrix}, \qquad \overline{y}_2 = \begin{bmatrix} 3 \\ -1 \\ 1 \end{bmatrix}, \qquad \overline{y}_3 = \begin{bmatrix} -1 \\ -3 \\ -2 \end{bmatrix}. \tag{12.41}$$

But we note that $3\overline{y}_1 \equiv \overline{y}_2 \mod 3$ and $4\overline{y}_1 \equiv \overline{y}_3 \mod 3$ so that we only have one solution. As a result, the figure 8 knot has 25 colorings or $Col_{R_5}(K_8) = 25$.

The relations in $Q(K_{4.63})$ from Figure 12.4b are:

$$d \triangleright d = a, \qquad a \triangleright c = b, \tag{12.42}$$

$$b \triangleright d = c, \qquad c \triangleright b = d. \tag{12.43}$$

These relations reduce to three equations.

$$a \triangleright c = b \qquad b \triangleright a = c \qquad c \triangleright b = a. \tag{12.44}$$

From this set of equations, we obtain

$$A(K_{4.62}) = \begin{bmatrix} -1 & -1 & 2 \\ -1 & 2 & -1 \\ 2 & -1 & -1 \end{bmatrix}. \tag{12.45}$$

We find the solutions to $A(K_{4.62})_{4,4}\overline{y}_i = \pm 3\overline{e}_i$ are

$$\overline{y}_1 = \begin{bmatrix} 2 \\ 1 \end{bmatrix}, \qquad \overline{y}_2 = \begin{bmatrix} 1 \\ -1 \end{bmatrix}. \tag{12.46}$$

Solutions to $A(K_{4.62})\overline{x} \equiv \overline{0} \mod 3$ have the form

$$c_1 \begin{bmatrix} 2 \\ 1 \\ 0 \end{bmatrix} + c_2 \begin{bmatrix} 1 \\ 1 \\ 1 \end{bmatrix} \tag{12.47}$$

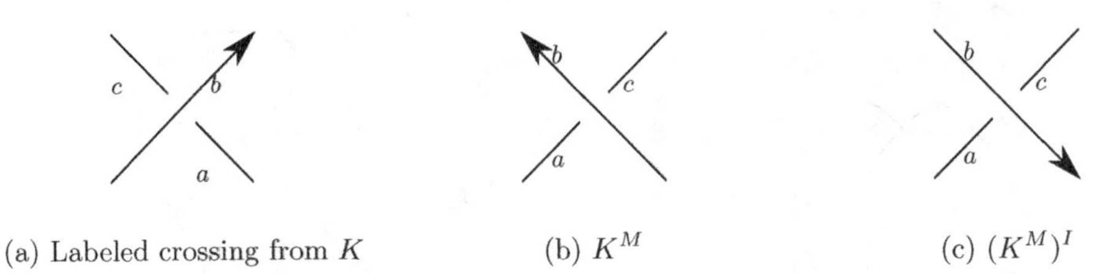

(a) Labeled crossing from K (b) K^M (c) $(K^M)^I$

FIGURE 12.5: Labeled crossings

where $c_i \in \mathbb{Z}_3$. Hence, $K_{4.62}$ has a total of 9 colorings in R_3.

In Chapter 6, we examined the symmetries of a virtual link diagram K. A summary is shown in Table 12.1. We proved that $(K^M)^S = K^F$ as diagrams, but several questions remained. We mark the remaining questions in Table 12.1.

Symbol	Description	Questions
K^I	The inverse of K	
K^S	The switch of K	Is K^M equivalent to K^S ?
K^M	The mirror of K	Is K^M equivalent to K^S ?
K^F	The flip of K	Is K equivalent to K^F ?

TABLE 12.1: Symmetries of K

There is a relationship between $det(K)$ and $det((K^M)^I)$.

Theorem 12.8. *For all virtual knot diagrams K, $det(K) = det((K^M)^I)$.*

Proof. The quandle equations obtained from the crossings in the diagrams K and $(K^M)^I$ are equivalent. All three diagrams have the same set of generators. The quandle relations determined by the crossings:

$$\begin{aligned} K &: a \triangleright b = c, \\ K^M &: c \triangleright b = a, \\ (K^M)^I &: a \triangleright b = c. \end{aligned} \tag{12.48}$$

Hence, K and $(K^M)^I$ have the same set of generators and relations and $det(K) = det((K^M)^I)$. □

Consider our earlier goal: distinguishing K and K^F. We examine a labeled Kishino knot K_K and its flip in Figure 12.6. The quandle $Q(K_K)$ can only be labeled with one element

(a) Labeled K_K (b) Labeled K_K^F

FIGURE 12.6: Comparing $Q(K_K)$ and $Q(K_K^F)$

and $Col_{R_3}(K_K) = 3$. The quandle associated with the flip of Kishino's knot is non-trivial:

$Q(K_K^F) = \{a, b, c, d | a \triangleright c = b, a \triangleright c = d, c \triangleright a = b, c \triangleright a = d\}$. Calculation shows that $Col_{R_3}(K_K^F) = 9$.

We've proved the following statement.

Theorem 12.9. *There exists a virtual knot diagram K such that $Col_X(K) \neq Col_X(K^F)$. That is, there exists a virtual knot diagram where $K \nsim K^F$.*

The edges of the diagrams do not correspond so we cannot give a general proof about the relationship between a knot and its flip. We pursue this in the next chapter when we introduce biquandles. A non-classical knot is said to be **pseudo classical** if $K \sim K^F$.

Theorem 12.10. *For all classical knot diagrams K, $K \sim K^F$.*

Proof. We sketch the proof of this statement. A classical knot K represents a mapping of the circle into three dimensional space. The knot can be physically rotated in three dimensional space and the result projected onto the plane to obtain K^F. Then there is a sequence of Reidemeister moves representing this rotation, so that $K \sim K^F$. □

Recall that the f-polynomial does not detect the virtualization move. Here, we introduce a diagrammatic move that the quandle does not detect. The T move (first introduced by L. H. Kauffman) is defined on two strands in the diagram and is shown in Figure 12.7. The orientation on strand a is fixed.

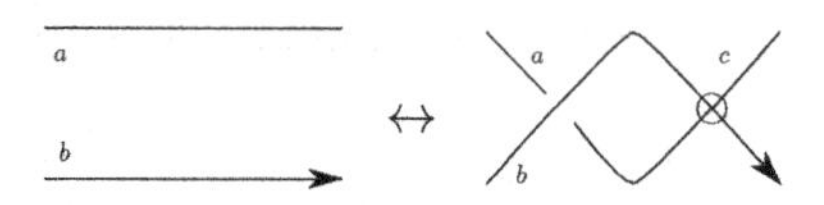

FIGURE 12.7: The T move

Theorem 12.11. *For all virtual knot diagrams K, the diagram K can be labeled with R_n if and only if $T^2(K)$ can be labeled with R_n.*

Proof. Let K be an oriented virtual knot diagram with a region of the diagram that appears as below.

(12.49)

The two parallel edges in K are replaced with the tangle below, creating the diagram $T^2(K)$. The diagram $T^2(K)$ is identical to K outside this region.

(12.50)

Examining the labels on $T^2(K)$, we see that

$$d \triangleright b = c, \qquad c \triangleright b = a, \tag{12.51}$$

which reduces to $(d \triangleright b) \triangleright b = a$. In the dihedral quandle, R_n,

$$a \equiv (d \triangleright b) \triangleright b \qquad \equiv 2b - (2b - d) \mod n \equiv d, \tag{12.52}$$

so that $a \equiv d \mod n$. □

Exercises

1. For the virtual knot diagrams in Figure 12.8, compute det(K).
2. For the virtual knot diagrams in Figure 12.8, use linear combinations to describe the possible labelings of the diagram.
3. For the virtual knot diagrams in Figure 12.9, compute $\det(K)$.
4. Prove that the equations obtained from the contra-oriented Reidemeister II move do not change the GCD of the minors of $A(K)$.

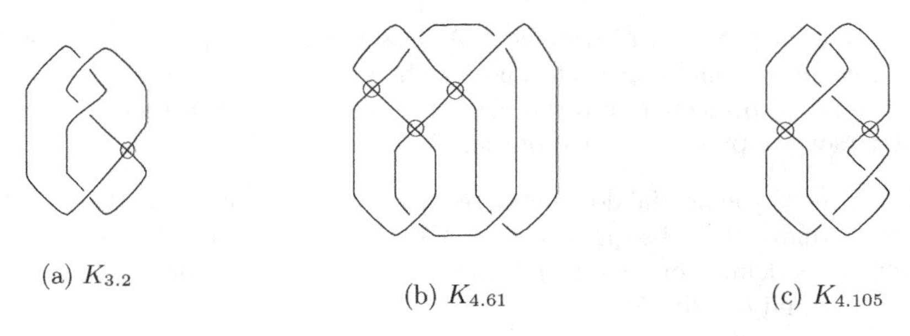

(a) $K_{3.2}$ (b) $K_{4.61}$ (c) $K_{4.105}$

FIGURE 12.8: Compute $\det(K)$

(a) K (b) L

FIGURE 12.9: Compute $\det(K)$

12.3 THE ALEXANDER POLYNOMIAL

Using the same approach, we compute the Alexander polynomial of an oriented virtual knot diagram K. The Alexander polynomial, $\Delta_K(t)$, is an element of $\mathbb{Z}[t, t^{-1}]$. The notation $\mathbb{Z}[t, t^{-1}]$ denotes the set of all Laurent polynomials with coefficients in $\mathbb{Z}$. The Laurent polynomials have the form

$$f(t) = \sum_{i=-n}^{m} c_i t^i. \tag{12.53}$$

If $f(t)$ and $g(t)$ are in $\mathbb{Z}[t, t^{-1}]$ then the product $f(t)g(t)$ is also a Laurent polynomial. However, $\frac{f}{g}$ may not be an element of $\mathbb{Z}[t, t^{-1}]$.

The quandle relation $a \triangleright b = c$ is defined as $ta + (1-t)b = c$. We obtain a system of equations from the quandle relation defined by each classical crossing. From the set of equations, we construct a matrix $Al(K)$ with coefficients in $\mathbb{Z}[t, t^{-1}]$.

The columns of $Al(K)$ sum to zero, so that $det(Al(K)) = 0$ and $Al(K)$ is non-invertible. As before, the null space contains vectors of the form: $c\bar{1}$. We want to determine if there are any non-trivial solutions modulo some polynomial in $\mathbb{Z}[t, t^{-1}]$.

Calculating these polynomial solutions is tricky; so we obtain the invariant from the first part of this computation: determining the greatest common divisor of the minors of $Al(K)$. The **Alexander polynomial** of K, $\Delta_K(t)$, is defined to be the greatest common divisor of the minors of $Al(K)$:

$$\Delta_K(t) = GCD\{det(Al(K)_{i,j}) | 1 \leq i \leq n, 1 \leq j \leq n\}. \quad (12.54)$$

For the trefoil T, we compute $\Delta_T(t)$. Recall the quandle relation $a \triangleright c = b, b \triangleright a = c$, and $c \triangleright b = a$ to produce the matrix:

$$Al(T) = \begin{bmatrix} t & -1 & 1-t \\ 1-t & -t & -1 \\ -1 & 1-t & t \end{bmatrix}. \quad (12.55)$$

All the minors of this matrix have the form $\pm(1 - t + t^2)$ and $\Delta_K(t) = t^{-1} - 1 + t^1$.

We prove that $\Delta_K(t)$ is a knot invariant using the same approach used to show that $\det(K)$ is a knot invariant.

Lemma 12.12. *For all oriented virtual knot diagrams K and K', if $K \sim K'$ then $Al(K)$ and $Al(K')$ are related by:*

1. *Augmentation*
2. *Row addition*
3. *Row exchange*
4. *Row multiplication: scalar multiplying rows by $\pm t^i$.*

Proof. The proof uses the same arguments as in Lemma 12.4. □

Theorem 12.13. *For all oriented virtual knot diagrams K and K', if $K \sim K'$ then $\Delta_K(t) = \pm t^i \Delta_K(t)$.*

Proof. The matrices $Al(K)$ and $Al(K')$ are related by the operations given in Lemma 12.12. Only the operation row multiplication affects the GCD of the minors. Row multiplication changes the minors by a multiple of $\pm t^i$. □

Exercises

1. Compute $\Delta_K(t)$ for the knot diagrams in Figure 12.10.
2. Prove that for virtual knot diagrams K, $\Delta_K(t) = \Delta_{K^I}(t)$.
3. Prove that for all virtual knot diagrams K, $\Delta_K(-1) = det(K)$.
4. Prove for all classical knot diagrams K, $\Delta_K(1) = \pm 1$.

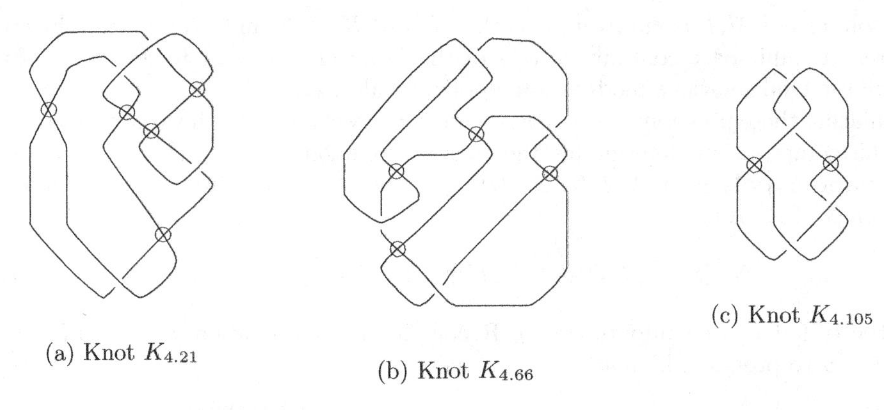

(a) Knot $K_{4.21}$

(b) Knot $K_{4.66}$

(c) Knot $K_{4.105}$

FIGURE 12.10: Compute the Alexander polynomial

12.4 THE FUNDAMENTAL GROUP

We define the fundamental group of a diagram K, which is denoted as $\pi_1(K)$. The fundamental group is an important invariant of classical knots. In addition to the quandle structure, the fundamental group also has the algebraic structure called a group. A **group** is a set G with an operation $*$ that satisfies the following axioms.

1. For all a, b in G, $a * b$ is an element of G (closure).
2. For all a, b, c in G, $(a * b) * c = a * (b * c)$ (associativity).
3. For all a in G, there exists an element e such that $a * e = e * a = a$ (identity element).
4. For all a in G, there exists an element a^{-1} such that $a * a^{-1} = a^{-1} * a = e$ (inverse element).

A group can be defined using a set of generators and relations, just as we defined quandles.

We define the fundamental group using the quandle structure and then verify that it satisfies the quandle axioms. For a labeled virtual knot diagram K, the elements of $\pi_1(K)$ are finite products of the edge labels and their inverses. The group operation $*$ in $\pi_1(K)$ is multiplication. The quandle relation is defined as a product of generators and their inverses:

$$a \triangleright b = b^{-1}ab. \tag{12.56}$$

This operation is called **conjugation** and $b^{-1}ab$ is called a **conjugate**. This type of quandle is called a **conjugation quandle**.

Reviewing our three quandle axioms, we calculate that

1. For all $a \in Q$, $a \triangleright a = a$ since $a^{-1}aa = a$.
2. For all $b, c \in Q$, there exists a unique a such that $a \triangleright b = c$. The unique solution is $a = bcb^{-1}$ and $b^{-1}ab = c$.
3. For all $a, b, c \in Q$, the third quandle axiom is also satisfied. We calculate that

$$(a \triangleright b) \triangleright c = c^{-1}(b^{-1}ab)c. \tag{12.57}$$

Next,

$$(a \triangleright c) \triangleright (b \triangleright c) = (c^{-1}bc)^{-1}(c^{-1}ac)(c^{-1}bc) = c^{-1}(b^{-1}ab)c. \tag{12.58}$$

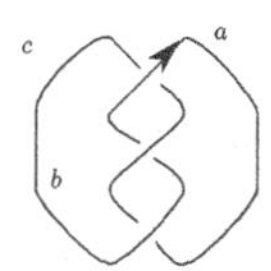

FIGURE 12.11: Labeling for fundamental group

Theorem 12.14. *For all oriented virtual knot diagrams K and K', if $K \sim K'$ then $\pi_1(K) \cong \pi_1(K')$.*

Proof. This follows immediately from the fact that $Q(K) \cong Q(K')$. □

We calculate the fundamental group of the labeled trefoil shown in Figure 12.11. The trefoil has three quandle relations:

$$a \triangleright b = c \qquad b \triangleright c = a \qquad c \triangleright a = b. \tag{12.59}$$

Rewriting the relations from 12.59 in terms of the fundamental group, we obtain

$$b^{-1}ab = c \qquad c^{-1}bc = a \qquad a^{-1}ca = b. \tag{12.60}$$

We can then eliminate the variable c to obtain the equations

$$b^{-1}a^{-1}bab = a \qquad a^{-1}b^{-1}aba = b \tag{12.61}$$

which reduce to $bab = aba$. Then,

$$\pi_1(T) = \{a, b | bab = aba\}. \tag{12.62}$$

Classical links and knots can also be viewed as the complement of a physical knot or link in three-dimensional space. The fundamental group describes all possible loops in this three-dimensional space.

Exercises

1. Compute the fundamental group of the knot diagrams shown in Figure 12.10.
2. Define $a \triangleright b$ as $b^{-n}ab^n$. Prove this operation satisfies all three quandle axioms.

12.5 OPEN PROBLEMS AND PROJECTS

Open problems

1. Given a family of knots, determine what finite quandles can non-trivially color the knot.
2. If we know that a knot diagram can be non-trivially colored with quandle X, does that imply that any other quandles can not be used to color K? Does it imply that other quandles can be used to color K?
3. The following statement only holds for classical knots. Let K and L be classical knot diagrams. Then $Col_{R_n}(K\sharp L) = \frac{1}{n}\mathrm{Col}_{R_n}(K)\mathrm{Col}_{R_n}(L)$. What is its extension to virtual knot diagrams?

Projects

1. For the knot table in the appendix, determine (if possible) the minimum value of n that results in a non-trivial coloring.
2. Find colorings of virtual knot diagrams that minimize the number of colors used. Refer to Louis H. Kauffman and Pedro Lopes. On the minimum number of colors for knots. *Advances in Applied Mathematics*, 40(1):36–53, 2008.
3. Find a virtual knot diagram and prove that it is non-classical by comparing $Col_X(K)$ and $Col_X(K^F)$.

Further Reading

1. Sang Youl Lee. Genera and periodicity of virtual knots and links. *J. Knot Theory Ramifications*, 21(4):15,1250037, 2012
2. Seiichi Kamada. Knot invariants derived from quandles and racks. *Invariants of knots and 3-manifolds (Kyoto, 2001)*, 4(July 2002):103–117, 2002
3. J. Scott Carter. A survey of quandle ideas. In *Introductory lectures on knot theory*, volume 46 of *Ser. Knots Everything*, pages 22–53. World Sci. Publ., Hackensack, NJ, 2012
4. R. H. Fox. A quick trip through knot theory. *Topology of 3-manifolds and related topics (Proc. The Univ. of Georgia Institute, 1961)*, pages 120–167, 1962
5. J. S. Carter, D. S. Silver, and S. G. Williams. Three dimensions of knot coloring. *The American Mathematical Monthly*, 121(6):506–514, 2014
6. Louis H. Kauffman and Pedro Lopes. On the minimum number of colors for knots. *Advances in Applied Mathematics*, 40(1):36–53, 2008
7. Yongju Bae. Coloring link diagrams by Alexander quandles. *J. Knot Theory Ramifications*, 21(10):1250094, September 2012

CHAPTER 13

Biquandles

In this chapter, we study the biquandle. We construct a biquandle associated with a virtual link diagram and prove that this biquandle is invariant under the diagrammatic moves. Then, we use the biquandle to construct a system of equations and an associated matrix equation of the form $A\overline{x} = \overline{0}$. The determinant of this matrix is an invariant of virtual links up to a factor of the form $\pm t^i s^j$. We also examine whether or not this structure differentiates between the different symmetries of K.

The original application of the biquandle to virtual links occurs in the paper "Bi-oriented Quantum Algebras, and a Generalized Alexander Polynomial for Virtual Links" by L. Kauffman and D. Radford. This paper was published in the book *Diagrammatic morphisms and applications* in 2003. However, the definition of the biquandle predates the virtual links and its original definition is in an unpublished manuscript by J. Scott Carter and M. Saito. The biquandle structure provides many different invariants of virtual links. In particular, S. Nelson has written many papers on the topic of biquandles jointly with undergraduates. Undergraduate-friendly papers are listed in the section on further reading and provide an excellent starting point for undergraduate research.

13.1 THE BIQUANDLE STRUCTURE

A **biquandle** is a set X with a map $B : X \times X \to X \times X$ that satisfies four conditions. The map B can be defined as the product of two operations $L : X \times X \to X$ and $R : X \times X \to X$ such that $B(a, b) = (L(a, b), R(a, b))$. The map B satisfies the following four conditions.

1. B is an invertible.
2. $(B \times I)(I \times B)(B \times I) = (I \times B)(B \times I)(I \times B)$.
3. For all $a, b \in X$, there exists an $x \in X$ such that
$$a = L(b, x) \qquad b = L^{-1}(a, R(b, x)) \qquad x = R^{-1}(a, R(b, x)). \tag{13.1}$$
For all $a, b \in X$, there exists an $x \in X$ such that
$$a = R^{-1}(x, b) \qquad x = L(L^{-1}(x, b), a) \qquad b = R(L^{-1}(x, b), a). \tag{13.2}$$
4. For all $a \in X$, there exists an x such that $(a, x) = B(a, x)$. Note that since B is invertible then $B^{-1}(a, x) = (a, x)$.

The last three conditions above can be rewritten as follows.

1. There exist operations $L^{-1} : X \times X \to X$ and $R^{-1} : X \times X \to X$ such that

$$L^{-1}(L(a,b), R(a,b)) = a \tag{13.3}$$
$$R^{-1}(L(a,b), R(a,b)) = b. \tag{13.4}$$

2. The operations L and R also have the following structure:

$$L(a, L(b,c)) = L(L(a,b), L(R(a,b),c)) \tag{13.5}$$
$$R(R(a,b),c) = R(R(a, L(b,c)), R(b,c)) \tag{13.6}$$
$$R(L(a,b), L(R(a,b),c)) = L(R(a, L(b,c)), R(b,c)). \tag{13.7}$$

3. For all $a \in X$, there exists x such that $L(a,x) = a$ and $R(a,x) = x$.

The set X can be defined using a generating set $S = \{x_1, x_2, \ldots, x_n\}$. The set of **biquandle words**, $W(S)$, is recursively defined. If a is an element of S then a is an element of $W(S)$ and if a and b are elements of $W(S)$ then $L(a,b), R(a,b), L^{-1}(a,b)$, and $R^{-1}(a,b)$ are elements of $W(S)$. The set X is the set of **free biquandle words** which are equivalence classes of $W(S)$ determined by the biquandle conditions.

For an oriented virtual knot diagram K, we define the knot biquandle of K. In this context, the **edges** of the diagram begin and end at classical crossings. Label the edges of the diagram K. The edge labels form a set of generators, S. The operators $L : X \times X \to X$ and $R : L : X \times X \to X$ are associated with a positive crossing of a link diagram as shown in Figure 13.1a. The operators L^{-1} and R^{-1} are associated with a negative crossing as shown in Figure 13.1b. The **knot biquandle** of K, $BQ(K)$, is the set of equivalence

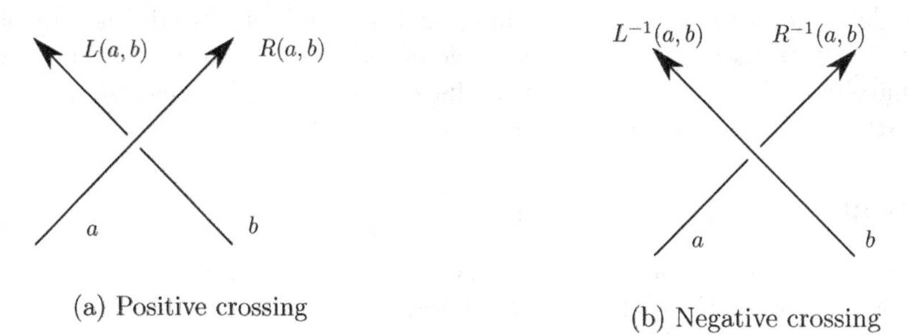

(a) Positive crossing

(b) Negative crossing

FIGURE 13.1: Biquandle labeled crossings

classes of words in $W(S)$ determined by the biquandle conditions and the relations given by the crossings in K.

As an example, we construct the biquandle of the oriented virtual trefoil diagram using the labeling in Figure 13.2. The generating set S is $\{a,b,c,d\}$. From the diagram of the oriented virtual trefoil, VT, we obtain the relations

$$L(d,b) = a, \qquad R(d,b) = c, \tag{13.8}$$
$$L(a,c) = b, \qquad R(a,c) = d. \tag{13.9}$$

Then $BQ(VT) = \{a,b,c,d | B(d,b) = (a,c), B(a,c) = (b,d)\}$. Note that the relation $B^{-1}(c,d) = (a,b)$ can be rewritten as the equivalent relation $B(a,b) = (c,d)$.

The biquandle of a knot diagram is invariant under the diagrammatic moves. For two equivalent oriented virtual knot diagrams K and K', the elements of $BQ(K)$ have

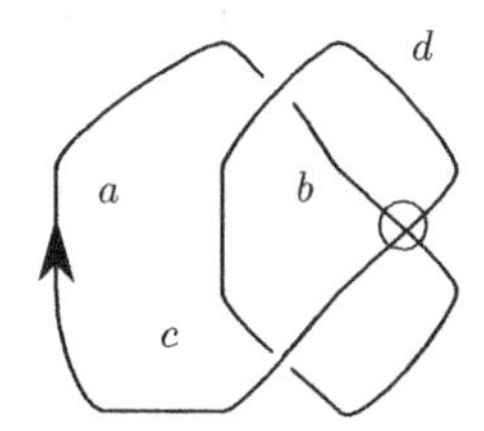

FIGURE 13.2: Biquandle labeled virtual trefoil

a one-to-one and onto correspondence with the elements of $BQ(K')$, which is denoted as $BQ(K) \cong BQ(K')$. We prove this correspondence holds for knot diagrams related by a single diagrammatic move. Consequently, if $K \sim K'$ then $BQ(K) \cong BQ(K')$.

Theorem 13.1. *For all oriented virtual knot diagrams K and K', if $K \sim K'$ then $BQ(K) \cong BQ(K')$.*

Proof. We prove that if K and K' are related by a single diagrammatic move, then $BQ(K)$ and $BQ(K')$ represent the same set of equivalence classes. Let K denote the left hand side of a diagrammatic move and let K' denote the right hand side.

Consider Reidemeister I in Figure 13.3. On the left hand side, in $BQ(K)$, there is a

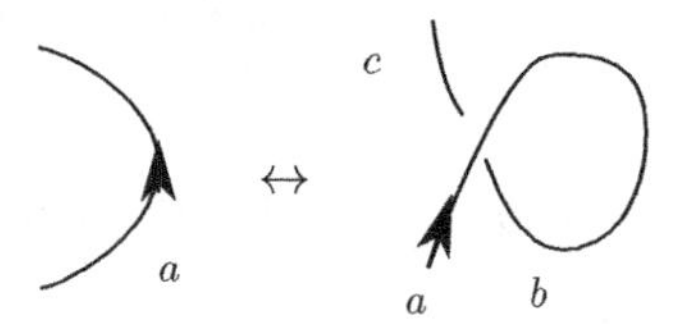

FIGURE 13.3: Reidemeister I move—Biquandle labeling

single element a. On the right hand side, in $BQ(K')$, there are the elements a, b, and c. The relations $b = R(a, b)$ and $c = L(a, b)$ are obtained from the crossing. Then, $c = a$ and there exists an element b of $BQ(K)$ such that $b = R(a, b)$.

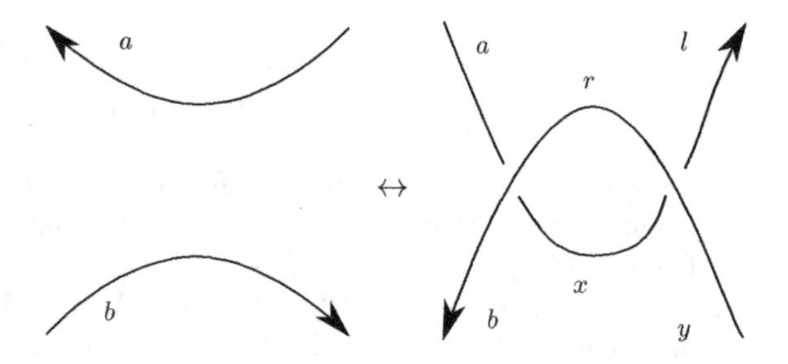

FIGURE 13.4: Contra-oriented Reidemeister II move—Biquandle labeling

The contra-oriented Reidemeister II move is illustrated in Figure 13.4. We prove that

$a = l$ and $b = y$. From the right hand side of the move (the diagram in Figure 13.4 K'), we obtain the following equations

$$a = L(b, x), \qquad r = R(b, x), \tag{13.10}$$
$$y = L^{-1}(l, r), \qquad x = R^{-1}(l, r).$$

From the biquandle conditions,

$$x = R^{-1}(l, R(b, x)), \qquad b = L^{-1}(l, R(b, x)). \tag{13.11}$$

Combining Equations 13.10 and 13.11, we obtain $b = y$. Next, since $b = L^{-1}(l, r)$ and $x = R^{-1}(l, r)$ then $L(b, x) = l$ by invertibility. Since $L(b, x) = a$ also, then $l = a$.

We leave the co-oriented case of the Reidemeister II move as an exercise.

The Reidemeister III move is shown in Figure 13.5.

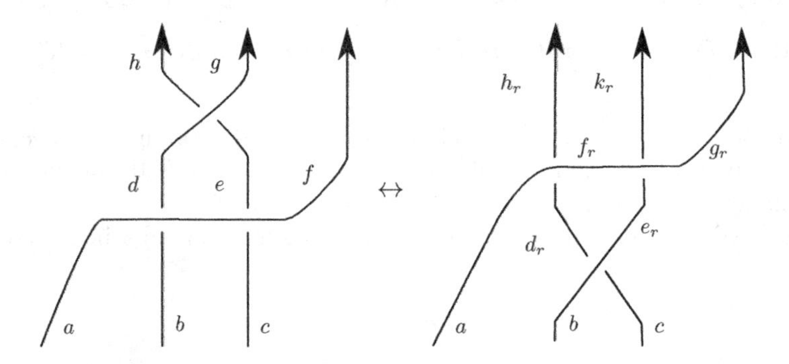

FIGURE 13.5: Biquandle labeled Reidemeister III move

From the left hand side of Figure 13.5, we obtain the following equations:

$$e = L(a, b) \qquad f = R(a, b), \tag{13.12}$$
$$d = L(R(a, b), c) \qquad g = R(R(a, b), c), \tag{13.13}$$
$$h = L(L(a, b), L(R(a, b), c)) \qquad k = R(L(a, b), L(R(a, b), c)). \tag{13.14}$$

From the right hand side of Figure 13.5, we obtain

$$d_r = L(b, c) \qquad e_r = R(b, c), \tag{13.15}$$
$$f_r = R(a, L(b, c)) \qquad g_r = R(R(a, L(b, c)), R(b, c)), \tag{13.16}$$
$$h_r = L(a, L(b, c)) \qquad k_r = L(R(a, L(b, c)), R(b, c)). \tag{13.17}$$

The intermediate labels are elements of both $BQ(K)$ and $BQ(K')$ by the recursive definition of $W(S)$. We want to prove that the labels on the endpoints are equivalent. We calculate that $h = h_r$ from the biquandle axiom $L(L(a, b), L(R(a, b), c)) = L(a, L(b, c))$. Next, the relation $R(L(a, b), L(R(a, b), c)) = L(R(a, L(b, c)), R(b, c))$ indicates that $k = k_r$. Finally, from the condition that $R(R(a, b), c) = R(R(a, L(b, c)), R(b, c))$, we conclude that $g = g_r$. Only one version of the Reidemeister III move needs to be considered, since the other possible Reidemeister III moves can be derived from this version of the Reidemeister III move. □

(a) Crossing from K (b) Crossing from $(K^M)^I$

FIGURE 13.6: Crossings in K and $(K^M)^I$

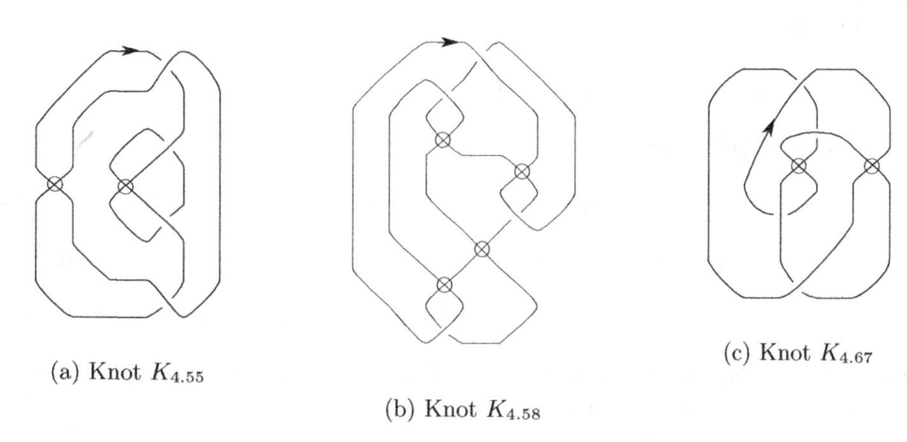

(a) Knot $K_{4.55}$ (b) Knot $K_{4.58}$ (c) Knot $K_{4.67}$

FIGURE 13.7: Compute $BQ(K)$

What information about the knot does the biquandle distinguish? We return to the virtual knot diagram K and its symmetries. In the earlier chapter, we weren't able to compare certain diagrams since the generating elements (determined by the edges) were not identical. With the biquandle, we obtain the same set of edges for each of the following diagrams: K^I, K^F, K^M, and K^S. Let's compare the relations obtained from a single crossing.

Theorem 13.2. *For all oriented virtual knot diagrams K, $BQ(K) \cong BQ((K^M)^I)$.*

Proof. We consider the labels on a classical crossing in K and its corresponding crossing in $(K^M)^I$. A classical crossing from K and the corresponding crossing in $(K^M)^I$ is illustrated in Figure 13.6. K and $(K^M)^I$ share the same set of generators, so we need only compare the relations obtained from each crossing. In K,

$$B(a, b) = (c, d). \tag{13.18}$$

In $(K^M)^I$, $B^{-1}(c, d) = (a, b)$. After rewriting, we see that $BQ(K)$ and $BQ((K^M)^I)$ are determined by the same set of generators and relations. □

Exercises

1. Compute $BQ(K)$ for the diagrams in Figure 13.7.
2. Prove that $BQ(K)$ is invariant under the co-oriented Reidemeister II move.
3. Compare the crossing relations between $BQ(K^M)$ and $BQ(K^S)$.
4. Compare the crossing relations between $BQ(K^F)$ and $BQ(K)$.

13.2 THE GENERALIZED ALEXANDER POLYNOMIAL

We constuct a biquandle X where B is defined by the equations

$$B(a,b) = (xa + yb, qa + zb). \tag{13.19}$$

The biquandle structure places restrictions on the values of x,y, q, and z. Using Equation 13.19,

$$L(a, L(b,c)) = L(L(a,b), L(R(a,b),c)). \tag{13.20}$$

We expand to obtain

$$L(a, xb + yc) = L(xa + yb, L(qa + zb, c)). \tag{13.21}$$

Expanding again, we obtain

$$xa + xyb + y^2c = L(xa + yb, xqa + xzb + yc). \tag{13.22}$$

Expanding for the final time, we obtain an equality with the variables x, y, q, and z

$$xa + xyb + y^2c = x^2a + xyb + xyqa + xyzb + y^2c. \tag{13.23}$$

The equality holds true for all a, b, and c in X. Collecting the coefficients of a, b, and c, we obtain three equations that the variables must satisfy:

$$x - x^2 - xyq = 0, \tag{13.24}$$
$$xyz = 0, \tag{13.25}$$
$$y^2 = y^2. \tag{13.26}$$

In Equation 13.24, if $x \neq 0$ then we conclude that $x = 1-qy$ and $z = 0$. This is the only solution that also satisfies the remaining two equations: $R(R(a,b),c) = R(R(a, L(b,c)), R(b,c))$, and $R(L(a,b), L(R(a,b),c)) = L(R(a, L(b,c)), R(b,c))$.

Relabeling the variables so that $q = s$ and $y = t$, we determine that

$$L(a,b) = tb + (1-st)a \tag{13.27}$$
$$R(a,b) = sa. \tag{13.28}$$

Then, B and B^{-1} can be written in matrix form

$$B = \begin{bmatrix} (1-st) & t \\ s & 0 \end{bmatrix}, \qquad B^{-1} = \begin{bmatrix} 0 & \frac{1}{s} \\ \frac{1}{t} & (1 - \frac{1}{st}) \end{bmatrix}. \tag{13.29}$$

We can rewrite the inverses of L and R

$$L^{-1}(a,b) = \frac{1}{s}b \tag{13.30}$$
$$R^{-1}(a,b) = \frac{1}{t}a + \left(1 - \frac{1}{st}\right)b. \tag{13.31}$$

For an oriented virtual knot diagram K, if B and B^{-1} are defined as these matrices, $BQ(K)$ determines a system of equations. The system of equations has coefficients in $\mathbb{Z}[t, t^{-1}, s, s^{-1}]$—the set of Laurent polynomials in the variables t and s. We investigate two questions:

1. What happens to the set of equations under the Reidemeister moves?

2. Can we find solutions to this system of equations?

Recall that the solution set to $A\overline{x} = \overline{0}$ is determined by the determinant of the matrix, so we focus on the determinant instead of the solution set.

We let $ABM(K)$ denote the matrix obtained from the system of equations, which we will refer to as the **Alexander biquandle matrix**. The **generalized Alexander polynomial** of K is the determinant of $ABM(K')$ and is denoted as $AB_K(s,t)$.

Theorem 13.3. *For oriented virtual knot diagrams K and K', if $K \sim K'$ then $AB_K(s,t) = \pm t^i s^j AB_{K'}(s,t)$.*

Proof. Let K denote the diagram on the left hand side of the Reidemeister moves and let K' represent the right hand side. The Reidemeister I move introduces the variables c and b as shown in Figure 13.3. The equations determined by this new crossing are

$$c = tb + (1 - st)a \tag{13.32}$$

$$b = sa. \tag{13.33}$$

Now

$$ABM(K') = \begin{bmatrix} & & ABM(K)^* & & 0 & * \\ & & & & \vdots & \vdots \\ & & & & 0 & * \\ 0 & \dots & 0 & (1-st) & t & -1 \\ 0 & \dots & 0 & s & -1 & 0 \end{bmatrix} \tag{13.34}$$

where $ABM(K)^*$ is related to $ABM(K)$ by row addition to $ABM(K')$. Then, $\det(ABM(K')) = -\det(ABM(K))$.

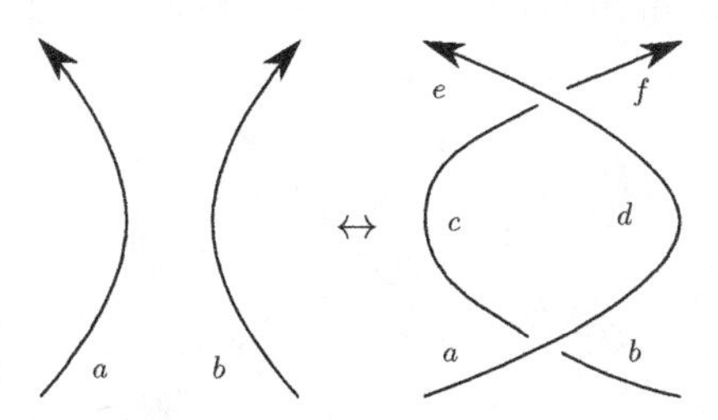

FIGURE 13.8: Co-oriented Reidemeister II move—Biquandle labeling

We consider the co-oriented Reidemeister II move. The right hand side (as shown in Figure 13.8) introduced the new variables c, d, e, and f and the new equations

$$d = sa, \qquad c = tb + (1 - st)a, \tag{13.35}$$

$$e = \frac{1}{s}d \qquad f, = \frac{1}{t}c + (1 - \frac{1}{st})d. \tag{13.36}$$

We construct the matrix $ABM(K')$:

$$ABM(K') = \left[\begin{array}{ccccccccc} & & & & & 0 & 0 & * & * \\ & & ABM(K)^* & & & \vdots & \vdots & \vdots & \vdots \\ & & & & & 0 & 0 & * & * \\ 0 & \dots & 0 & s & 0 & 0 & -1 & 0 & 0 \\ 0 & \dots & 0 & 0 & 0 & 0 & \frac{1}{s} & -1 & 0 \\ 0 & \dots & 0 & (1-st) & t & -1 & 0 & 0 & 0 \\ 0 & \dots & 0 & 0 & 0 & \frac{1}{t} & (1-\frac{1}{st}) & 0 & -1 \end{array}\right]. \tag{13.37}$$

After row operations, we obtain:

$$ABM(K')* = \left[\begin{array}{ccccccccc} & & & & & 0 & 0 & 0 & 0 \\ & & ABM(K) & & & \vdots & \vdots & \vdots & \vdots \\ & & & & & 0 & 0 & 0 & 0 \\ 0 & \dots & 0 & s & 0 & 0 & -1 & 0 & 0 \\ 0 & \dots & 0 & 1 & 0 & 0 & 0 & -1 & 0 \\ 0 & \dots & 0 & 1-st & t & -1 & 0 & 0 & 0 \\ 0 & \dots & 0 & 0 & 1 & 0 & 0 & 0 & -1 \end{array}\right]. \tag{13.38}$$

Then, applying co-factor expansion to Equation 13.38,

$$\det(ABM(K')*) = \det(ABM(K)). \tag{13.39}$$

Because $ABM(K')*$ and $ABM(K')$ are related by row operations, $\det(ABM(K')) = \pm t^i s^j (\det(ABM(K))$.

A similar result can be obtained for the Reidemeister III move. □

For an oriented knot diagram K, the generalized Alexander polynomial, $AB_K(s,t)$, is the determinant of $ABM(K)$. For two equivalent oriented knot diagrams K and K', the determinants of $ABM(K)$ and $ABM(K')$ differ only by a factor of the form $\pm t^i s^j$. We conclude that $AB_K(s,t)$ is an invariant of oriented knots.

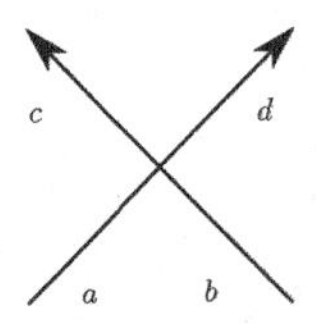

FIGURE 13.9: Labeled classical crossing

As an example, we compute $\det(ABM(K_{3.1})$. Begin by labeling the edges of the diagram as shown in Figure 13.10. From the diagram, we obtain the equations:

$$B(b,d) = (c,a), \qquad B(a,e) = (d,f), \qquad B(a,f) = (e,b).$$

After rewriting, we obtain:

$$td + (1-st)b = c, \qquad sb = a, \tag{13.40}$$

$$te + (1-st)a = f, \qquad sa = f, \tag{13.41}$$

$$tf + (1-st)c = e, \qquad sc = b. \tag{13.42}$$

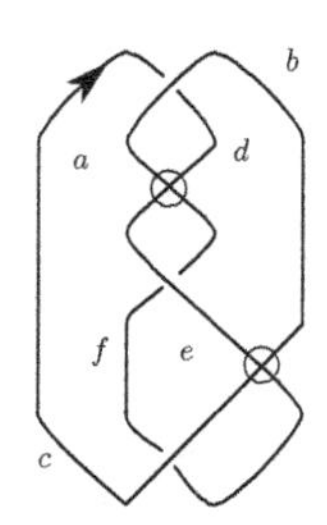

FIGURE 13.10: Labeled $K_{3.1}$

The system of equations results in the matrix:

$$ABM(K_{3.1}) = \begin{bmatrix} 0 & 1-st & -1 & t & 0 & 0 \\ -1 & s & 0 & 0 & 0 & 0 \\ 1-st & 0 & 0 & -1 & t & 0 \\ s & 0 & 0 & 0 & 0 & -1 \\ 0 & 0 & 1-st & 0 & -1 & t \\ 0 & -1 & s & 0 & 0 & 0 \end{bmatrix}. \tag{13.43}$$

Then $\det(ABM(K_{3.1}) = 1 - s - t^2 + s^3t^2 + st^3 - s^3t^3$.

We now consider when $\det(ABM(K)) = 0$. (The system of equations $ABM(K)\overline{x} = \overline{0}$ has a non-trivial solution when $\det(ABM(K)) = 0$.) Consider a single classical crossing labeled as shown in Figure 13.9. If this crossing is positively signed, we obtain the equations:

$$d = sa \tag{13.44}$$
$$c = tb + (1-st)a. \tag{13.45}$$

Converting to matrix format:

$$\begin{bmatrix} s & 0 & 0 & -1 \\ (1-st) & t & -1 & 0 \end{bmatrix} \begin{bmatrix} a \\ b \\ c \\ d \end{bmatrix} = \begin{bmatrix} 0 \\ 0 \\ 0 \\ 0 \end{bmatrix}. \tag{13.46}$$

If we multiply the columns corresponding to the left hand side of the crossing (edges a and c) by s^i and the columns corresponding to the right side (edges b and d) by s^{i+1}, we obtain the matrix

$$\begin{bmatrix} s^i s & 0 & 0 & -s^{i+1} \\ s^{i+1}(1-st) & s^{i+1}t & -s^i & 0 \end{bmatrix}. \tag{13.47}$$

The sum of the columns of the matrix in Equation 13.47 is the zero vector. If the crossing is negatively signed, we obtain the equations:

$$c = \frac{1}{t}b, \tag{13.48}$$
$$d = \frac{1}{t}a + (1 - \frac{1}{st})b. \tag{13.49}$$

Converting to matrix form:

$$\begin{bmatrix} 0 & \frac{1}{s} & -1 & 0 \\ \frac{1}{t} & (1-\frac{1}{st}) & 0 & -1 \end{bmatrix}. \tag{13.50}$$

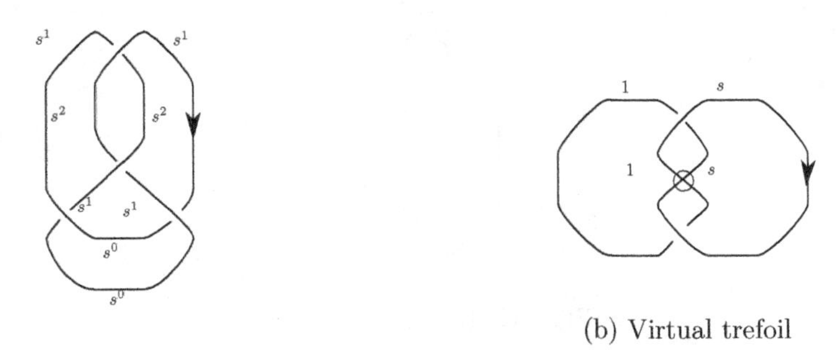

(b) Virtual trefoil

(a) Figure 8 knot

FIGURE 13.11: Labeling edges with factors s^i

If we multiply the left hand columns (a and c) of the matrix in Equation 13.50 by s^i and the right hand columns (b and d) by s^{i+1}, we obtain the matrix in Equation 13.51. In Equation 13.51, the sum of the matrice's columns is the zero vector.

$$\begin{bmatrix} 0 & \frac{s^{i+1}}{s} & -s^i & 0 \\ \frac{s^i}{t} & s^{i+1}(1-\frac{1}{st}) & 0 & -s^{i+1} \end{bmatrix}. \tag{13.51}$$

For an oriented virtual knot diagram K, if we can multiply the columns of $ABM(K)$ by factors of the form s^i so that the resulting columns sum to the zero vector then the columns of $ABM(K)$ are linearly dependent and $\det(ABM(K)) = 0$. If these factors exist, we can denote the columns of $ABM(K)$ as C_e (where e is the corresponding edge label), let $\sigma(e)$ denote the integer exponent for each edge label, and

$$\sum_e s^{\sigma(e)} C_e = \overline{0}. \tag{13.52}$$

The edges of the figure 8 knot, K_8, shown in Figure 13.11a, are labeled with the factors in Equation 13.52 . However, the virtual trefoil has no such labeling, as shown in Figure 13.11b.

To determine if a virtual link diagram K has such a labeling, we convert the diagram into a directed graph with $2n$ vertices and $3n$ directed edges as shown in Figure 13.12. We call this new diagram G_K.

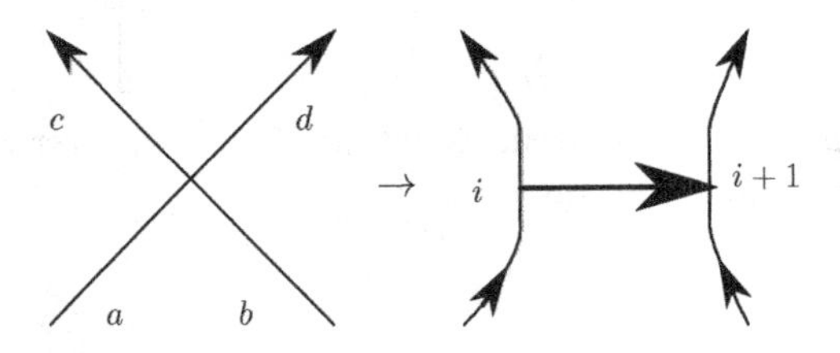

FIGURE 13.12: Link diagram conversion

For an edge e in G_K, $i(e)$ denotes the initial point and $t(e)$ denotes the terminal point. There are two types of edges in G_K: edges from the original link diagram and **inserted**

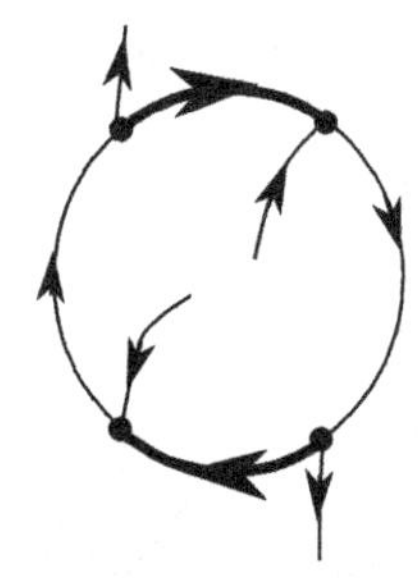

FIGURE 13.13: Cycle with inserted edges

edges that are added during the conversion process. In G_K, we can construct sequences of directed adjacent edges: $e_1, e_2, \ldots, e_n$, where $t(e_i) = i(e_{i+1})$. A sequence of edges that does not repeat any edges or vertices and $t(e_n) = i(e_1)$ is called a **cycle**. We consider cycles $e_1, e_2, \ldots, e_n$ with the following property: where at least one edge is an inserted edge.

If the knot diagram K has a labeling then we can label the vertices of the directed graph with i and $i+1$. Following a sequence of edges in the direction of orientation, the vertex labels never decrease. Suppose that G_K has a cycle that contains inserted edges. The graph G_K is finite and this cycle can only contain a finite number of inserted edges. On a cycle with inserted edges, the vertex labels will decrease following the direction of orientation. For example, a cycle with n inserted edges has a vertex that is labeled both s^i and s^{n+i}. This contradicts the fact that K had an edge labeling.

Theorem 13.4. *For all classical knot diagrams D, the columns of $ABM(K)$ are linearly dependent and* $\det(ABM(D)) = 0$.

Proof. Let D be a classical knot diagram. If G_D contains a cycle with an inserted edge, this cycle forms a simple, closed curve in the plane.

Each non-inserted edge in G_D is part of a cycle that does not contain inserted edges.

We consider a cycle with inserted edges. in Figure 13.13. The thickened edges represent inserted edges while the thin edges are oriented edges from the original knot diagram. The outward pointing edges are each part of a cycle. But a cycle containing these edges must cross the cycle containing the inserted edges with a virtual crossing. This contradicts the fact that D did not contain virtual crossings. G_D does not contain any cycles with inserted edges and we are able to label the edges of D and, correspondingly, the columns of $ABM(D)$ with factors of the form s^i. The columns of $ABM(K)$ are linearly dependent and $\det(ABM(K)) = 0$. □

Note that there are virtual knots K with $AB_K(s,t) = 0$, which do not necessarily satisfy this condition.

Exercises

1. Construct $ABM(K)$ for the diagrams in Figure 13.14.
2. Compute the $AB_K(s,t)$ for the diagrams in Figure 13.14.
3. Verify that $B^{-1}(B(a,b)) = (a,b)$.
4. Prove that if $s = 1$, we obtain a system of equations equivalent to the system produced by the Alexander quandle.

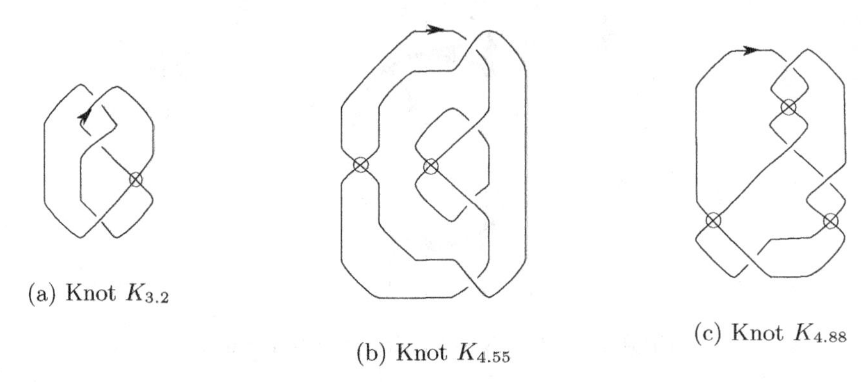

(a) Knot $K_{3.2}$

(b) Knot $K_{4.55}$

(c) Knot $K_{4.88}$

FIGURE 13.14: Compute $AB_K(s,t)$

13.3 OPEN PROBLEMS AND PROJECTS

Open problems

1. Does the abstract biquandle differentiate between all diagrams K and $(K^M)^I$?
2. For every non-trivial virtual knot diagram, is there a finite biquandle that distinguishes the knot from the unknot?
3. What criterion can we place on virtual knots K and L to obtain results about the biquandle structure of $K \sharp L$?
4. What information do we obtain from the minors? Can you find a restriction on the diagram so that all minors give the same polynomial up to $\pm t^i s^j$?

Projects

1. Evaluate the generalized Alexander polynomial for different values of s and t. What can you prove? Can you compute the Alexander polynomial from minors of the matrix $ABM(K)$?
2. Read about other finite biquandle structures.
3. Compute the generalized Alexander polynomial for families of knots.

Further Reading

1. A. S. Crans, Allison Henrich, and Sam Nelson. Polynomial knot and link invariants from the virtual biquandle. *J. Knot Theory Ramifications*, 22(4):1–12, 2013
2. Jorg Sawollek. On Alexander-Conway polynomials for virtual knots and links. *arXiv preprint math/9912173*, 1999
3. Louis H. Kauffman and David Radford. Bi-oriented quantum algebras, and a generalized Alexander polynomial for virtual links. *Diagrammatic morphisms and applications (San Francisco, CA, 2000)*, 318:113–140, 2003
4. Naoko Kamada and Seiichi Kamada. Biquandles with structures related to virtual links and twisted links. *J. Knot Theory Ramifications*, 21(13):1–15, 2012

5. M. Elhamdadi and S. Nelson. *Quandles: an introduction to the algebra of knots.* American Mathematical Society, 2015

CHAPTER 14

Gauss diagrams

Gauss diagrams offer an alternative viewpoint to virtual knot diagrams. Gauss diagrams look nothing like link diagrams **but** there is a one-to-one correspondence between virtual knots and equivalence classes of Gauss diagrams. Equivalence classes of Gauss diagrams are based on Gauss diagram analogs of the Reidemeister moves. In this chapter, we introduce Gauss words and show how to construct a Gauss word from an oriented virtual knot diagram. We then introduce Gauss diagrams (constructed from a Gauss word) and introduce several virtual link invariants that are easily computed from the Gauss diagram but not from the virtual knot diagram.

14.1 GAUSS WORDS AND DIAGRAMS

A **Gauss word** is a decorated list of symbols where each symbol occurs twice. The decorations include a $\pm$ sign and a bar symbol. To obtain a Gauss word from an oriented knot diagram, select a base point on the diagram and assign a label to each crossing. We then record crossing information as we traverse the virtual knot diagram from the base point in the direction of the orientation. The information recorded includes the label and sign of the crossing. We mark the traversal of an overpass by placing a bar on the crossing label. For example, if we traverse the overpass of a positive crossing labeled a, we record $\overline{a}+$. Traversing the negatively signed underpass of the crossing b, we record $b-$. We compute the Gauss words of the two labeled diagrams in Figure 14.1. The Gauss word obtained from

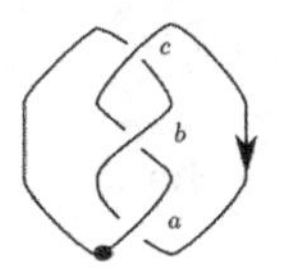

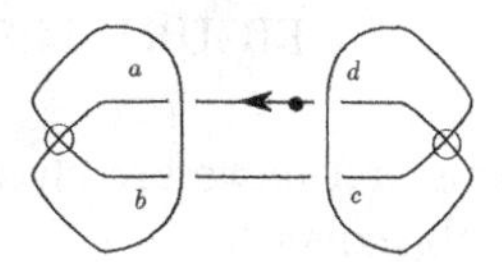

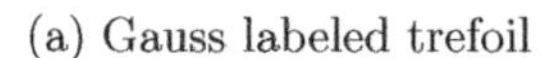

(a) Gauss labeled trefoil

(b) Gauss labeled Kishino's knot

FIGURE 14.1: Gauss labeled diagrams

the labeled trefoil diagram in Figure 14.1a is

$$\overline{a}+b+\overline{c}+a+\overline{b}+c+. \tag{14.1}$$

The Gauss word obtained from the oriented diagram of Kishino's knot (in Figure 14.1b) is

$$a+\overline{b}-\overline{a}+b-\overline{c}-d+c-\overline{d}+. \tag{14.2}$$

Changes in the position of the base point result in a cyclical shift of the Gauss word. The virtual crossings are completely ignored when the crossing information is transcribed.

Consider the Gauss diagram analogs of the diagrammatic moves. As an example, we perform a Reidemeister II move on the oriented trefoil diagram from Figure 14.1a. The result is the oriented virtual knot diagram in Figure 14.2. The Gauss word obtained from the diagram is given in Equation 14.3.

$$d - e + \overline{a} + b + \overline{c} + a + \overline{b} + c + \overline{d} - \overline{e} + . \tag{14.3}$$

In comparison to Equation 14.1, the Reidemeister II move resulted in the addition of two

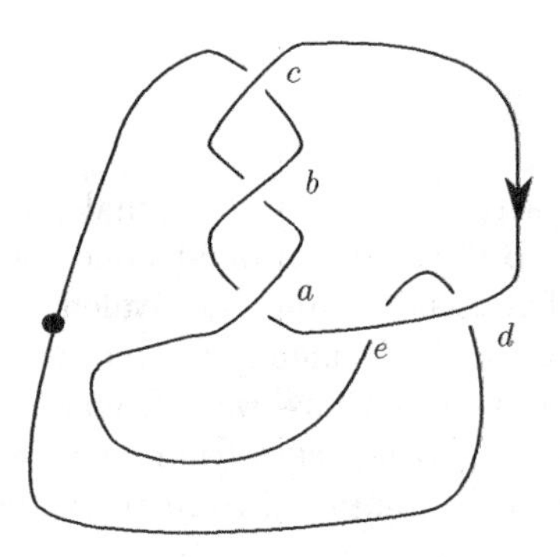

FIGURE 14.2: Trefoil after Reidemeister II move

sub-sequences to the Gauss word. The sequence $d - e+$ was added to the beginning and the sequence $\overline{d} - \overline{e}+$ was added to the end.

We describe how each classical Reidemeister move alters a Gauss word. The virtual Reidemeister moves change only virtual crossings and do not affect the Gauss word. Let S_1, S_2, S_3 denote sub-sequences contained in a Gauss word. The left and right hand side

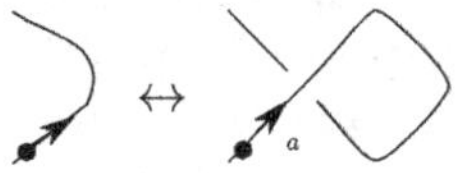

FIGURE 14.3: Gauss labeled Reidemeister I move

of a Reidemeister I move are shown in Figure 14.3. The move changes the Gauss word by inserting a sub-sequence:

$$S_1 \longleftrightarrow \overline{a} + a + S_1. \tag{14.4}$$

The sub-sequence $\overline{a} + a+$ is added to the word by the Reidemeister I move.

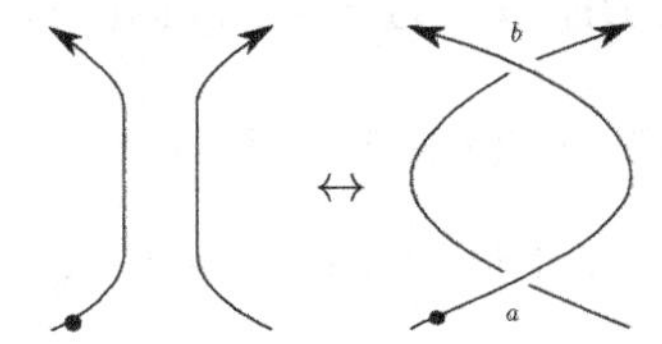

FIGURE 14.4: Gauss labeled co-oriented Reidemeister II move

In Figure 14.4, compare the left and right hand sides of a co-oriented Reidemeister II move. The corresponding Gauss words are

$$S_1S_2 \longleftrightarrow \overline{a}+\overline{b}-S_1a+b-S_2. \tag{14.5}$$

The Reidemeister II moves change the Gauss word by inserting two sub-sequences $\overline{a}+\overline{b}-$ and $a+b-$.

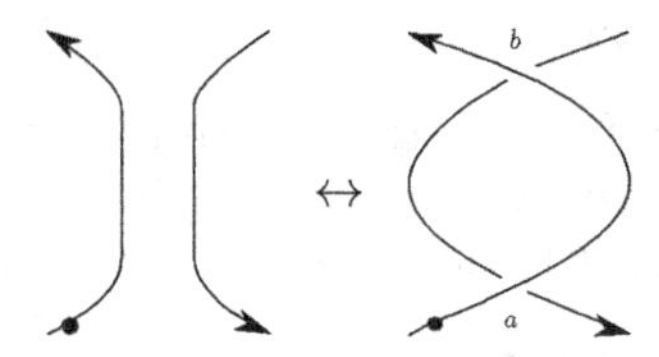

FIGURE 14.5: Gauss labeled contra-oriented Reidemeister II move

The left and right hand sides of a contra-oriented Reidemeister II move are shown in Figure 14.5. The corresponding Gauss words are:

$$S_1S_2 \longleftrightarrow \overline{a}+\overline{b}-S_1b-a+S_2. \tag{14.6}$$

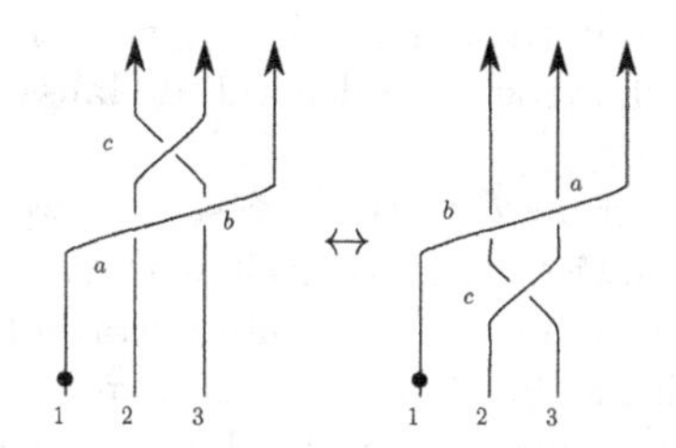

FIGURE 14.6: Gauss labeled Reidemeister III move

The left and right hand side of a Reidemeister III move are shown in Figure 14.6. The Gauss words obtained from the left and right hand side of a Reidemeister III move are

$$\overline{a}+\overline{b}+S_1a+\overline{c}+S_2b+c+S_3 \longleftrightarrow \overline{b}+\overline{a}+S_1\overline{c}+a+S_2c+b+S_3. \tag{14.7}$$

Unlike the other moves, this move does not involve the insertion or deletion of a sub-sequence. Instead, three sub-sequences are modified. For example, $\bar{a}+\bar{b}+$ becomes $\bar{b}+\bar{a}+$.

There are different choices of orientation and sign for each of these moves. In the Reidemeister I move, the crossing can be either positively signed or negatively signed.

Theorem 14.1. *Equivalence classes of Gauss words modulo the Reidemeister move analogs and cyclic permutation are in one-to-one correspondence with the set of oriented virtual knots.*

Proof. The analogs are constructed directly from the Reidemeister moves. If a sequence of Reidemeister moves relates the diagrams K and K', we can construct a sequence of analogous modifications to the Gauss words. □

From this theorem, the following corollary about crossing number is clear.

Corollary 14.2. *For all classical knots K, $\mathsf{c}(K) = \mathsf{cc}(K)$.*

We can recover an oriented virtual knot diagram from the Gauss word by the following algorithm. Draw a base point on the plane. We construct a closed curve consisting of $2n$ segments.

1. Examine the 1st symbol in the word; draw a line segment (this is curve 1) from the basepoint. Choose a point on the line segment and label it with the crossing information on the first symbol.
2. Examine the n-th symbol. If this is the first time this symbol is encountered, continue your curve with a line segment that does not intersect the existing curve. Mark the midpoint of the line segment to decorate with crossing information.
3. If this is the second time that the symbol is encountered, add a curve segment that passes through the marked point that was created the first time the segment was encountered. Avoid all existing classical crossings and marked points. If the curve intersects another line segment and creates a crossing, the crossing is virtual.
4. After the last symbol, connect the first and last curve segments so that the curve is closed.

We are now ready to define a Gauss diagram. A **Gauss diagram** is a counterclockwise oriented circle with $2n$ marked points. Pairs of marked points with the same symbol are connected by a **chord** contained in the circle. Chords with an orientation and sign are called **arrows**. A Gauss diagram with arrows is a decorated Gauss diagram. The set of decorated Gauss diagrams is denoted as $\mathcal{G}$.

Given an oriented knot diagram D with n crossings, we construct its Gauss diagram, $G(D)$. Construct a counterclockwise oriented circle with $2n$ evenly spaced points. Label the $2n$ points sequentially with the symbols in the Gauss word of D. Then connect each pair of points with the same label with a chord. The arc is then marked with the sign of the label and then directed from the label with the bar to the label without the bar. In practice, the chords are drawn with a slight arch to avoid triple (or higher) points of intersection.

Recall the Gauss word corresponding to trefoil, T,

$$\bar{a} + b + \bar{c} + a + \bar{b} + c + . \tag{14.8}$$

The decorated Gauss diagram, $G(T)$, appears in Figure 14.7a.

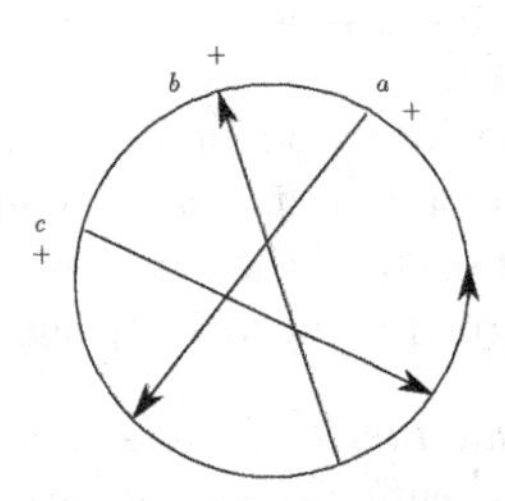

(a) Gauss diagram for T

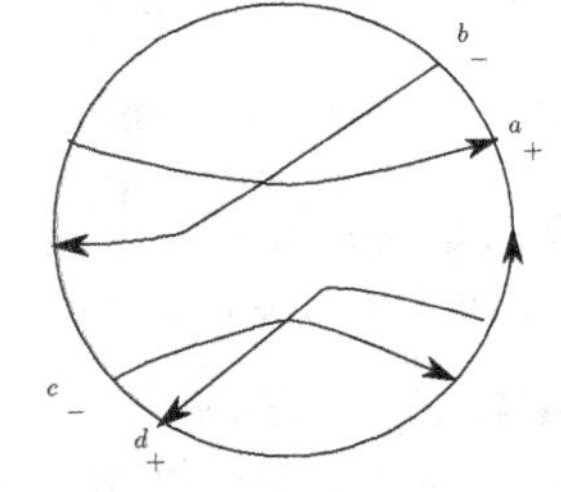

(b) Gauss diagram for Kishino's knot

FIGURE 14.7: Example Gauss diagrams

Recall the Gauss word corresponding to Kishino's knot (Equation 14.2). The Gauss diagram corresponding to Kishino's knot is in Figure 14.7b.

We describe the analogs of the Reidemeister moves for Gauss diagrams. In the Gauss diagram analogs of the Reidemeister moves, we draw the oriented circle using solid arcs and dotted arcs. On the solid arcs, we see exactly the arrows involved in a single Reidemeister move. On the dotted arcs, we have no knowledge of the arrows or their endpoints. The dotted arcs act as placeholders for the remainder of the diagram. The Reidemeister I move analog introduces or removes a single, isolated chord with either sign as shown in Figure 14.8.

There are two analogs of the Reidemeister II move. Both moves introduce or remove two oppositely signed chords as shown in Figure 14.9. Note that there are no chords between the endpoints of the two chords.

There are two analogs of the Reidemeister III moves based on the orientation and ordering of the edges in the Reidemeister III moves. Both analogs involve triangular shaped configurations of chords as shown in Figure 14.10. (Be careful that you have correctly signed and oriented the moves!) O. Ostlund proved which analogs of the Reidemeister moves are sufficient to define equivalence classes of decorated Gauss diagrams in his 2004 paper, "A diagrammatic approach to link invariants of finite degree," which was published in *Mathematica Scandinavica.*

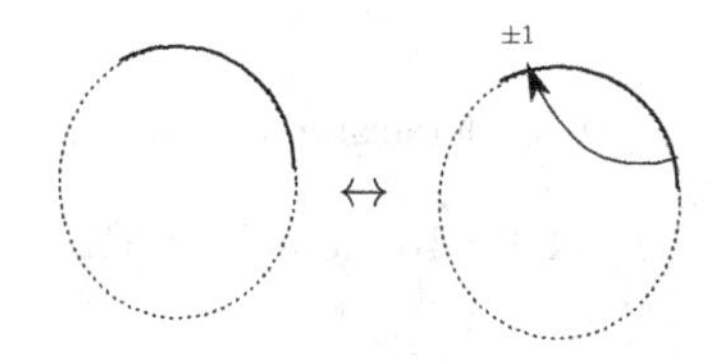

FIGURE 14.8: Reidemeister I analog change

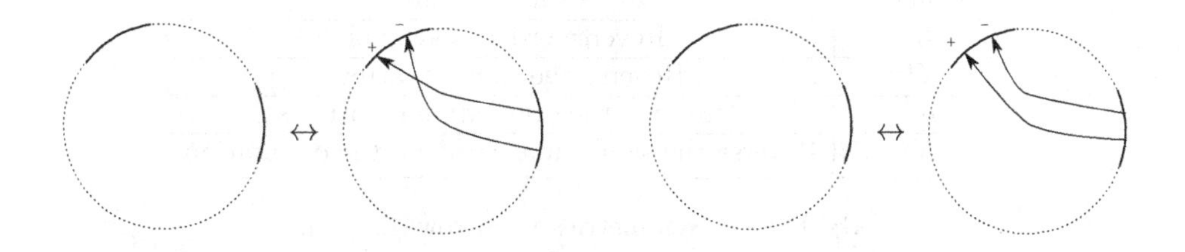

(a) Co-oriented Reidemeister II analog (b) Contra-oriented Reidemeister II analog

FIGURE 14.9: Reidemeister II analogs

An **arrow diagram** is an equivalence class of decorated Gauss diagrams related by a finite sequence of the Reidemeister move analogs. We have the following theorem:

Theorem 14.3. *Oriented virtual knots are in one-to-one correspondence with arrow diagrams.*

Proof. There is a bijective map from the set of knots to the set of arrow diagrams. We can construct this map by using the constructions outlined above. □

We use arrow diagrams instead of oriented knot diagrams because some invariants are

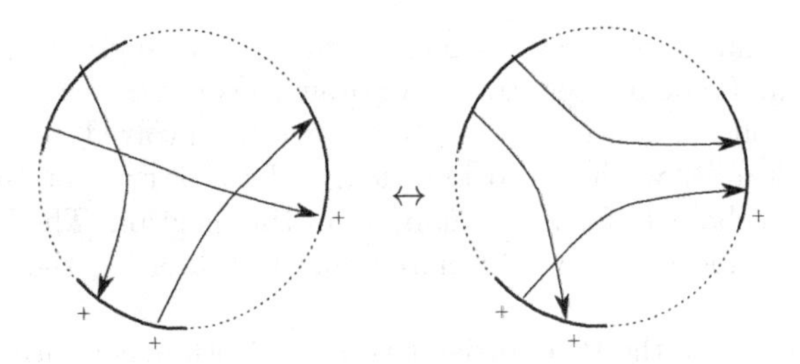

(a) Reidemeister III analog

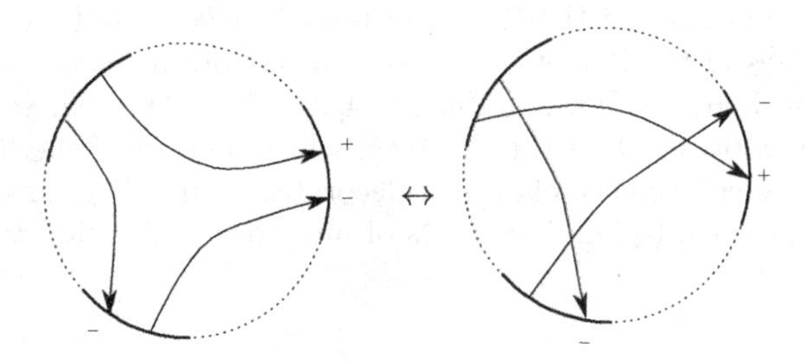

(b) Reidemeister III analog

FIGURE 14.10: Reidemeister III analogs

much easier to compute arrow diagrams! For example, constructing symmetries of an oriented knot diagram becomes very simple. We outline how to construct symmetries using arrow diagrams.

Symmetry	Construction
K^I	Reverse order of symbols
K^M	Reverse the signs on arrows
K^F	Reverse the orientation on arrows
K^S	Reverse the signs on arrows and the orientation

TABLE 14.1: Symmetries via arrow diagrams

From this point of view, it is easy to prove the following theorem about $(K^S)^M$ and K^F.

Theorem 14.4. *For all oriented virtual knot diagrams D, $(D^S)^M \sim D^F$.*

Proof. We begin with $G(D)$. Then we reverse the signs and orientation on all arrows to obtain $G(K^S)$. Next, we reverse all the signs on all arrows to obtain $G((D^S)^M)$. But this is exactly the diagram $G(D^F)$. □

We can also easily construct analogs of the virtualization moves.

Exercises

1. Construct Gauss words for the oriented versions of the knot diagrams shown in Figure 14.11.

Virtualization	Construction
Sign virtualization	Reverse sign on arrow
Way virtualization	Reverse the orientation on arrow

TABLE 14.2: Virtualization in arrow diagrams

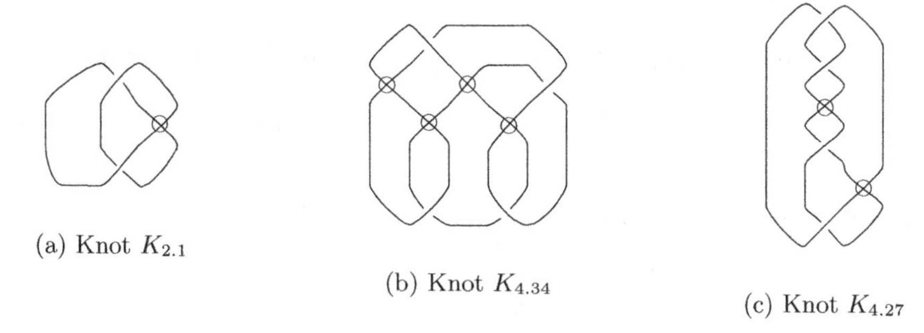

(a) Knot $K_{2.1}$

(b) Knot $K_{4.34}$

(c) Knot $K_{4.27}$

FIGURE 14.11: Compute the Gauss word

2. Construct a knot diagram corresponding to $\overline{a}_+ b_+ \overline{c}_- d_- \overline{b}_+ a_+ \overline{d}_- c_-$.
3. Construct a knot diagram corresponding to $\overline{a}_- \overline{b}_- c_- \overline{d}_- a_- b_- \overline{c}_- d_-$.
4. Verify Table 14.2 for the virtual figure 8 knot (with base point and orientation) in Figure 14.12.
5. Construct oriented knot diagrams for the Gauss diagrams shown in Figure 14.13.
6. Construct decorated Gauss diagrams for the virtual knots in Figure 14.11.
7. List all possible Gauss word analogs for the Reidemeister II and III moves.

14.2 PARITY AND PARITY INVARIANTS

Arrow diagrams can be used to compute the **parity** of individual crossings in an oriented knot diagram. Parity can either be used to construct an independent invariant of the knot or to amplify an invariant. Parity was first introduced by Vassily O. Manturov in the early 2000s; one of his papers on this topic is included in the further reading list.

Let $\mathcal{A}(G)$ denote the set of arrows in a fixed Gauss diagram G. We define a function $p : \mathcal{A} \to \mathbb{Z}_2$. For an arrow $a \in \mathcal{A}$, let N_A denote the set of chords that intersect a. The

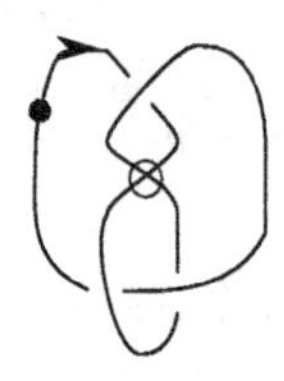

FIGURE 14.12: Compute the Gauss diagrams of the symmetries

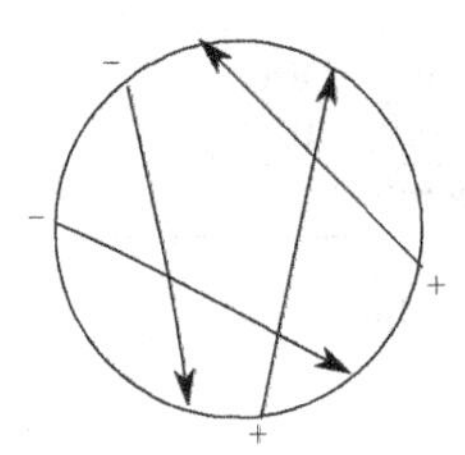

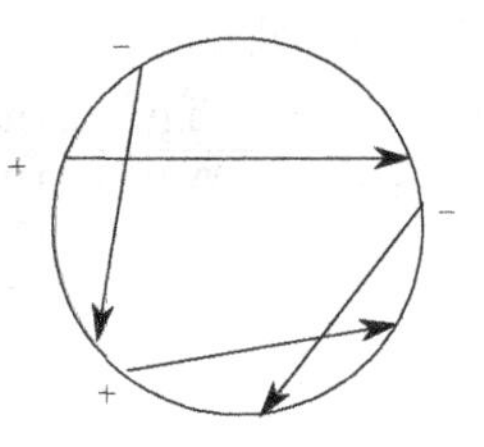

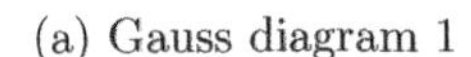

(a) Gauss diagram 1

(b) Gauss diagram 2

FIGURE 14.13: Construct the knot diagram

notation $|N_a|$ denotes the number of elements in the set. Then

$$p(a) = |N_a| \mod 2. \tag{14.9}$$

The value of $p(a)$ is the parity of the arrow and the parity of the crossings in the oriented virtual knot diagram corresponding to G. Let G and G' be equivalent decorated Gauss diagrams related by a single Reidemeister move analog. For an arrow a in $\mathcal{A}(G)$ with a corresponding arrow a' in $\mathcal{A}(G')$, we prove that $p(a) = p(a')$.

1. A Reidemeister I analog introduces an isolated arrow. Let a be an arrow in G not involved in the Reidemeister I move. Then G' contains a corresponding arrow a' and $|N_a| = |N_{a'}|$. (The arrow introduced by the Reidemeister I move has parity zero.)

2. The Reidemeister II move introduces (or removes) two arrows of the same parity. Let G denote the side of the Reidemeister move with no arrows. The diagram G' contains the two arrows introduced by the Reidemeister II move, c_1 and c_2. For an arrow a' in G' (with corresponding arrow a in G), either the arrows c_1 and c_2 both intersect an arrow a' or neither chord intersects a'. In the first case, $|N_{a'}| = |N_a| + 2$ so that $p(a) = p(a')$. If c_1 and c_2 do not intersect a' then $|N_{a'}| = |N_a|$.

3. Given a and corresponding arrow a', the Reidemeister III move does not change the number of arrows that intersect an arrow so that $|N_a| = |N_{a'}|$.

The number of arrows with parity 1 changes under the Reidemeister analogs. We define an invariant of arrow diagrams $\mathcal{A}_1 : \mathcal{G} \to \mathbb{Z}$. For a decorated Gauss diagram G,

$$\mathcal{A}_1(G) = \sum_{a \in \mathcal{A}(G)} sgn(a)p(a). \tag{14.10}$$

The **parity** of the diagram G, $\mathcal{A}_1(G)$, is the number of crossings with parity 1 counted with sign.

Theorem 14.5. *For all decorated Gauss diagrams G and G', if G and G' are equivalent then $\mathcal{A}_1(G) = \mathcal{A}_1(G')$.*

Proof. The Reidemeister I analog introduces (or removes) arrows that have parity zero. These arrows do not change the parity of other arrows in the diagram.

The Reidemeister II analog introduces (or removes) two arrows that have identical intersection sets, and are oppositely signed. The parity of other arrows is unchanged by the Reidemeister II move. As a result, $\mathcal{A}_1(G) = \mathcal{A}_1(G')$.

In the Reidemeister III analog, consider an arrow c in G and the corresponding arrow c' in G'. If c is not part of the Reidemeister III move, then $p(c) = p(c')$. If c and c' are involved in the Reidemeister III move, we note that $p(c) = p(c')$ (although the intersection sets change). □

Parity can be used to determine that non-classical virtual knots are not the unknot by determining that the genus of a virtual knot is greater than one.

Lemma 14.6. *For all decorated Gauss diagrams G, if G' is a decorated Gauss diagram related to G by reversing the orientation of a single arrow and changing the sign of the arrow then* $\mathcal{A}_1(G) = \mathcal{A}_1(G') \mod 2$.

Proof. Let c be an arrow in the diagram G. Suppose that G' is obtained from G by flipping the direction and sign of the arrow c. For every arrow a in G, there is a corresponding arrow a' in G'. We note that $p(a) = p(a')$. Now since the sign of c changes, we see that if $\mathcal{A}_1(G) = p + 1$ then $\mathcal{A}_1(G') = p - 1$. □

Theorem 14.7. *For all oriented, classical knot diagrams D, for all crossings c in D, $p(c) = 0$ and $\mathcal{A}_1(G(D)) = 0$.*

Proof. Let D be a classical knot diagram and let $G(D)$ be the corresponding arrow digaram. Suppose that $p(c) = 1$ for some crossing c. If we vertically smooth the corresponding crossing in D then the resulting smoothed diagram has two components. Arrows (in $G(D)$) that intersect the arrow c indicate crossings that contain edges in both components of the smoothed diagram. However, two closed curves in the plane intersect an even number of times by the Jordan curve theorem, so that $p(c) = 0$. This contradicts our assumption that $p(c) = 1$. As a result, $p(c) = 0$ and $\mathcal{A}_1(G(D)) = 0$. □

Corollary 14.8. *For all oriented, virtual knot diagrams D, $\mathcal{A}_1(G(D)) \neq 0$ then D is not classical and $D \not\sim U$.*

We use the parity of a crossing to amplify the bracket polynomial. We denote the **parity bracket** of an oriented virtual knot diagram K as $\langle K \rangle_p$. This modification of the bracket polynomial was first introduced by Vassily O. Manturov. The resolution of a crossing is based on its parity. For a crossing c, if $p(c) = 1$ then we rigidly fuse the crossing. Otherwise, we apply the usual skein relation expansion. We illustrate the two possible types of expansions for a crossing in Figures 14.14 and 14.15.

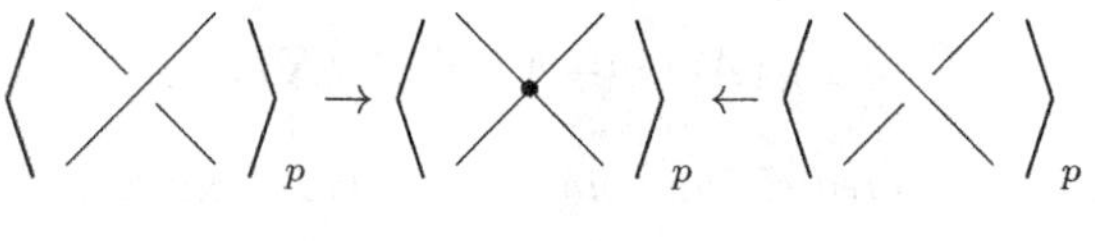

FIGURE 14.14: Parity bracket expansion for crossings with parity 1

The states of the parity bracket polynomial consist of four valent graphs (each vertex is adjacent to 4 edges) and closed loops that possibly contain virtual crossings. We evaluate the states as follows:

1. $\langle \bigcirc \rangle_p = 1$.

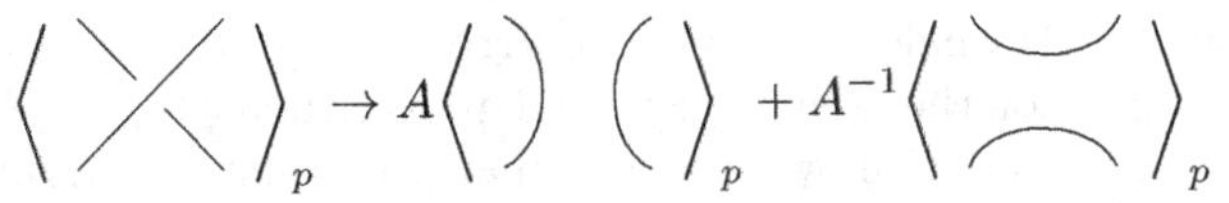

(a) Positive crossing

(b) Negative crossing

FIGURE 14.15: Parity bracket expansion for crossings with parity 0

2. $\langle \bigcirc \cup K \rangle_p = (-A^2 - A^2)\langle K \rangle_p.$

3. We reduce the graph using the following relation:

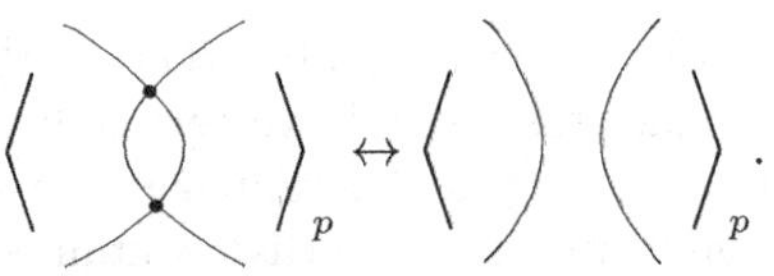

$$(14.11)$$

The parity bracket is a formal sum of 4-valent graphs with coefficients in $\mathbb{Z}[A, A^{-1}]$ (Laurent polynomials with coefficients in $\mathbb{Z}$). We recall from Theorem 14.7 that if K is a classical knot diagram then $p(c) = 0$ for all crossings in the diagram K. This means that if K is a classical diagram then $\langle K \rangle = \langle K \rangle_p$. Theorem 14.7 is not true for virtual knot diagrams. For example, every crossing in Kishino's knot, K_K, has parity 1. In the expansion, we rigidly fuse each crossing in the diagram and the formal sum consists of a single 4-valent graph.

$$\langle K_K \rangle_p = \qquad (14.12)$$

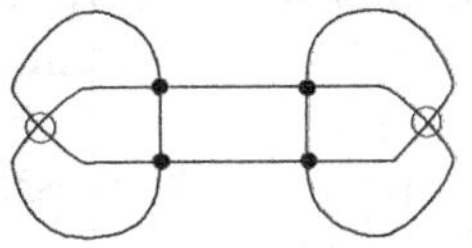

The **normalized parity bracket polynomial** of K is defined as

$$P_K(A) = (-A^{-3})^{w(K)}\langle K \rangle_p. \qquad (14.13)$$

Theorem 14.9. *For all oriented virtual knot diagrams K and K', if $K \sim K'$ then $P_K(A) = P_{K'}(A)$.*

Proof. A crossing introduced by a Reidemeister I move has parity 0 and the usual expansion is applied to this crossing.

We consider a Reidemeister II move. Both crossings in a Reidemeister II move have the same parity. We compute the expansion of a Reidemeister II move where both crossings have parity 1:

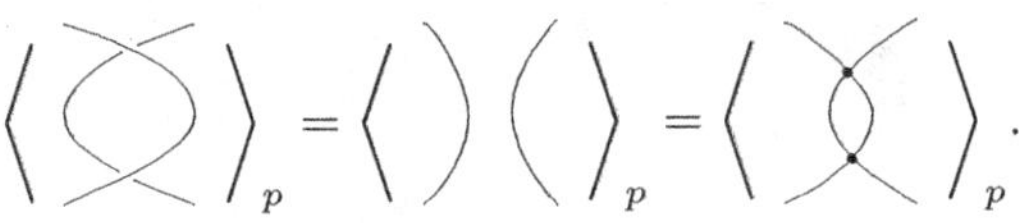

$$(14.14)$$

In the case of the Reidemeister III move, two of the crossings have parity 1 or none of the crossings have parity 1. If none of the crossings have parity 1 the usual expansion applies to the move. We focus on one of the three cases where two of the crossings have parity 1:

$$\langle \quad \rangle_p = A\langle \quad \rangle_p + A^{-1}\langle \quad \rangle_p \tag{14.15}$$

$$= A\langle \quad \rangle_p + A^{-1}\langle \quad \rangle_p \tag{14.16}$$

$$= A\langle \quad \rangle_p + A^{-1}\langle \quad \rangle_p \tag{14.17}$$

$$= \langle \quad \rangle_p . \tag{14.18}$$

The remaining two cases are left as an exercise. □

Consider double way virtualization, an example of which is shown in shown in Figure 14.16.

Theorem 14.10. *The normalized parity bracket is not invariant under the double way virtualization.*

Proof. The proof is left as an exercise. □

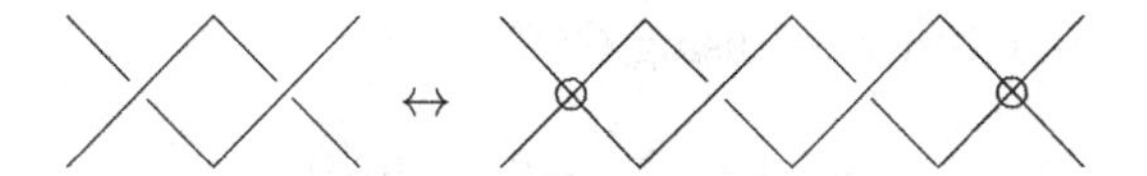

FIGURE 14.16: Double way virtualization

Exercises

1. Compute $\mathcal{A}_1(G(K))$ for the knots in Figure 14.17.
2. Compute the normalized parity bracket $P_K(A)$ for the knots shown in Figure 14.17.
3. Prove the remaining cases of the Reidemeister III move.
4. Prove Theorem 14.10.

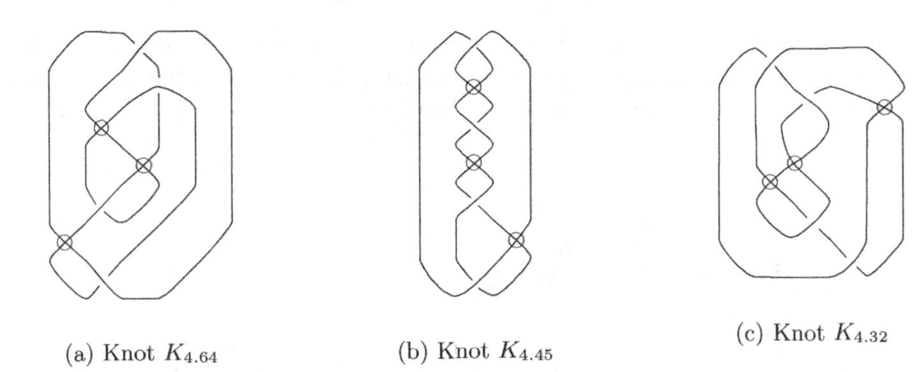

(a) Knot $K_{4.64}$ (b) Knot $K_{4.45}$ (c) Knot $K_{4.32}$

FIGURE 14.17: Compute the normalized parity bracket

14.3 CROSSING WEIGHT NUMBER

For an oriented virtual knot diagram K, the crossing weight numbers $\mathcal{C}_a(K)$ can be easily computed from the corresponding Gauss diagram $G(K)$. Let K be an oriented virtual knot diagram and let $G(K)$ denote the corresponding Gauss diagram. We abuse notation by letting c denote a classical crossing in the diagram K and the corresponding arrow in $G(K)$. Let $\mathcal{A}(K)$ denote the set of arrows in $G(K)$. Let N_c denote the set of arrows that the arrow c intersects. If $a \in N_c$, determine $int_c(a)$ as illustrated in Figure 14.18.

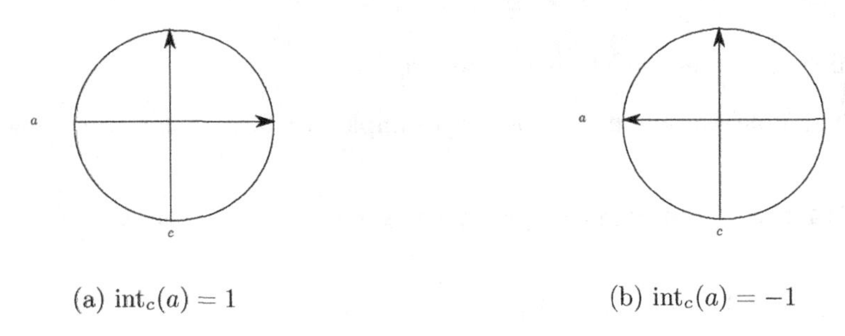

(a) $\text{int}_c(a) = 1$ (b) $\text{int}_c(a) = -1$

FIGURE 14.18: Evaluating $\text{int}_c(a)$

Now, the **weight** of c is defined using $G(K)$.

$$\mathbf{w}(c) = \sum_{a \in N_c} sgn(a) int_c(a). \tag{14.19}$$

Let $\mathbf{W}_a(G)$ denote the set of chords of weight a. Then,

$$\mathbf{C}_a(G(K)) = \sum_{c \in \mathbf{W}_a(G)} \text{sgn}(c). \tag{14.20}$$

We realize that $\mathbf{C}_a(G(K)) = \mathcal{C}_a(K)$.

Theorem 14.11. *For all equivalent oriented virtual knot diagrams K and K', if $a \neq 0$, then $\mathbf{C}_a(G(K)) = \mathbf{C}_a(G(K'))$. The crossing weight number is an invariant of oriented virtual knots.*

Remark 14.1. *The definition of the weight of c is defined so that $\mathbf{C}_a(G(K))$ and $\mathcal{C}_a(K)$ are equal. Usually, the definition is reversed.*

Using the diagram $G(K)$, we can also quickly compute the crossing weights of K^I, K^S, K^M, and K^F. We can sometimes determine if switching a single crossing changes the equivalence class of a diagram K. Let G be a Gauss diagram with an arrow c. Let G' be the diagram obtained from G by changing the direction and sign of the arrow c, forming the arrow c'. We use a to denote all other arrows in G and a' to denote the corresponding arrows in G'. We have the following theorem:

Theorem 14.12. *Let G be a Gauss diagram and let G' be a Gauss diagram obtained from G by changing the sign and direction of arrow c to form c'. Then*

1. $\mathrm{w}(c) = -\mathrm{w}(c')$
2. *For all $a \neq c$ such that $a \in \mathcal{A}(G)$ and the corresponding $a' \in \mathcal{A}(G)$ then* $\mathrm{w}(a) = \mathrm{w}(a')$.

Proof. Note that $\mathrm{int}_c(a) = -\mathrm{int}_{c'}(a')$. The results immediately follow. □

We apply this result to obtain the following theorem.

Theorem 14.13. *For all non-zero integers a, $\mathbf{C}_a(K)$ is a finite type invariant of degree* ≤ 1.

Proof. Let K be an oriented virtual knot diagram. Without loss of generality, assume that K has two crossings of weight a, say x and y with positive sign. We let K_{-+} denote the oriented virtual knot diagram obtained by switching the crossing x, K_{+-} denote the oriented virtual knot diagram obtained by switching the crossing y, and K_{--} denote the oriented virtual knot diagram obtained by switching the crossings x and y.

We denote the Gauss diagram of K as G_{++}. Let G_{-+} denote $G(K_{-+})$, G_{+-} denote $G(K_{+-})$, and G_{--} denote $G(K_{--})$.

Suppose that $\mathbf{C}_a(G_{++}) = k$. Then by Theorem 14.13,

$$\mathbf{C}_a(G_{-+}) = k - 1 \qquad \mathbf{C}_a(G_{+-}) = k - 1 \qquad \mathbf{C}_a(G_{--}) = k - 2. \tag{14.21}$$

Note that $\mathbf{C}_a(G_{++}) - \mathbf{C}_a(G_{-+}) - \mathbf{C}_a(G_{+-}) + \mathbf{C}_a(G_{--}) = 0$. However, $\mathbf{C}_a(G_{++}) - \mathbf{C}_a(G_{-+}) \neq 0$. We conclude that $\mathbf{C}_a(K)$ is a finite type invariant of degree ≤ 1. □

Exercises

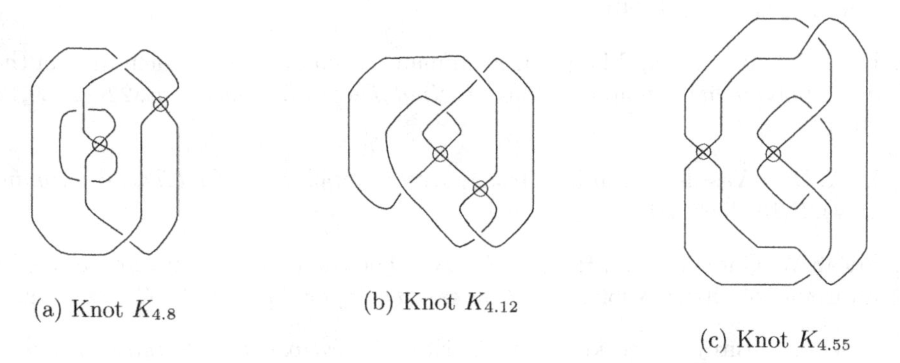

(a) Knot $K_{4.8}$

(b) Knot $K_{4.12}$

(c) Knot $K_{4.55}$

FIGURE 14.19: Compute the a weights for each knot

1. Determine all non-zero values of $\mathbf{C}_a(K)$ for the diagrams in Figure 14.19.
2. For the positive virtual trefoil, VT, verify that $\mathbf{C}_a(G(VT)) = \mathcal{C}_a(VT)$ for all non-zero a.

14.4 OPEN PROBLEMS AND PROJECTS

Open problems

1. If a crossing in a diagram K is cosmetic, then changing the over/underpassing information of the crossing does not change the equivalence class of K. We would like to prove that with only a few exceptions, crossings are not cosmetic.
2. Choose subsets of arrows in Gauss diagrams and construct finite type invariants of higher degree.
3. Find a relationship between virtual finite type invariants and classical finite type invariants.
4. Use parity (or crossing weights) to amplify other knot invariants.

Projects

1. Find families of virtual knot diagrams which are non-classical and $\mathbf{C}_a(K) = 0$ for all $a \neq 0$.
2. Construct knot diagrams that are detected by the parity bracket but have $\mathbf{C}_a(K) = 0$ for all $a \neq 0$.
3. Compute the values of finite type invariants arising from the arrow polynomial and compare to the crossing weights.
4. Read about cosmetic crossings.

Further Reading

1. S. Chmutov, S. Duzhin, and J. Mostovoy. *Introduction to Vassiliev knot invariants.* Cambridge University Press, Cambridge, 2012
2. Aaron Kaestner and Louis H Kauffman. Parity biquandles. *Knots in Poland. III. Part 1*, 100:131–151, 2014
3. Lena C. Folwaczny and Louis H. Kauffman. A linking number definition of the affine index polynomial and applications. *J. Knot Theory Ramifications*, 22(12):30,1341004, 2013
4. H. A. Dye. Vassiliev invariants from parity mappings. *J. Knot Theory Ramifications*, 22(4):21,1340008, 2013
5. Micah W. Chrisman and Heather A. Dye. The three loop isotopy and framed isotopy invariants of virtual knots. *Topology and its Applications*, 173:107–134, August 2014
6. V. O. Manturov. Free knots and parity. In *Introductory lectures on knot theory*, volume 46 of *Ser. Knots Everything*, pages 321–345. World Sci. Publ., Hackensack, NJ, 2012
7. Louis H. Kauffman. Introduction to virtual knot theory. *J. Knot Theory Ramifications*, 21(13):37,1240007, 2012

CHAPTER 15

Applications

Two applications of virtual knot theory include quantum computation and textiles. Other applications of knot theory include chemistry and biology, with a specific emphasis on DNA recombination. Several references are given in the further readings sections on these topics.

15.1 QUANTUM COMPUTATION

Classical computers use a bit to encode information. A bit stores either a 0 or 1 and data is encoded in strings of 0's and 1's. Logical gates act on these strings of data—taking two bits as input and producing a single bit as output. Clearly, what we think of as a "computer computation" requires thousands of bits and operations. To perform a task on a computer, we need to know how many bits are needed to store the data and how many operations (such as addition or multiplication) are required to perform the computation. The number of operations required to perform the task is referred to as the computational complexity. To put this in perspective, currently, a computation on x bits requires a number of operations that can be expressed as a polynomial in the variable x.

For example, let's consider matrix multiplication of an $n \times n$ matrix. In an $n \times n$ matrix, there are n^2 entries. How many operations does it take to multiply two $n \times n$ matrices? We let A,B, and C denote $n \times n$ matrices and suppose that $AB = C$. Recall that the entry in the i-th row and j-th column of C is computed as follows

$$C_{ij} = \sum_{k=1}^{n} A_{ik} B_{kj}. \tag{15.1}$$

The computation of C_{ij} in Equation 15.1 requires n multiplication operations and $n-1$ addition operations. A single entry in the matrix C requires roughly $2n$ operations. There are n^2 entries; computing the entire matrix C requires roughly $2n^3$ operations. Notice that $2x^3$ is a polynomial, so we say that we can compute matrix multiplication in polynomial time.

Now consider the bracket polynomial. A virtual link diagram D, with n classical crossings, has 2^n states. Recall the definition of the bracket polynomial:

$$\langle D \rangle = \sum_{s \in S} A^{\alpha(s)-\beta(s)} d^{|s|}. \tag{15.2}$$

There are 2^n states and each state must be evaluated (count the number of total loops) and then the values are added. Taking this simplistic viewpoint, each state requires two operations: evaluation and addition to a running total. For a virtual link diagram with n classical crossings, that is $(2^n) \times 2$ total operations. The total number of computations is an exponential expression based on the number of classical crossings.

Algorithms or computations that require an exponential number of calculations are challenging, if not impossible to compute. Their rapid growth in size means that it is very easy to run out of computer memory.

Quantum computation is fundamentally different from classical computation. The basic unit of data is a **qubit**—a "quantum bit". A qubit is neither a 0 or a 1, but a **superposition** of the two classical states. Qubits are expressed as linear combinations:

$$|\psi\rangle = \alpha|0\rangle + \beta|1\rangle \tag{15.3}$$

where α and β are complex numbers. The notation $|\psi\rangle$ is called a ket and is part of Dirac notation. When measured, the qubit collapses to a 0 with probability $|\alpha|^2$ and state 1 with probability $|\beta|^2$. Until the qubit is measured, the qubit carries information regarding both bits. Returning to the bracket polynomial, a type A smoothing could be encoded as a 0 and a type B smoothing could be encoded as a 1. For a virtual knot diagram with n classical crossing, all 2^n states could be encoded using n qubits, instead of $n(2^n)$ classical bits.

Two famous quantum algorithms are Shor's Algorithm (factoring a number into primes) and Grover's algorithm (sorting an unstructured list). Both of these quantum algorithms are "faster" than their classical counterparts. One reason that knot theorists are interested in quantum algorithms is because there is also a quantum algorithm for the computation of the f-polynomial.

Another reason for knot theorists' interest in quantum computation is one of the models of quantum computation. Theoretical models of quantum computation involve three main steps:

1. Initialize the system (construct qubits)
2. Apply logical gates (operations)
3. Measure the outcome (read the results)

The first step (system initialization) consists of choosing a linear combination of qubit vectors of length n with entries in $\mathbb{Z}_2$ and complex coefficients. The logical gates (operations) are simulated via matrix multiplication using unitary matrices. (We let A^* denote the conjugate transpose of a matrix. Then a unitary matrix satisfies the equation $(A^*)^T = A^{-1}$ or that $A(A^*)^T = I = (A^*)^T A$.) One model of quantum computation is topological quantum computation. This model involves anyons—particles that are arranged in a line in a 2-D plane. Through some mechanism, these particles move around each other in the plane. Adding the dimension of time, a braid emerges from this scenario. The particles involved in this scenario are called non-abelian anyons.

By braid, we mean an $n-n$ tangle with some specific restrictions. The **n-strand braid group** B_n has the set of generators $\{\sigma_1, \sigma_2, \dots \sigma_{n-1}\}$. The braid group also contains the n-strand identity braid as shown in Figure 15.1. The inverse of σ_i is obtained by switching the positive crossing to a negative crossing. The elements of B_n consist of all $n-n$ tangles formed by concatenation of the generators. The generators satisfy the following identities

$$\sigma_i \sigma_j = \sigma_j \sigma_i \text{ for all } |i-j| > 1, \tag{15.4}$$

$$\sigma_i \sigma_{i\pm 1} \sigma_i = \sigma_{i\pm 1} \sigma_i \sigma_{i\pm 1}. \tag{15.5}$$

An example of concatenation is shown in Figure 15.2. The braid group can be mapped

FIGURE 15.1: Braid generators

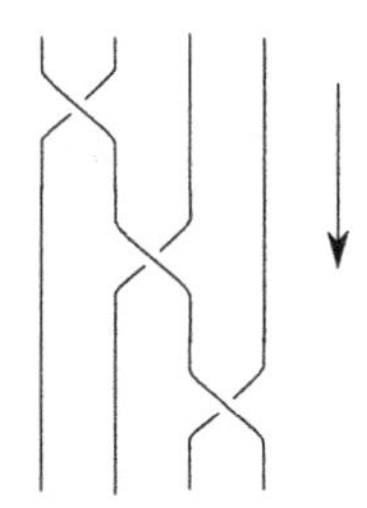

FIGURE 15.2: Braid: $\sigma_1\sigma_2\sigma_3$

onto a set of $2n \times 2n$ matrices: $\phi : B_n \to SL_n(\mathbb{C})$. These matrices respect the Reidemeister moves and, when unitary, can also be viewed as simulated logical gates in our simulated quantum computer. Research has shown these braid operators are universal, meaning that the entire set of logical gates can be constructed using a small set of matrices.

Exercises

1. For the braid group B_4, verify that $\sigma_1\sigma_3 = \sigma_3\sigma_1$.

2. For the braid group B_4, construct the diagram $\sigma_1\sigma_2\sigma_1$. Which Reidemeister move does this braid correspond to?

15.2 TEXTILES

Traditional woven textiles have a warp and weft thread structure—the warp threads extend across the width of the fabric while the weft threads run the length of the fabric. The interactions between the warp and weft threads are regular—repeating along both the length and width of the fabric. Using mathematical terminology, **fabric** is a doubly periodic oriented plane knot diagram. We consider several examples of woven fabric in Figure 15.3.

The fact that the fabric can be described using a small region of the doubly periodic pattern leads to the following definition from Vassiliev. A **fabric kernel** is a $n - m$ tangle with no virtual crossings as shown in Figure 15.4. Given an $n-m$ tangle L, we can construct a fabric kernel as shown in Figure 15.5.

The fabric kernel can be viewed in three distinct contexts: as a link on a torus, as a classical link (where components X and Y mark the frame of the torus), or as a virtual link diagram. Each possiblity is shown in Figure 15.6.

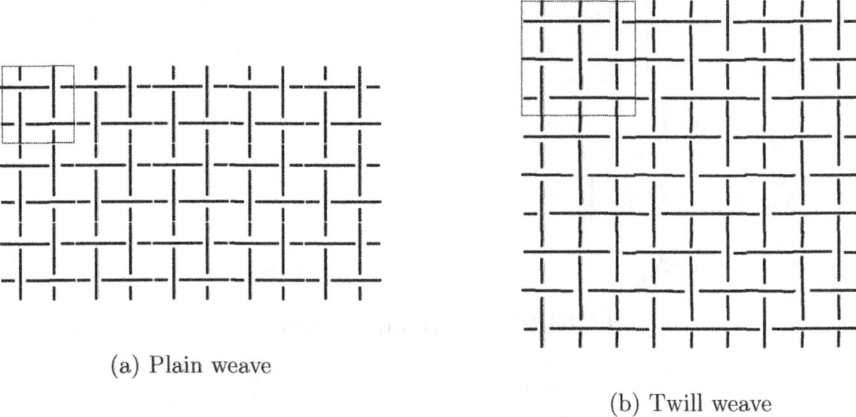

(a) Plain weave

(b) Twill weave

FIGURE 15.3: Woven fabrics

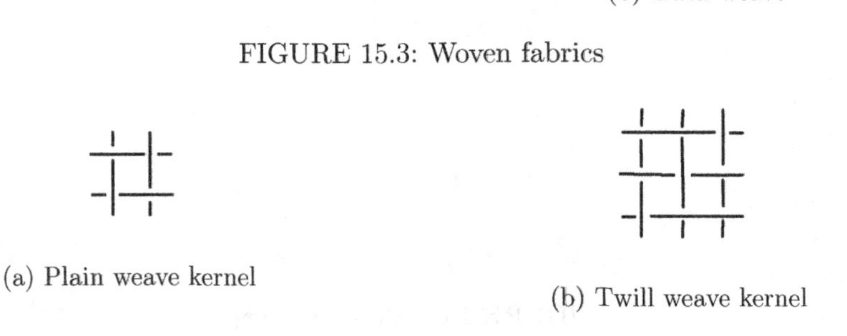

(a) Plain weave kernel

(b) Twill weave kernel

FIGURE 15.4: Woven kernels

We can ask the following questions about fabric kernels.

1. When do two different fabric kernels result in the same fabric?
2. Does a fabric kernel result in fabric that contains closed components (like chain mail, as shown in Figure 15.7d)?
3. Can knot invariants be used to identify different fabrics?

The following invariants can be applied to the fabric kernel and can be used to describe or differentiate between different fabric kernels. We apply these invariant to the kernel of Figure 15.7a, which is shown in Figure 15.8.

1. The **crossing number** of the kernel, L, denoted $\mathcal{C}(L)$ counts the number of crossings in the kernel. There are 2 crossings in Figure 15.7a.
2. The **number of components** that result when the edges of the fabric kernel are identified to form a torus is also an invariant. In Figure 15.7a, two components are formed when the edges are identified.
3. The linking number of the components is also an invariant. We order the components of the fabric kernel after identifying the edges. Then, the linking number of the fabric kernel is defined as
$$\mathcal{L}(L) = \sum_{i<j} |l(K_i, K_j)| \tag{15.6}$$
where $l(K_i, K_j)$ is the sum of the signs of the crossings where components i and j meet. There are two components in Figure 15.7a and $\mathcal{L}(L) = 0$.

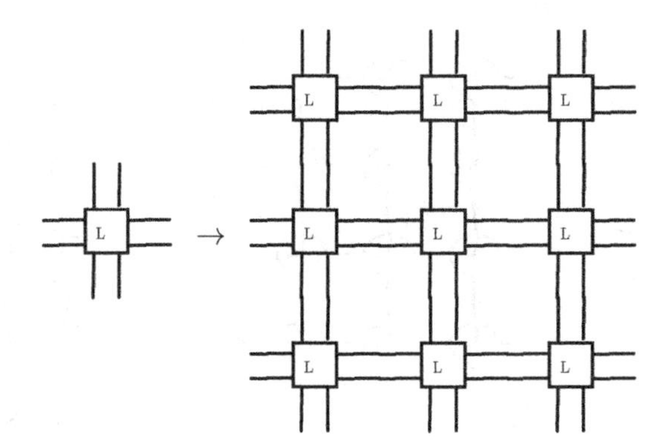

FIGURE 15.5: Fabric from the kernel L

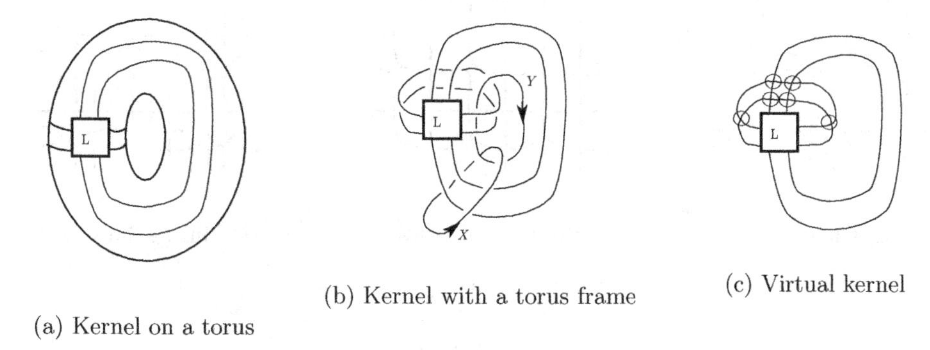

(a) Kernel on a torus

(b) Kernel with a torus frame

(c) Virtual kernel

FIGURE 15.6: Kernel visualization

4. The **axial type** of each strand can also be determined by computing the repeat of an individual strand across the surface of the kernel. The repeat is then expressed in terms of X and Y, where X denotes the oriented, horizontal boundary of the fabric kernel and Y denotes the oriented, vertical boundary of the fabric kernel as shown in Figure 15.8. Tracing the components, we see that the axial type of one component is $X - Y$ and the other component is $X + Y$. Traditional fabric is either knitted (constructed from one continuous fiber) or woven (constructed from warp and weft fibers). The **number of axial components** identifies the fabric type—knitted fabric has one axial fiber and woven fabric has two axial fibers (warp and weft). However, new industrial processes and 3-D printing open up the possibility of multi-axial fabrics and fabrics that include closed components. Additional fabric structures are shown in Figure 15.7. In particular, chain mail fabrics (which contain closed components) have axial types 0, as the component bounds a closed region.

More traditional knot invariants can also be applied as in H. R. Morton and S. Grishanov's paper "Doubly periodic textile patterns" published on *arxiv.org*.

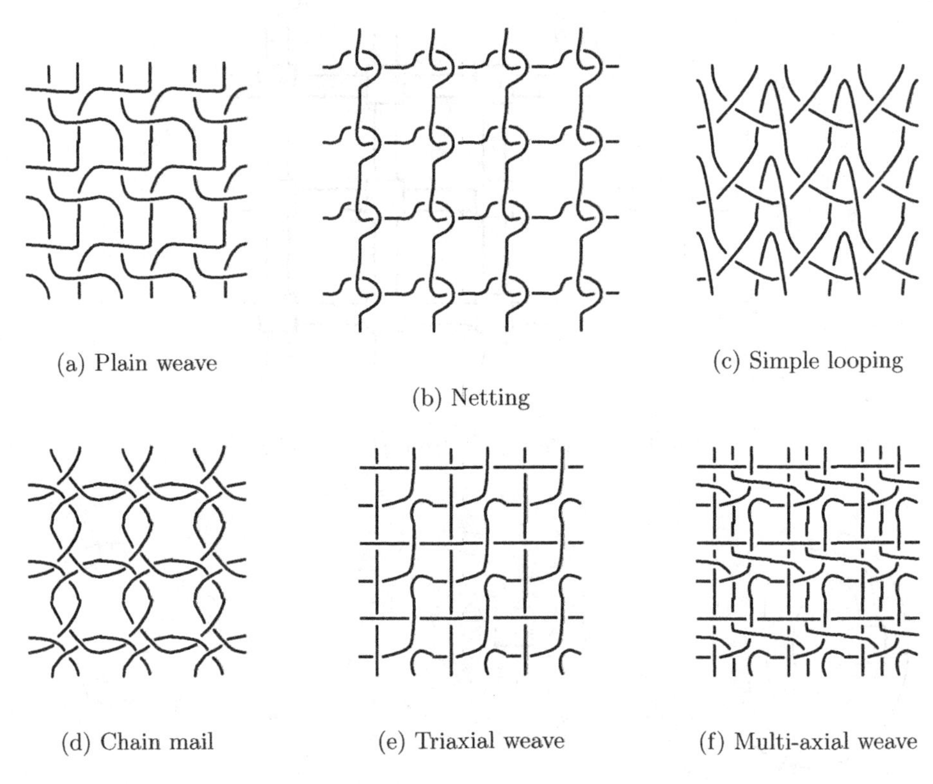

(a) Plain weave

(b) Netting

(c) Simple looping

(d) Chain mail

(e) Triaxial weave

(f) Multi-axial weave

FIGURE 15.7: Different fabrics

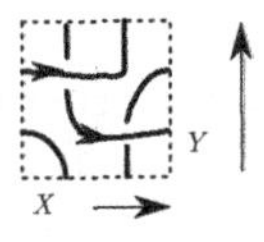

FIGURE 15.8: Plain weave tile

Exercises

1. For the kernels of the fabric diagrams in Figures 15.7b–15.7f compute: the crossing number, the number of components, the linking number, and the axial types.

15.3 OPEN PROBLEMS AND PROJECTS

Projects

1. Read and present on mathematical biology and knot theory's role.

2. Read about quantum algorithms for the Jones polynomial.

3. Read about mosaic knots.

4. Read about textiles, look up common fabric types, and construct fabric kernels.

Further Reading

1. S. J. Lomonaco and L. H. Kauffman. Quantum knots and mosaics. *Quantum Information Processing*, pages 1–32, 2008

2. Anirban Pathak. *Elements of quantum computation and quantum communication*. CRC Press, 2013

3. Dorothy Buck and Erica Flapan, editors. *Applications of knot theory*, volume 66 of *Proceedings of Symposia in Applied Mathematics*. American Mathematical Society, Providence, RI, 2009

4. De Witt Sumners. Lifting the curtain: using topology to probe the hidden action of enzymes. *Match*, (34):51–76, 1996

5. S. Grishanov, V. Meshkov, and A. Omelchenko. A topological study of textile structures. Part I: An introduction to topological methods. *Textile Research Journal*, 79(8):702–713, April 2009

6. S. Grishanov, V. Meshkov, and A. Omelchenko. A topological study of textile structures. Part II: Topological invariants in application to textile structures. *Textile Research Journal*, 79(9):822–836, May 2009

APPENDIX A

Tables

In this chapter, the knot tables are based on Jeremy Green's table of virtual knots.

The knot invariants listed in the table are the (un-normalized) arrow polynomial and the generalized Alexander polynomial. From the arrow polynomial, the bracket polynomial can be recovered by evaluating each K_i term at 1.

A.1 KNOT TABLES

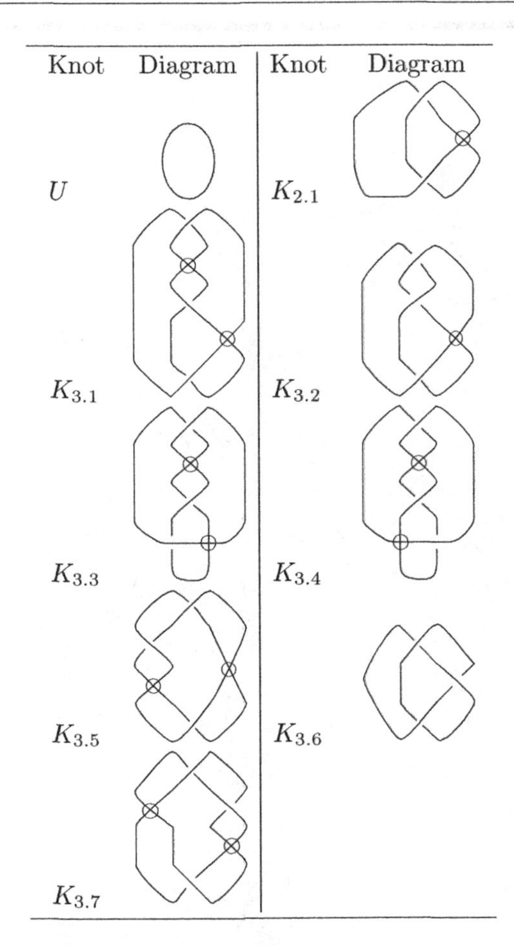

TABLE A.1: Knots with 3 or fewer crossings

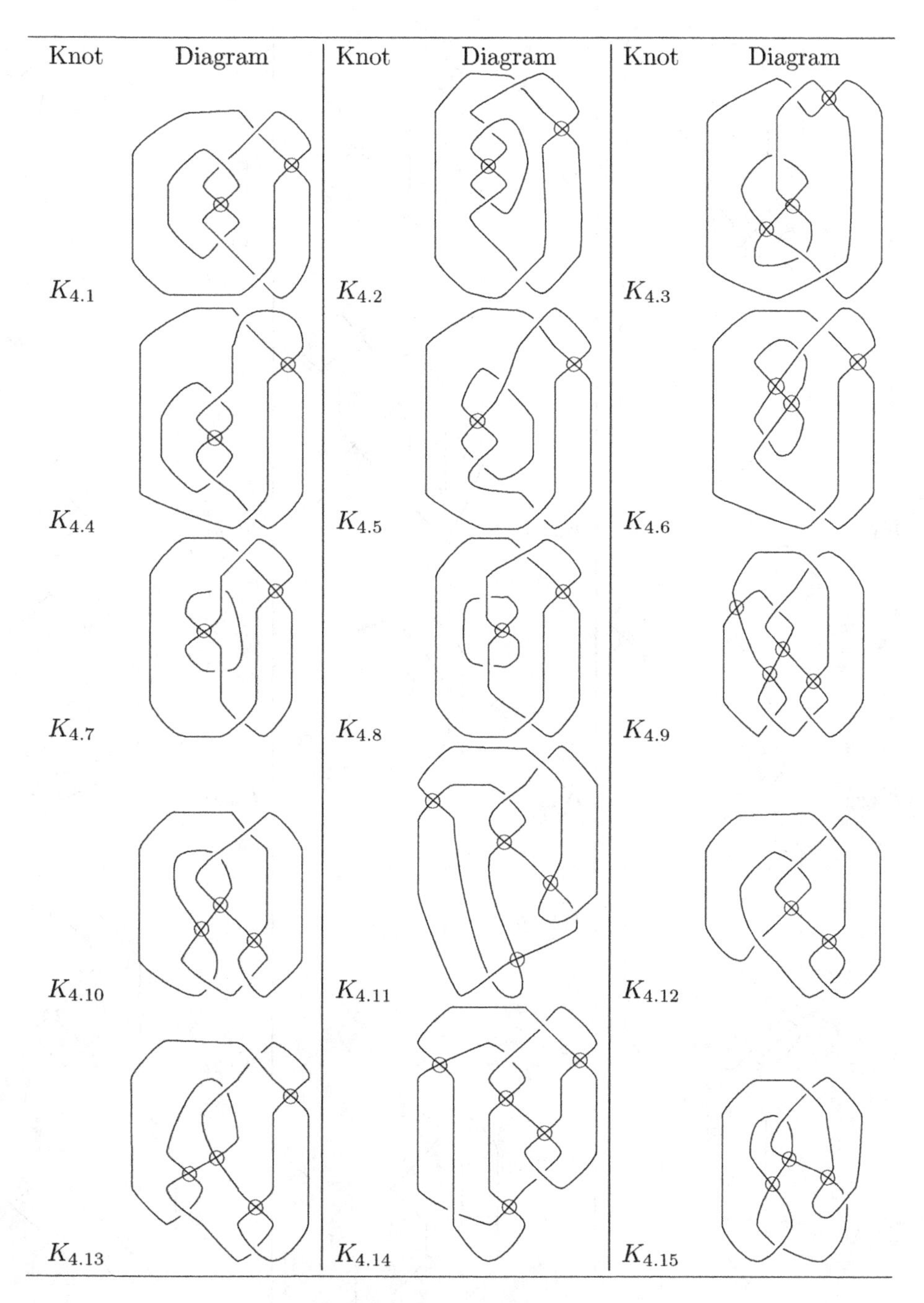

TABLE A.2: Knots $K_{4.1}$–$K_{4.15}$

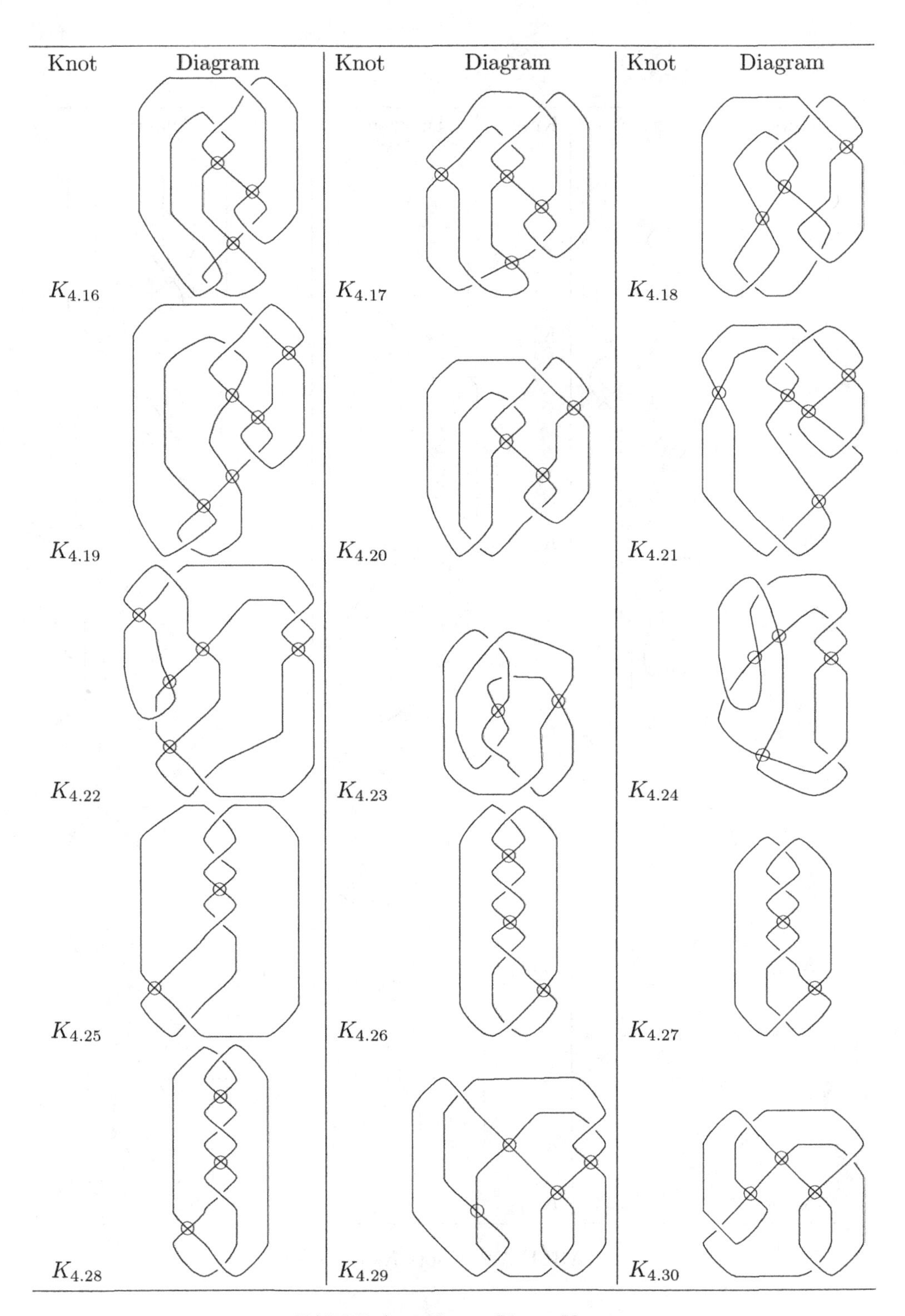

TABLE A.3: Knots $K_{4.16}$–$K_{4.30}$

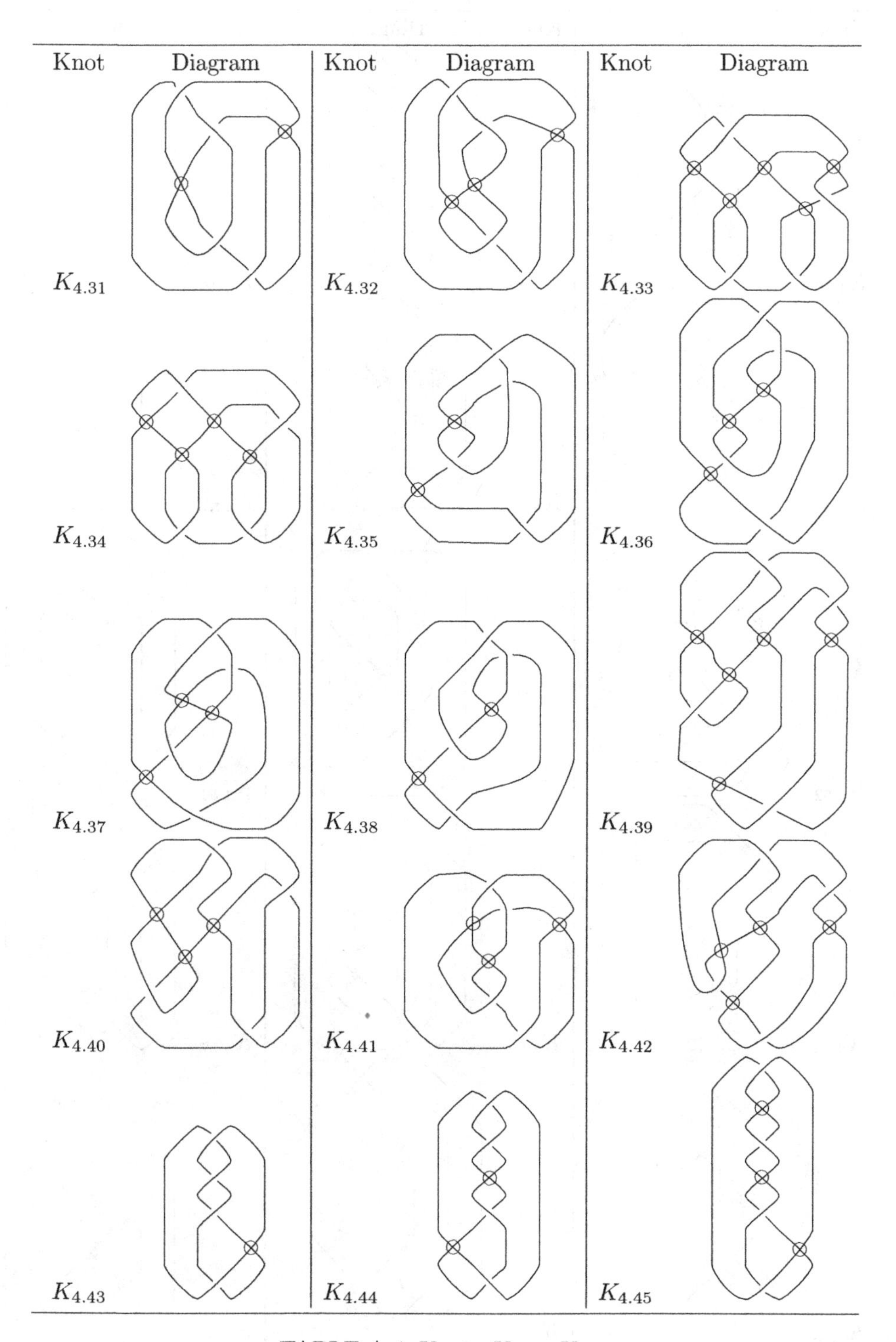

TABLE A.4: Knots $K_{4.31}$–$K_{4.45}$

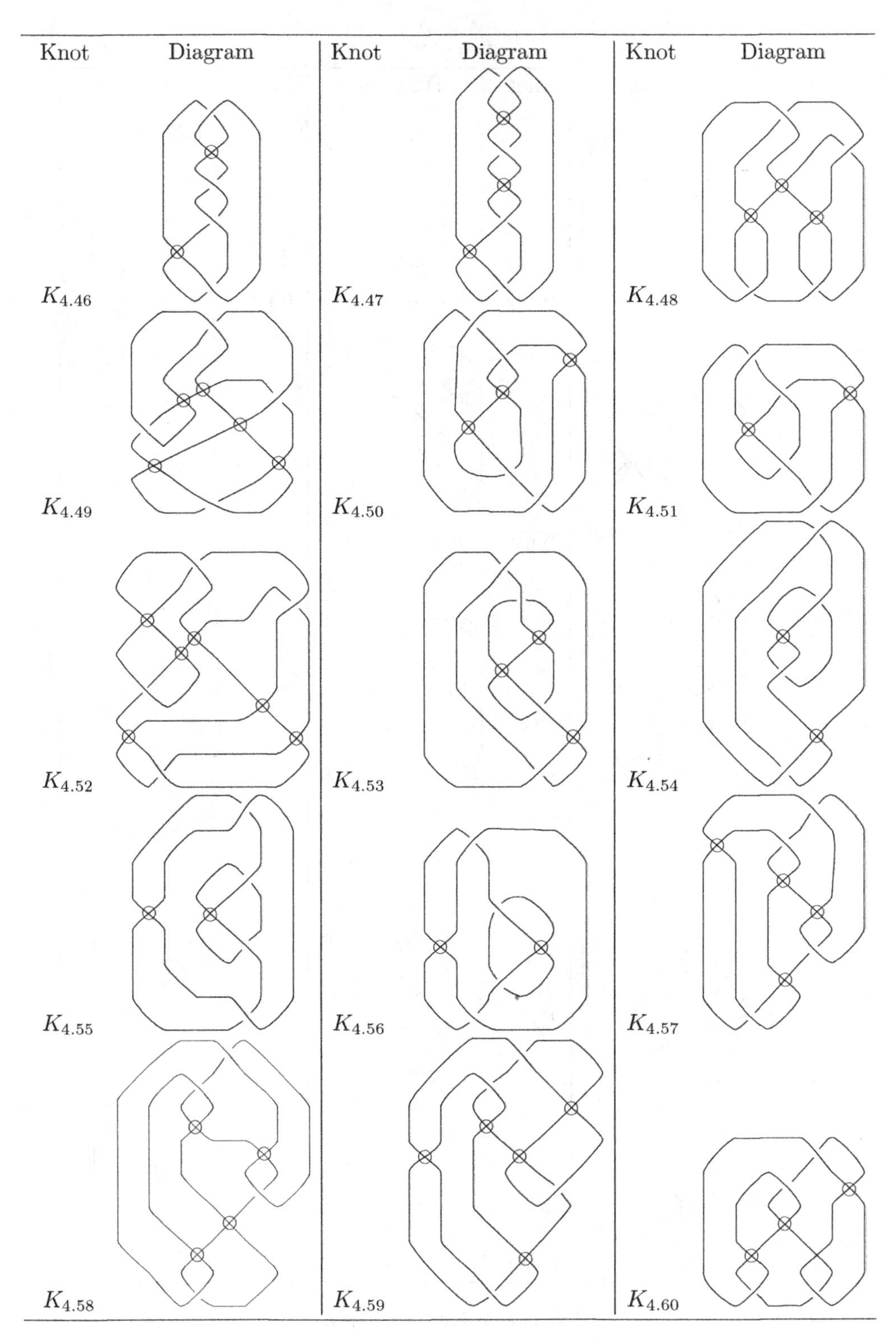

TABLE A.5: Knots $K_{4.46}$–$K_{4.60}$

TABLE A.6: Knots $K_{4.61}$–$K_{4.75}$

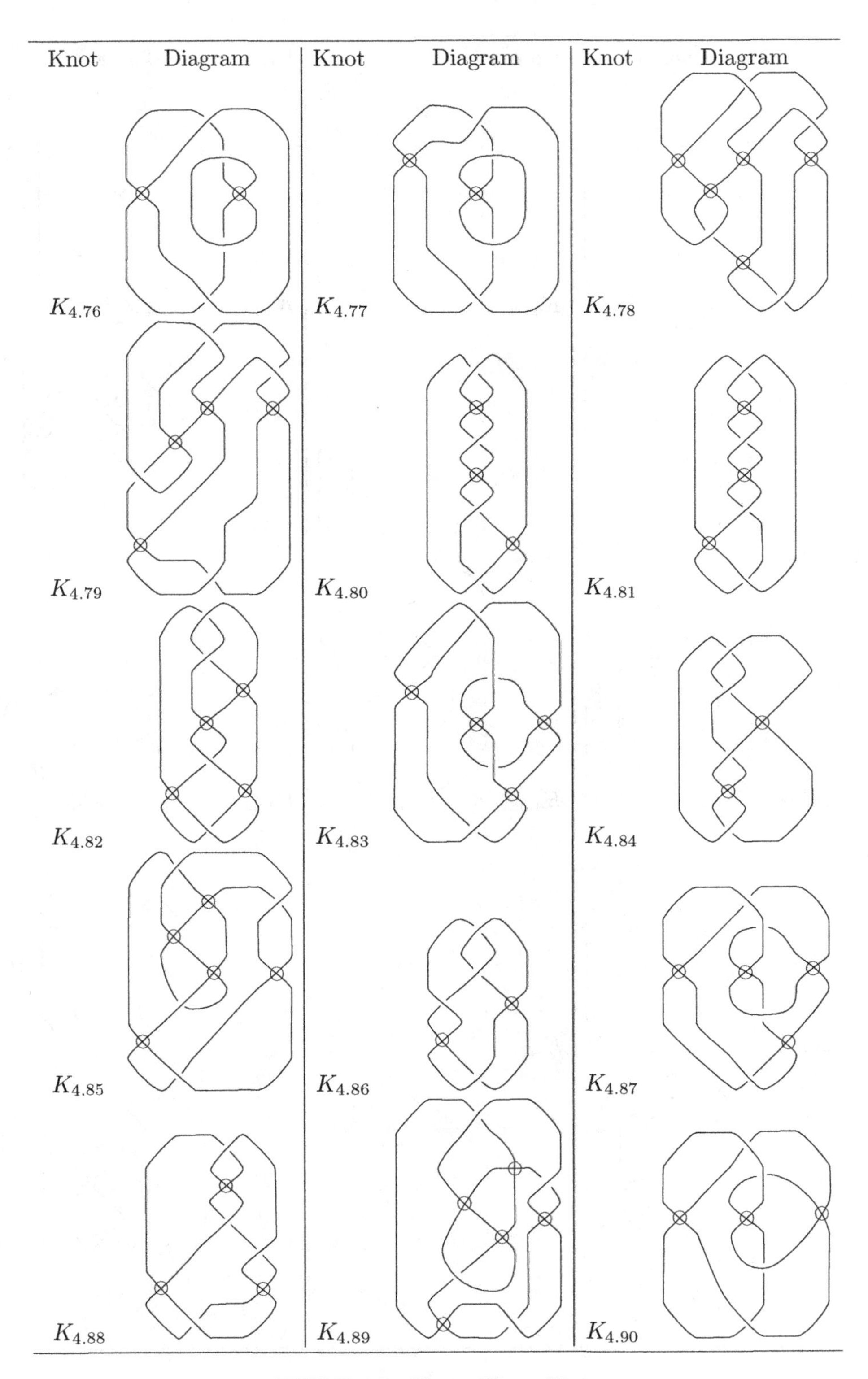

TABLE A.7: Knots $K_{4.76}$–$K_{4.90}$

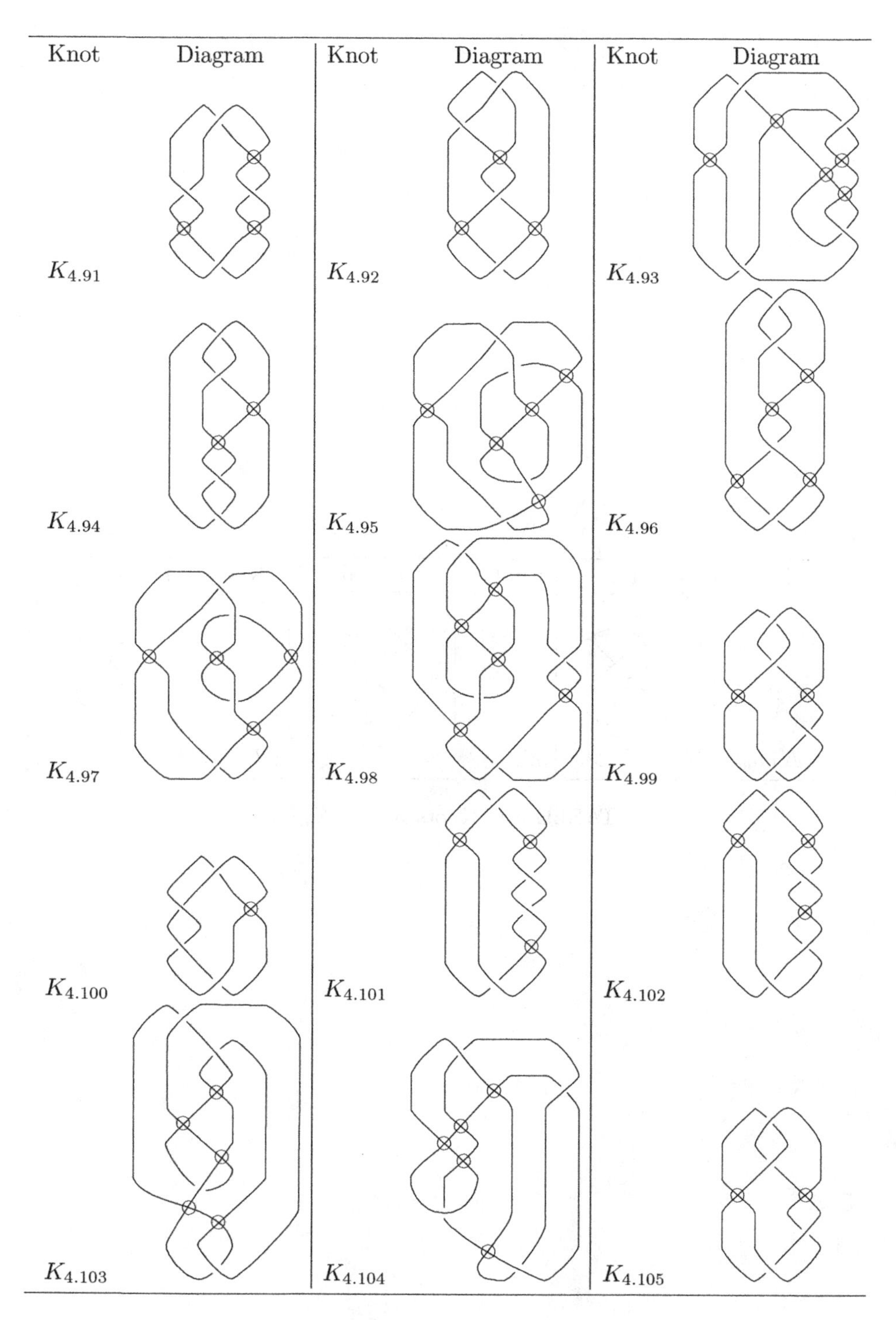

TABLE A.8: Knots $K_{4.91}$–$K_{4.105}$

Knot	Diagram	Knot	Diagram	Knot	Diagram
$K_{4.106}$		$K_{4.107}$		$K_{4.108}$	

TABLE A.9: Knots $K_{4.106}$–$K_{4.108}$

A.2 KNOT INVARIANTS

Knot	Arrow Polynomial	v(K)	g(K)	Generalized Alexander
4.01	$A^8K_1^2 - 3K_1^2 + 2 - 2A^4K_1^2 + 2K_2^1 - 2A^2K_1 + 2A^{-2}K_1 + A^{-4}$	2	1	$1 - 2s + s^2 - 2t + 2st + 2s^2t - 2s^3t + t^2 + 2st^2 - 4s^2t^2 + s^4t^2 - 2st^3 + 2s^3t^3 + s^2t^4 - s^4t^4$
4.02	$-A^6K_1 - A^4K_1^2 + 2K_2 + 3 - 2K_1^2 + A^2K_1 + A^{-2}K_1 - A^{-4}K_1^2 - A^{-6}K_1$	2	1	$s - s^2 + t - 2st - s^2t + 2s^3t - t^2 - st^2 + 4s^2t^2 - s^3t^2 - s^4t^2 + 2st^3 - s^2t^3 - 2s^3t^3 + s^4t^3 - s^2t^4 + s^3t^4$
4.03	$A^8K_1^2 - A^4 - K_1^2 + 1 - 2A^2K_1^1 - A^4K_2 + K_2 + 2A^{-2}K_1^1 + A^{-4}$	2	1	$-1 + s + 2t - st - s^2t - t^2 - 2st^2 + 2s^2t^2 + s^3t^2 + 2st^3 - s^3t^3 - s^4t^3 - s^2t^4 + s^4t^4$
4.04	$A^2 - A^4K_1^1 - 2A^2K_1^2 - 2A^{-2}K_1^2 + A^{-2}K_2 + A^2K_2 + 2A^{-2} + 1K_1$	2	1	$-s + s^2 + 2st - 2s^3t - st^2 - 2s^2t^2 + 2s^3t^2 + s^4t^2 + s^2t^3 - s^4t^3$
4.05	$-A^4K_1^1 + A^2 - 2A^2K_1^2 - 2A^{-2}K_1^2 + A^{-2}K_2 + A^2K_2 + K_1 + 2A^{-2}$	2	1	$-t + s^2t + t^2 + 2st^2 - 2s^2t^2 - s^3t^2 - 2st^3 + 2s^3t^3 + s^2t^4 - s^3t^4$
4.06	$-A^6K_1 - A^4K_1^2 + K_2 + A^{-4}K_1^2 - A^{-4}K_2 + A^2K_1 + 2 + A^{-2}K_1 - A^{-4} - A^{-6}K_1$	2	1	$-s + s^2 + st + s^2t - 2s^3t - 2s^2t^2 + s^3t^2 + s^4t^2 + s^3t^3 - s^4t^3$
4.07	$A^8K_1^2 - 3K_1^2 + 2K_2^1 - 2A^4K_1^2 + 2 - 2A^2K_1^1 + 2A^{-2}K_1^1 + A^{-4}$	2	1	$-1 + s + t - s^2t - st^2 + s^3t^2 + s^2t^3 - s^4t^3 - s^3t^4 + s^4t^4$
4.08	$-A^6K_1 - A^4K_1^2 + 3 + 2K_2 - 2K_1^2 + A^2K_1 + A^{-2}K_1 - A^{-4}K_1^2 - A^{-6}K_1$	2	1	0
4.09	$-A^4 - A^2K_1 + A^{-2}K_1 + A^{-4}$	1	1	$-1 + s + t + st - 2s^2t - 2st^2 + 2s^3t^2 + 2s^2t^3 - s^3t^3 - s^4t^3 - s^3t^4 + s^4t^4$
4.10	$-A^6 - A^4K_1 + 2A^2 + 2K_1 + A^{-2} - A^2K_1^2 - A^{-2}K_1^2 + A^{-2}K_2 - A^{-4}K_1$	2	1	$t - s^2t - t^2 - st^2 + s^2t^2 + s^3t^2 + st^3 - s^3t^3$
4.11	$-A^{-2}K_1^2 + A^2 - A^4K_1 - A^2K_1^2 + A^{-2}K_2 + K_1 + A^{-2}$	2	1	$s - s^2 - t - 2st + s^2t + 2s^3t + t^2 + 2st^2 + s^2t^2 - 3s^3t^2 - s^4t^2 - st^3 - s^2t^3 + s^3t^3 + s^4t^3$
4.12	$-A^6K_1 - A^4K_2 + A^2K_1 + 1 + 2K_2 - A^{-4}K_2 + A^{-2}K_1 - A^{-6}K_1$	2	1	$-s + s^2 + t + s^2t - 2s^3t - t^2 - st^2 + s^3t^2 + s^4t^2 + 2st^3 - s^2t^3 - s^4t^3 - s^2t^4 + s^3t^4$
4.13	$A^4 - A^{-4}K_1^2 + 1 - A^4K_1^2 - 2K_1^2 + 2K_2 + A^{-4}$	2	1	$t - st - t^2 - st^2 + 2s^2t^2 + 2st^3 - s^2t^3 - s^3t^3 - s^2t^4 + s^3t^4$
4.14	$-A^6K_1 - 2A^{-4}K_1^2 + A^2K_1 + 2K_2 + 2 - 2K_1^2 + A^{-4}$	2	1	$-s + s^2 - t + st + s^2t - s^3t + t^2 + st^2 - 2s^2t^2 - st^3 + s^3t^3$
4.15	$K_2 - A^4K_1^2 - K_1^2 + 1 - A^2K_1 + A^{-2}K_1 + A^{-4}$	2	1	$1 - s - st + s^2t - t^2 + st^2 + s^2t^2 - s^3t^2 + st^3 - 2s^2t^3 + s^4t^3 + s^3t^4 - s^4t^4$

TABLE A.10: Knots 1–15

Knot	Arrow Polynomial	v(K)	g(K)	Generalized Alexander
4.16	$-A^6 - A^4K_1 + A^2 + 2K_1 + A^{-2} - A^{-4}K_1$	1	1	0
4.17	$-A^6K_1 - A^4K_2 + A^2K_1 + 2 + 2K_2 - A^{-4}K_1^2 - K_1^2 + A^{-2}K_1 - A^{-6}K_1$	2	1	$-st + s^2t + s^2t^2 - s^3t^2 + st^3 - s^2t^3 - s^2t^4 + s^3t^4$
4.18	$-2A^{-2}K_1^2 - A^4K_1 + A^2K_2 - 2A^2K_1^2 + 2A^{-2} + A^2 + K_1 + A^{-2}K_2$	2	1	$-st + s^2t + st^2 - s^3t^2 - s^2t^3 + s^3t^3$
4.19	$A^4 + K_2 + 1 - A^4K_1^2 - K_1^2$	2	1	$s - s^2 - s^2t + s^3t - st^2 + s^2t^2 + s^2t^3 - s^3t^3$
4.20	$-A^6K_1 + K_2 + A^2K_1 + 2 - A^{-4}K_1^2 - K_1^2$	2	1	$t - st - t^2 + s^2t^2 + st^3 - s^2t^3$
4.21	$A^2 + A^2K_2 - A^4K_1 - A^2K_1^2 - 2A^{-2}K_1^2 + 2A^{-2} + 2K_1 + A^{-2}K_2 - A^{-4}K_1 - A^{-6}K_1^2$	2	1	$s - s^2 + st - 2s^2t + s^3t - st^2 + s^3t^2 - st^3 + 2s^2t^3 - s^3t^3 + s^2t^4 - s^3t^4$
4.22	$-A^6K_1 + A^2K_1 + K_2 + 2 - A^2K_1K_2 - A^{-2}K_1K_2 - K_1^2 - A^{-4}K_1^2 + A^2K_1^3 + 2A^{-2}K_1^3 + A^{-6}K_1^3 - A^{-2}K_1 - A^{-6}K_1$	3	2	$-st + s^2t - t^2 + st^2 + s^2t^2 - s^3t^2 + t^3 + st^3 - 2s^2t^3 - st^4 + s^3t^4$
4.23	$A^8K_1 - 2A^4K_1 + 2A^{-2} - A^2K_1^2 - A^{-2}K_1^2 + K_1 + A^{-2}K_2$	2	1	$-s + s^2 + s^2t - s^3t + st^2 - s^2t^2 - s^2t^3 + s^3t^3$
4.24	$A^2 - 2A^4K_1 + A^4K_1^3 + 2K_1^3 + A^{-4}K_1^3 - A^{-4}K_1 + A^{-2} + K_1 + A^{-2}K_2 - 1K_1K_2 - A^{-4}K_1K_2 - A^{-2}K_1^2 - A^{-6}K_1^2$	3	2	$s - s^2 + t - 2s^2t + s^3t - 2st^2 + s^2t^2 + s^3t^2 - t^3 + 2s^2t^3 - s^3t^3 + st^4 - s^3t^4$
4.25	$-A^6K_1 - A^{-2}K_1 + A^{-6}K_1 + A^2K_1 - A^{-4} + 1 + A^{-8}$	1	1	$1 - s - t + st - s^2t^2 + s^3t^2 + s^2t^3 - s^3t^3$
4.26	$A^2K_1 - A^{-2}K_1K_2 - A^2K_1K_2 + 1 + A^{-2}K_3$	3	2	$-t + st - s^2t^2 + s^3t^2 + t^3 + s^2t^3 - s^3t^3 - s^4t^3 - st^4 + s^4t^4$
4.27	$A^{-6} - A^{-2} + A^2 - A^{-4}K_1 + A^{-8}K_1$	1	1	$-1 + s + t - s^2t - st^2 + s^2t^2$
4.28	$A^2 + K_3 + K_1 - A^{-4}K_1K_2 - K_1K_2$	3	2	$-1 + s + s^2t^2 - s^3t^2 + t^3 - s^2t^3 - st^4 + s^3t^4$
4.29	$1 - A^4K_1^2 - K_1^2 + K_2 - A^2K_1 + A^{-2}K_1 + A^{-4}$	2	1	$1 - 2s + s^2 + 2s^2t - 2s^3t - t^2 + st^2 - s^2t^2 + s^4t^2 + st^3 - 2s^2t^3 + s^3t^3 + s^3t^4 - s^4t^4$
4.30	$-2A^{-2}K_1^2 + A^2K_2 + A^2 - 2A^2K_1^2 + A^{-2}K_2 - A^4K_1 + 2A^{-2} + K_1$	2	1	$-t + 2st - s^2t + t^2 - 2s^2t^2 + s^3t^2 - 2st^3 + 2s^2t^3 + s^2t^4 - s^3t^4$

TABLE A.11: Knots 16–30

Knot	Arrow Polynomial	v(K)	g(K)	Generalized Alexander
4.31	$-A^6K_1^2 - 2A^2K_1^2 + 2A^2K_2 - A^{-2}K_1^2 - A^4K_1 + 2K_1 + 2A^{-2} + A^2 - A^{-4}K_1$	2	1	$-s + s^2 + st + s^2t - 2s^3t - 2s^2t^2 + s^3t^2 + s^4t^2 + s^3t^3 - s^4t^3$
4.32	$-A^6K_1 + A^2K_1 + 3 - A^4K_1^2 - K_1^2 - A^{-4} + K_2 + A^{-2}K_1 - A^{-6}K_1$	2	1	$t - s^2t - t^2 - st^2 + s^2t^2 + s^3t^2 + st^3 - s^3t^3$
4.33	$-A^4K_1 + K_1 + A^{-2}$	1	1	$-st + s^2t + st^2 - s^3t^2 - s^2t^3 + s^3t^3$
4.34	$-A^6K_1 + A^2K_1 + 2 - A^{-4}K_1^2 + K_2 - K_1^2$	2	1	$t - st - t^2 + s^2t^2 + st^3 - s^2t^3$
4.35	$-K_1^2 - A^{-4}K_1^2 + 2 + A^{-4}K_2$	2	1	$-s + s^2 + s^2t - s^3t + st^2 - s^2t^2 - s^2t^3 + s^3t^3$
4.36	$A^2 + 2K_1 - A^4K_1 - A^{-4}K_1 + A^{-2}K_2 - A^{-6}K_2$	2	1	$s - s^2 + st - 2s^2t + s^3t - st^2 + s^3t^2 - st^3 + 2s^2t^3 - s^3t^3 + s^2t^4 - s^3t^4$
4.37	$K_2 - A^4K_2 - A^2K_1 + A^{-2}K_1 + A^{-4}$	2	1	$-1 + s - s^2t + s^3t + t^2 - s^4t^2 - st^3 + s^2t^3 - s^3t^4 + s^4t^4$
4.38	$A^2K_2 - A^4K_1 - A^2K_1^2 - A^{-2}K_1^2 + 2A^{-2} + K_1$	2	1	$-s^2t + s^3t + s^2t^2 - s^4t^2 - s^3t^3 + s^4t^3$
4.39	$A^2K_2 - A^4K_1K_2 - K_1K_2 - A^2K_1^2 - A^{-2}K_1^2 - A^4K_1 + 2A^{-2} - A^{-4}K_1 + A^4K_1^3 + 2K_1^3 + A^{-4}K_1^3$	3	2	$-st + s^2t - t^2 + st^2 + s^2t^2 - s^3t^2 + t^3 + st^3 - 2s^2t^3 - st^4 + s^3t^4$
4.40	$-A^6K_1 - A^{-4} + 2 + A^2K_1$	1	1	$-st + s^2t + st^2 - s^3t^2 - s^2t^3 + s^3t^3$
4.41	$A^8K_1 - 2A^4K_1 - A^2 + K_1 + 2A^{-2}$	1	1	0
4.42	$-A^6K_1 + A^6K_1^3 + 2A^2K_1^3 + A^{-2}K_1^3 - A^2K_1 + 2 - A^2K_1K_2 - A^{-2}K_1K_2 - K_1^2 - A^{-4}K_1^2 + A^{-4}K_2$	2	2	$-t + st + st^2 - s^2t^2 + t^3 - st^3 - st^4 + s^2t^4$
4.43	$-A^6K_1 - A^{-2}K_1 + A^{-6}K_1 + A^2K_1 - A^{-4} + 1 + A^{-8}$	1	1	$-2st + 2s^2t + 2st^2 - 2s^3t^2 - 2s^2t^3 + 2s^3t^3$
4.44	$-A^4K_1 + A^{-2} + K_1$	1	1	$1 - s - t + s^2t + st^2 - s^2t^2$
4.45	$-K_1K_2 - A^4K_1K_2 + K_1 + K_3 + A^{-2}$	3	1	$1 - s - st + s^2t + st^2 - s^3t^2 - t^3 - s^2t^3 + s^3t^3 + s^4t^3 + st^4 - s^4t^4$

TABLE A.12: Knots 30–45

Knot	Arrow Polynomial	v(K)	g(K)	Generalized Alexander
4.46	1	0	0	$1-s-t-st+2s^2t+2st^2-s^2t^2-s^3t^2-s^2t^3+s^3t^3$
4.47	$A^2K_3^1+1-A^{-2}K_1K_2-A^2K_1K_2+A^{-2}K_1$	3	2	$t-s^2t-st^2+s^3t^2-t^3+s^2t^3+st^4-s^3t^4$
4.48	$A^4-2A^4K_1^2-2K_1^2+1+2K_2-A^2K_1+A^{-2}K_1+A^{-4}$	2	1	$-1+s+t-st-s^2t+s^3t-st^2+2s^2t^2-s^4t^2+st^3-s^3t^3-s^2t^4+s^4t^4$
4.49	$A^2K_2-A^{-2}K_1^2-A^2K_1^2-A^4K_1+K_1+2A^{-2}$	2	1	$s-s^2-st+s^3t+s^2t^2-s^3t^2$
4.50	$-A^6K_1^2-A^2K_1^2+A^2+2K_1+A^2K_2-A^4K_1-A^{-4}K_1+A^{-2}$	2	1	$t-s^2t-t^2-st^2+s^2t^2+s^3t^2+st^3-s^3t^3$
4.51	$-A^6K_1+A^2K_1+3-A^4K_1^2-2K_1^2+2K_2-A^{-4}K_1^2+A^{-2}K_1-A^{-6}K_1$	2	1	$s-s^2-st-s^2t+2s^3t+2s^2t^2-s^3t^2-s^4t^2-s^3t^3+s^4t^3$
4.52	$-A^6K_1+A^2K_1+2-A^{-4}$	1	1	$-st+s^2t+st^2-s^3t^2-s^2t^3+s^3t^3$
4.53	$A^8-2A^4-2A^2K_1+1+2A^{-2}K_1+A^{-4}$	1	1	$1-s-t+s^2t+st^2-s^3t^2-s^2t^3+s^4t^3+s^3t^4-s^4t^4$
4.54	$-A^6-A^4K_1+A^6K_1^2+A^{-2}-A^2K_2+A^2+K_1-A^{-2}K_1^2+A^{-2}K_2$	2	1	$s-s^2-2st+2s^3t+st^2+2s^2t^2-2s^3t^2-s^4t^2-s^2t^3+s^4t^3$
4.55	$A^4+2K_2+1-A^4K_1^2-2K_1^2-A^{-4}K_1^2+A^{-4}$	2	1	0
4.56	$A^4+1-A^4K_1^2+2K_2-2K_1^2-A^{-4}K_1^2+A^{-4}$	2	1	0
4.57	$-A^6K_2-A^4K_1+2A^2K_2+2K_1+2A^{-2}-A^2K_1^2-A^{-2}K_1^2-A^{-4}K_1$	2	1	$st-s^2t-s^2t^2+s^3t^2-st^3+s^2t^3+s^2t^4-s^3t^4$
4.58	$-A^6K_1+A^2K_1+3-A^4+A^{-2}K_1-A^{-4}-A^{-6}K_1$	1	1	0
4.59	$A^4K_2-A^{-4}K_1^2-A^4K_1^2-2K_1^2+3+A^{-4}K_2$	2	1	0
4.60	$-A^6K_1-2A^{-4}K_1^2+A^2K_1+3+K_2-2K_1^2+A^{-4}K_2$	2	1	$-st+s^2t+st^2-s^3t^2-s^2t^3+s^3t^3$

TABLE A.13: Knots 46–60

Knot	Arrow Polynomial	v(K)	g(K)	Generalized Alexander
4.61	$1-A^4-A^2K_1+A^{-2}K_1+A^{-4}$	1	1	$1-s-t-st+2s^2t+2st^2-2s^3t^2-2s^2t^3+s^3t^3+s^4t^3+s^3t^4-s^4t^4$
4.62	$A^2-A^4K_1K_2-K_1K_2-A^2K_1^2-A^{-2}K_1^2-A^4K_1+A^{-2}K_2-A^{-4}K_1+A^{-2}+A^4K_1^3+2K_1^3+A^{-4}K_1^3$	2	1	$t+2st-2s^2t-s^3t-3st^2-s^2t^2+3s^3t^2+s^4t^2-t^3+3s^2t^3-s^3t^3-s^4t^3+st^4-s^3t^4$
4.63	$A^2-A^4K_1-A^2K_1^2-A^{-2}K_1^2+A^{-2}K_2+K_1+A^{-2}$	2	1	$-t-st+2s^2t+t^2+2st^2-s^2t^2-2s^3t^2-st^3-s^2t^3+2s^3t^3$
4.64	$-A^6K_1-A^{-4}K_2+K_2+1+A^2K_1$	2	1	$-t+s^3t+t^2+st^2-s^3t^2-s^4t^2-st^3+s^4t^3$
4.65	$A^8K_1-2A^4K_1+A^{-2}-A^2K_2+K_1+A^{-2}K_2$	2	1	$-s+s^2-st+2s^2t-s^3t+st^2-s^3t^2+st^3-2s^2t^3+s^3t^3-s^2t^4+s^3t^4$
4.66	$-A^6K_1+A^{-2}K_1^3+1+A^6K_1^3+2A^2K_1^3-A^{-4}K_1^2+K_2-A^2K_1-A^2K_1K_2-A^{-2}K_1K_2-K_1^2+A^{-4}$	3	2	$s-s^2+t-2s^2t+s^3t-2st^2+s^2t^2+s^3t^2-t^3+2s^2t^3-s^3t^3+st^4-s^3t^4$
4.67	$-A^{-4}K_1^2+1+K_2-K_1^2+A^{-4}$	2	1	$st-s^2t-s^2t^2+s^3t^2-st^3+s^2t^3+s^2t^4-s^3t^4$
4.68	$A^2-A^4K_1+A^{-2}+2K_1-A^{-4}K_1-A^{-6}$	1	1	0
4.69	$A^4K_2-2A^4K_1^2-2K_1^2+K_2+2-A^2K_1+A^{-2}K_1+A^{-4}$	2	1	$1-s-t-st+2s^2t+2st^2-2s^3t^2-2s^2t^3+s^3t^3+s^4t^3+s^3t^4-s^4t^4$
4.70	$-A^6K_1^2-A^2K_1^2+A^2K_2+2K_1-A^4K_1+A^2-A^{-4}K_1+A^{-2}$	2	1	$t-s^2t-t^2-st^2+s^2t^2+s^3t^2+st^3-s^3t^3$
4.71	$-A^6K_1+A^2K_1+2K_2-A^4K_1^2-2K_1^2+3-A^{-4}K_1^2+A^{-2}K_1-A^{-6}K_1$	2	1	0
4.72	1	0	0	0
4.73	$A^8K_2-2A^4K_2-2A^2K_1+K_2+2A^{-2}K_1+A^{-4}$	2	1	$-1+2s-s^2-2s^2t+2s^3t+t^2-s^4t^2-2st^3+2s^2t^3+s^2t^4-2s^3t^4+s^4t^4$
4.74	$-A^6K_2-A^4K_1+A^6K_1^2+2A^{-2}-A^2+A^2K_2+K_1-A^{-2}K_1^2$	2	1	$-s+s^2+2s^2t-2s^3t+st^2-2s^2t^2+s^4t^2-s^2t^3+2s^3t^3-s^4t^3$
4.75	$-A^6K_1-A^4+3+A^2K_1+A^{-2}K_1-A^{-4}-A^{-6}K_1$	1	1	0

TABLE A.14: Knots 61–75

Knot	Arrow Polynomial	v(K)	g(K)	Generalized Alexander
4.76	$A^4K_2 - A^4K_1^2 + 3 - 2K_1^2 - A^{-4}K_1^2 + A^{-4}K_2$	2	1	0
4.77	$A^4K_2 - 2K_1^2 - A^4K_1^2 + 3 - A^{-4}K_1^2 + A^{-4}K_2$	2	1	0
4.78	$-A^6K_1K_2 - A^2K_1K_2 - A^4K_1^2 - K_1^2 + A^{-2}K_1^3 + K_2 + 1 + A^6K_1^3 + 2A^2K_1^3 - 2A^2K_1 + A^{-4}$	3	2	$-1 + s - s^2t + s^3t + s^2t^2 - s^4t^2 + t^3 - s^3t^3 - st^4 + s^4t^4$
4.79	$A^8K_1^3 + 2A^4K_1^3 + K_1^3 - 3A^4K_1 + 2A^{-2} - A^4K_1K_2 - K_1K_2 - A^2K_1^2 - A^{-2}K_1^2 + K_1 + A^{-2}K_2$	3	2	$-t + st + st^2 - s^2t^2 + t^3 - st^3 - st^4 + s^2t^4$
4.80	$-A^6K_1K_2 - A^2K_1K_2 - A^{-2}K_1 + A^2K_1 + A^{-6}K_1 + A^2K_3 - A^{-4} + 1 + A^{-8}$	3	2	$-1 + s + t^3 - s^4t^3 - st^4 + s^4t^4$
4.81	$A^4K_3 - A^{-2} + A^2 + A^{-6} - A^4K_1K_2 - K_1K_2 - A^{-4}K_1 + K_1 + A^{-8}K_1$	3	2	$-t + st + t^3 - s^3t^3 - st^4 + s^3t^4$
4.82	$-A^6K_1 + A^{-4}K_1^2 - A^4K_1^2 + 2 - 2A^{-4} + A^2K_1 + A^{-8}$	2	1	$-1 + s^2 + s^2t - s^3t + t^2 + st^2 - s^2t^2 - s^3t^2 - st^3 - s^2t^3 + 2s^3t^3$
4.83	$-K_1K_2 - A^4K_1K_2 + K_1 + A^{-2} + K_3$	3	2	$st - s^3t + t^2 - st^2 - s^2t^2 + s^4t^2 - t^3 - st^3 + s^2t^3 + s^3t^3 + st^4 - s^4t^4$
4.84	$2 + A^2K_1 - A^{-4} - A^{-2}K_1 - K_1^2 + A^{-8}K_1^2$	2	1	$1 - s^2 - 2st + s^2t + s^3t - t^2 + st^2 + s^2t^2 - s^3t^2 + st^3 - s^2t^3$
4.85	$-A^2K_1^2 + A^2 + A^{-6}K_1^2$	2	1	$st - s^3t - s^2t^2 + s^4t^2 - st^3 + s^3t^3 + s^2t^4 - s^4t^4$
4.86	$A^8 - A^4 + 2 - A^{-4} - K_1^2 + A^{-8}K_1^2$	2	1	$1 - s^2 - st + s^3t - t^2 + s^2t^2 + st^3 - s^3t^3$
4.87	$-A^6K_1K_2 - A^2K_1K_2 - A^4K_1^2 - 2A^{-4} + A^2K_1 + 2 + A^2K_3 + A^{-4}K_1^2 + A^{-8}$	3	2	$1 - s^2 - st^2 + s^4t^2 - t^3 + s^2t^3 + st^4 - s^4t^4$
4.88	$2 + A^2K_1 + A^2K_3 - A^{-4} - A^2K_1K_2 - A^{-2}K_1K_2 - K_1^2 + A^{-8}K_1^2$	3	2	$t^2 - st^2 - t^3 + s^2t^3 + st^4 - s^2t^4$
4.89	$A^4 - 2A^4K_1^2 + 2A^{-4}K_1^2 - 2A^{-4} + 1 + A^{-8}$	2	1	$-1 + s^2 + t^2 - s^4t^2 - s^2t^4 + s^4t^4$
4.90	$A^8K_1^2 - A^4 - 2K_1^2 + 3 - A^{-4} + A^{-8}K_1^2$	2	1	0

TABLE A.15: Knots 76–90

Knot	Arrow Polynomial	v(K)	g(K)	Generalized Alexander
4.91	$-A^{10}K_1^3 - A^6K_1^3 + A^2K_1^3 + 2A^6K_1 + A^{-2}K_1^3 - 2A^2K_1 + A^{-4}$	3	1	$-1 + s^3 + s^2t - s^4t + st^2 - s^3t^2 + t^3 - s^2t^3 - st^4 + s^4t^4$
4.92	$-A^6K_3 - A^4 + 2 + A^2K_3 - A^{-4} + A^{-8}$	3	1	$-1 + s^3 + st - s^4t + t^3 - s^3t^3 - st^4 + s^4t^4$
4.93	$A^4K_3 + A^{-6}K_1^2 + A^2 - A^2K_1^2 - A^4K_1K_2 - K_1K_2 + K_1$	3	2	$st - s^2t - t^2 + s^3t^2 + t^3 - s^3t^3 - st^4 + s^2t^4$
4.94	$-A^6K_1 + 2 - A^4 + A^2K_1 - A^{-4} + A^{-8}$	1	1	$st - s^2t - st^2 + s^3t^2 + s^2t^3 - s^3t^3$
4.95	$-A^4K_3 + A^{-2} + K_3$	3	1	$-1 + s^3 + st - s^4t + t^3 - s^3t^3 - st^4 + s^4t^4$
4.96	$A^{-6}K_1^2 - A^2K_1^2 + A^2$	2	1	$-1 + s^2 + st - s^3t + t^2 - s^2t^2 - st^3 + s^3t^3$
4.97	$-A^2K_1K_2 - A^{-2}K_1K_2 + A^2K_3 + 1 + A^{-2}K_1$	3	1	$-1 + s^3 + 2st - s^3t - s^4t + t^2 - st^2 - s^2t^2 + s^4t^2 - st^3 + s^2t^3$
4.98	1	0	0	0
4.99	$A^8 - A^4 - A^{-4} + 1 + A^{-8}$	0	0	0
4.100	$-A^{10}K_1 + A^6K_1 - A^2K_1 + A^{-2}K_1 + A^{-4}$	1	1	$-1 + s + t - s^2t - st^2 + s^3t^2 + s^2t^3 - s^4t^3 - s^3t^4 + s^4t^4$
4.101	$A^8K_1 + K_1^3 - A^{-4}K_1 + A^{-2} - A^8K_1^3 - A^4K_1^3 + A^{-4}K_1^3$	3	1	$-s + s^3 - t - st + 3s^2t - s^4t + 3st^2 - 3s^3t^2 + t^3 - 3s^2t^3 + s^3t^3 + s^4t^3 - st^4 + s^3t^4$
4.102	$-A^6K_1 - A^{-2}K_1^3 + 1 + A^6K_1^3 + A^2K_1^3 - A^{-6}K_1^3 - A^2K_1 + A^{-6}K_1 + A^{-2}K_1$	3	1	$-s + s^3 - t + 2s^2t - s^4t + 2st^2 - 2s^3t^2 + t^3 - 2s^2t^3 + s^4t^3 - st^4 + s^3t^4$
4.103	$A^4K_1 + A^2 + K_3 - A^4K_1K_2 - K_1K_2 - A^2K_1^2 + A^{-6}K_1^2$	3	2	$-1 + s^3 + st - s^4t + t^2 - st^2 + s^2t^3 - s^3t^3 - s^2t^4 + s^4t^4$
4.104	$A^2K_3 - A^{-2}K_3 - A^{-4} + 1 + A^{-8}$	3	1	$-1 + s^3 + st - s^4t + t^3 - s^3t^3 - st^4 + s^4t^4$
4.105	$-A^4 + 1 + A^{-8}$	0	0	0
4.106	$A^2 - A^2K_1^2 + A^{-6}K_1^2$	2	1	$1 - s^2 - st + s^3t - t^2 + s^2t^2 + st^3 - s^3t^3$
4.107	1	0	0	$1 - s^2 - 2st + 2s^3t - t^2 + 2s^2t^2 - s^4t^2 + 2st^3 - 2s^3t^3 - s^2t^4 + s^4t^4$
4.108	$A^8 - A^4 - A^{-4} + 1 + A^{-8}$	0	0	0

TABLE A.16: Knots 90–108

References by chapter

B.1 Chapter 1

[1] Colin C Adams. *The knot book.* American Mathematical Society, Providence, RI, 2004.

[2] Steven A Bleiler. A note on unknotting number. *Math. Proc. Cambridge Philos. Soc.*, 96(3):469–471, 1984.

[3] Robert Bosch. Simple-closed-curve sculptures of knots and links. *J. Math. Arts*, 4(2):57–71, June 2010.

[4] Peter R Cromwell. *Knots and links.* Cambridge University Press, Cambridge, 2004.

[5] Peter R. Cromwell. The distribution of knot types in Celtic interlaced ornament. *J. Math. Arts*, 2(2):61–68, June 2008.

[6] Naoko Kamada and Seiichi Kamada. Abstract link diagrams and virtual knots. *J. Knot Theory Ramifications*, 9(1):93–106, 2000.

[7] Louis H Kauffman. Virtual knot theory. *European J. Combin.*, 20(7):663–690, 1999.

[8] J. Dennis Lawrence. *A Catalog of Special Plane Curves.* Dover, 1972.

[9] W B Raymond Lickorish. *An introduction to knot theory*, volume 175 of *Graduate Texts in Mathematics.* Springer-Verlag, New York, 1997.

[10] Charles Livingston. *Knot theory*, volume 24 of *Carus Mathematical Monographs.* Mathematical Association of America, Washington, DC, 1993.

[11] Sam Nelson. The combinatorial revolution in knot theory. *Notices Amer. Math. Soc.*, 58(11):1553–1561, 2011.

[12] Olof-Petter Östlund. Invariants of knot diagrams and relations among Feidemeister moves. *J. Knot Theory Ramifications*, 10(8):1215–1227, 2001.

[13] Dale Rolfsen. *Knots and links*, volume 7 of *Mathematics Lecture Series.* Publish or Perish, Inc., Houston, TX, 1990.

[14] Daniel Silver. Knot theory's odd origins. *American Scientist*, 94(2):158, 2006.

B.2 Chapter 2

[1] Zhiyun Cheng. A polynomial invariant of virtual knots. *Proc. Amer. Math. Soc.*, 142(2):713–725, 2014.

[2] Micah W. Chrisman and Heather A. Dye. The three loop isotopy and framed isotopy invariants of virtual knots. *Topology and its Applications*, 173:107–134, August 2014.

[3] H A Dye. Smoothed invariants. *J. Knot Theory Ramifications*, 21(13):17,1240003, 2012.

[4] H. A. Dye. Vassiliev invariants from parity mappings. *J. Knot Theory Ramifications*, 22(4):21,1340008, 2013.

[5] Lena C. Folwaczny and Louis H. Kauffman. A linking number definition of the affine index polynomial and applications. *J. Knot Theory Ramifications*, 22(12):30,1341004, 2013.

[6] Allison Henrich. A sequence of degree one Vassiliev invariants for virtual knots. *J. Knot Theory Ramifications*, 19(4):461–487, 2010.

[7] Louis H. Kauffman. A self-linking invariant of virtual knots. *Fund. Math.*, 184:135–158, 2004.

[8] Louis H. Kauffman. An affine index polynomial invariant of virtual knots. *J. Knot Theory Ramifications*, 22(4):30,1340007, 2013.

[9] W B Raymond Lickorish. *An introduction to knot theory*, volume 175 of *Graduate Texts in Mathematics*. Springer-Verlag, New York, 1997.

[10] V. O. Manturov. Free knots and parity. In *Introductory lectures on knot theory*, volume 46 of *Ser. Knots Everything*, pages 321–345. World Sci. Publ., Hackensack, NJ, 2012.

[11] Dale Rolfsen. *Knots and links*, volume 7 of *Mathematics Lecture Series*. Publish or Perish, Inc., Houston, TX, 1990.

B.3 Chapter 3

[1] Andrew Bartholomew, Roger Fenn, Naoko Kamada, and Seiichi Kamada. New invariants of long virtual knots. *Kobe J. Math.*, 27(1-2):21–33, 2010.

[2] Mario O. Bourgoin. Twisted link theory. *Algebr. Geom. Topol.*, 8(3):1249–1279, 2008.

[3] Roger Fenn, Richard Rimanyi, and Colin Rourke. The braid-permutation group. *Topology*, 36(1):123–135, January 1997.

[4] Ryo Hanaki. Pseudo diagrams of knots, links and spatial graphs. *Osaka J. Math.*, 47(3):863–883, 2010.

[5] A. Henrich, N. MacNaughton, S. Narayan, O. Pechenik, R. Silversmith, and J. Townsend. A midsummer knot's dream. *College Math. J.*, 42(2):126–134, 2011.

[6] Allison Henrich, Lee Johnson, Rebecca Hoberg, Elizabeth Minten, Slavik Jablan, and Ljiljana Radovic. The theory of pseudoknots. *J. Knot Theory Ramifications*, 22(7):21,1350032, 2013.

[7] Teruhisa Kadokami. Detecting non-triviality of virtual links. *J. Knot Theory Ramifications*, 12(06):781–803, September 2003.

[8] Teruhisa Kadokami. Some numerical invariants of flat virtual links are always realized by reduced diagrams. *Journal of Knot Theory and Its Ramifications*, 15(03):289–297, March 2006.

[9] Teruhisa Kadokami. Of 2-component flat virtual links with the supporting genus one. *J. Knot Theory Ramifications*, 17(5):633–647, 2008.

[10] Seiichi Kamada. Invariants of virtual braids and a remark on left stabilizations and virtual exchange moves. *Kobe J. Math.*, 21(1-2):33–49, 2004.

[11] Seiichi Kamada. Braid presentation of virtual knots and welded knots. *Osaka J. Math.*, 44(2):441–458, 2007.

[12] Taizo Kanenobu. Forbidden moves unknot a virtual knot. *J. Knot Theory Ramifications*, 10(1):89–96, 2001.

[13] Louis H. Kauffman. Virtual knot theory. *European J. Combin.*, 20(7):663–690, 1999.

[14] O. V. Manturov and V. O. Manturov. Free knots and groups. *J. Knot Theory Ramifications*, 19(2):181–186, 2010.

[15] Sam Nelson. Unknotting virtual knots with Gauss diagram forbidden moves. *J. Knot Theory Ramifications*, 10(6):931–935, 2001.

[16] Sam Nelson. Virtual crossing realization. *J. Knot Theory Ramifications*, 14(07):931–951, November 2005.

[17] Shin Satoh. Virtual knot presentation of ribbon torus-knots. *J. Knot Theory Ramifications*, 9(4):531–542, 2000.

[18] Vladimir Turaev. Lectures on topology of words. *Jpn. J. Math.*, 2(1):1–39, 2007.

[19] Vladimir Turaev. Topology of words. *Proc. Lond. Math. Soc. (3)*, 95(2):360–412, 2007.

[20] Manturov Vassily Olegovich. On free knots and links. 2012.

B.4 Chapter 4

[1] Tetsuya Abe, Ryo Hanaki, and Ryuji Higa. The unknotting number and band-unknotting number of a knot. *Osaka J. Math.*, 49(2):523–550, 2012.

[2] Colin Adams, Michelle Chu, Thomas Crawford, Stephanie Jensen, Kyler Siegel, and Liyang Zhang. Stick index of knots and links in the cubic lattice. *J. Knot Theory Ramifications*, 21(5):16,1250041, 2012.

[3] Colin C. Adams. *The knot book.* American Mathematical Society, Providence, RI, 2004.

[4] Denis Mikhailovich Afanasiev and Vassily Olegovich Manturov. On virtual crossing number estimates for virtual links. *J. Knot Theory Ramifications*, 18(6):1–16, 2009.

[5] Steven A. Bleiler. A note on unknotting number. *Math. Proc. Cambridge Philos. Soc.*, 96(3):469–471, 1984.

[6] Evarist Byberi and Vladimir Chernov. Virtual bridge number one knots. *Commun. Contemp. Math.*, 10(Suppl. 1):1013–1021, 2008.

[7] Alissa S. Crans, Sandy Ganzell, and Blake Mellor. The forbidden number of a knot. pages 1–14.

[8] Yuanan Diao, Claus Ernst, and Andrzej Stasiak. A partial ordering of knots and links through diagrammatic unknotting. *J. Knot Theory Ramifications*, 18(4):505–522, 2009.

[9] Ryo Hanaki. Pseudo diagrams of knots, links and spatial graphs. *Osaka J. Math.*, 47(3):863–883, 2010.

[10] Ryo Hanaki. Trivializing number of knots. *J. Math. Soc. Japan*, 66(2):435–447, 2014.

[11] Ryo Hanaki and Junsuke Kanadome. On an inequality between unknotting number and crossing number of links. *J. Knot Theory Ramifications*, 19(7):893–903, July 2010.

[12] A. Henrich, N. MacNaughton, S. Narayan, O. Pechenik, R. Silversmith, and J. Townsend. A midsummer knot's dream. *College Math. J.*, 42(2):126–134, 2011.

[13] A. Henrich, N. Macnaughton, S. Narayan, O. Pechenik, J. Townsend, Mikami Hirasawa, Naoko Kamada, and Seiichi Kamada. Classical and virtual pseudodiagram theory and new bounds on unknotting numbers and genus. *J. Knot Theory Ramifications*, 20(4):625–650, April 2011.

[14] Allison Henrich and Slavik Jablan. On the coloring of pseudoknots. *J. Knot Theory Ramifications*, 23(12):22,1450061, 2014.

[15] Mikami Hirasawa, Naoko Kamada, and Seiichi Kamada. Bridge presentations of virtual knots. *J. Knot Theory Ramifications*, 20(06):881–893, June 2011.

[16] S Jablan and R Sazdanovic. BJ-unknotting and unlinking numbers.

[17] Slavik Jablan and Radmila Sazdanovic. Unlinking number and unlinking gap. *J. Knot Theory Ramifications*, 16(10):1331–1355, 2007.

[18] Slavik V. Jablan. Unknotting number and infinity-unknotting number of a knot. *Filomat*, (12, part 1):113–120, 1998.

[19] Teruhisa Kadokami. The virtual crossing number of 2-component flat-virtual links with the supporting genus one. *J. Knot Theory Ramifications*, 17(5):633–647, 2008.

[20] Taizo Kanenobu and Yasuyuki Miyazawa. H(2)-unknotting number of a knot. *Commun. Math. Res.*, 25(5):433–460, 2009.

[21] Louis H. Kauffman. A self-linking invariant of virtual knots. *Fund. Math.*, 184:135–158, 2004.

[22] Vassily Olegovich Manturov. Virtual crossing numbers for virtual knots. *J. Knot Theory Ramifications*, 21(13):13,1240009, 2012.

[23] Yasutaka Nakanishi. Unknotting numbers and knot diagrams with the minimum crossings. *Math. Sem. Notes Kobe Univ.*, 11(2):257–258, 1983.

[24] Yasutaka Nakanishi. A note on unknotting number, II. *J. Knot Theory Ramifications*, 14(01):3–8, February 2005.

[25] Makoto Ozawa. Ascending number of knots and links. *J. Knot Theory Ramifications*, 19(1):15–25, January 2010.

[26] Migiwa Sakurai. An estimate of the unknotting numbers for virtual knots by forbidden moves. *J. Knot Theory Ramifications*, 22(3):10,1350009, March 2013.

[27] Shin Satoh. Virtual knot presentation of ribbon torus-knots. *J. Knot Theory Ramifications*, 9(4):531–542, 2000.

[28] Shin Satoh and Yumi Tomiyama. On the crossing numbers of a virtual knot. *Proceedings of the American Mathematical Society*, 140(1):367–376, 2012.

[29] Daniel S. Silver and Susan G. Williams. Virtual genus of satellite links. *J. Knot Theory Ramifications*, 22(3):4,1350008, March 2013.

B.5 Chapter 5

[1] Colin C. Adams. *The knot book.* American Mathematical Society, Providence, RI, 2004.

[2] J. H. Conway. An enumeration of knots and links, and some of their algebraic properties. In *Computational problems in abstract algebra (Proc. Conf., Oxford, 1967)*, pages 329–358. Pergamon, Oxford, 1970.

[3] Peter R. Cromwell. *Knots and links.* Cambridge University Press, Cambridge, 2004.

[4] Michael Eisermann and Christoph Lamm. Equivalence of symmetric union diagrams. *J. Knot Theory Ramifications*, 16(07):879–898, September 2007.

[5] C. Ernst. Tangle equations. *J. Knot Theory Ramifications*, 5(2):145–159, 1996.

[6] Allison Henrich and Slavik Jablan. On the coloring of pseudoknots. *J. Knot Theory Ramifications*, 23(12):22,1450061, 2014.

[7] Naoko Kamada. On the Jones polynomials of checkerboard colorable virtual links. *Osaka J. Math.*, 39(2):325–333, 2002.

[8] Louis H. Kauffman, Slavik Jablan, Ljiljana Radovic, and Radmila Sazdanovic. Reduced relative Tutte, Kauffman bracket and Jones polynomials of virtual link families. *J. Knot Theory Ramifications*, 22(4):1340003, April 2013.

[9] Louis H. Kauffman and Sofia Lambropoulou. On the classification of rational tangles. *Advances in Applied Mathematics*, 33(2):199–237, November 2004.

[10] Louis H. Kauffman and Sofia Lambropoulou. Tangles, rational knots and DNA. In *Lectures on topological fluid mechanics*, volume 1973 of *Lecture Notes in Math.*, pages 99–138. Springer, Berlin, 2009.

[11] Louis H. Kauffman and Pedro Lopes. The Teneva game. *J. Knot Theory Ramifications*, 21(14):1250125, 17, 2012.

[12] Daniel S. Silver and Susan G. Williams. Virtual tangles and a theorem of Krebes. *J. Knot Theory Ramifications*, 8(7):941–945, 1999.

[13] Daniel S. Silver and Susan G. Williams. Virtual genus of satellite links. *J. Knot Theory Ramifications*, 22(3):4,1350008, March 2013.

[14] Paul Zinn-Justin and Jean-Bernard Zuber. Matrix integrals and the generation and counting of virtual tangles and links. *J. Knot Theory Ramifications*, 13(3):325–355, 2004.

[15] Alexander Zupan. Properties of knots preserved by cabling. *Comm. Anal. Geom.*, 19(3):541–562, 2011.

B.6 Chapter 6

[1] Yongju Bae and Hugh R. Morton. The spread and extreme terms of Jones polynomials. *J. Knot Theory Ramifications*, 12(3):359–373, 2003.

[2] Peter R. Cromwell. *Knots and links.* Cambridge University Press, Cambridge, 2004.

[3] Heather A. Dye. Virtual knots undetected by 1- and 2-strand bracket polynomials. *Topology and its Applications*, 153:141–160, 2005.

[4] Shalom Eliahou, Louis H. Kauffman, and Morwen B. Thistlethwaite. Infinite families of links with trivial Jones polynomial. *Topology*, 42(1):155–169, January 2003.

[5] P. Freyd, D. Yetter, J. Hoste, W. B. R. Lickorish, K. Millett, and A. Ocneanu. A new polynomial invariant of knots and links. *Bull. Amer. Math. Soc. (N.S.)*, 12(2):239–246, April 1985.

[6] Masao Hara, Seiichi Tani, and Makoto Yamamoto. Degrees of the Jones polynomials of certain pretzel links. *J. Knot Theory Ramifications*, 9(7):907–916, 2000.

[7] Vaughan F. R. Jones. A polynomial invariant for knots via von Neumann algebras. *Bull. Amer. Math. Soc. (N.S.)*, 12(1):103–111, 1985.

[8] Naoko Kamada. On the Jones polynomials of checkerboard colorable virtual links. *Osaka J. Math.*, 39(2):325–333, 2002.

[9] Taizo Kanenobu. Infinitely many knots with the same polynomial invariant. *Proceedings of the American Mathematical Society*, 97(1):158–162, 1986.

[10] Taizo Kanenobu. The Homfly and the Kauffman bracket polynomials for the generalized mutant of a link. *Topology Appl.*, 8641(94):257–279, 1995.

[11] L. H. Kauffman. New invariants in the theory of knots. *Amer. Math. Monthly*, 95(3):195–242, 1988.

[12] Louis H. Kauffman. State models and the Jones polynomial. *Topology*, 26(3):395–407, 1987.

[13] Christopher King. A relationship between the Jones and Kauffman polynomials. *Transactions of the American Mathematical Society*, 329(1):307–323, 1992.

[14] Toshimasa Kishino and Shin Satoh. A note on non-classical virtual knots. *J. Knot Theory Ramifications*, 13(7):845–856, November 2004.

[15] Kunio Murasugi. Jones polynomials of alternating links. *Trans. Amer. Math. Soc.*, 295(1):147–174, 1986.

[16] Kunio Murasugi. Jones polynomials and classical conjectures in knot theory. *Topology*, 26(1):187–194, 1987.

[17] Morwen B. Thistlethwaite. A spanning tree expansion of the Jones polynomial. *Topology*, 26(3):297–309, 1987.

B.7 Chapter 7

[1] Peter Andrews. The classification of surfaces. *The American Mathematical Monthly*, 95(9):861–867, 1988.

[2] C. E. Burgess. Classification of surfaces. *The American Mathematical Monthly*, 92(5):349–354, 1985.

[3] J. Scott Carter, Seiichi Kamada, and Masahico Saito. Stable equivalence of knots on surfaces and virtual knot cobordisms. *J. Knot Theory Ramifications*, 11(3):1–12, 2008.

[4] Naoko Kamada and Seiichi Kamada. Abstract link diagrams and virtual knots. *J. Knot Theory Ramifications*, 9(1):93–106, 2000.

[5] Akio Kawauchi. The first Alexander Z[Z]-modules of surface-links and of virtual links. In *The Zieschang Gedenkschrift*, volume 14 of *Geom. Topol. Monogr.*, pages 353–371. Geom. Topol. Publ., Coventry, 2008.

[6] F. G. Korablev and S. V. Matveev. Reduction of knots in thickened surfaces and virtual knots. *Dokl. Akad. Nauk*, 437(6):748–750, 2011.

[7] Greg Kuperberg. What is a virtual link? *Algebr. Geom. Topol.*, 3:587–591 (electronic), 2003.

[8] Sang Youl Lee. Genera and periodicity of virtual knots and links. *J. Knot Theory Ramifications*, 21(4):15,1250037, 2012.

[9] Robert Messer and Philip Straffin. *Topology now!* Classroom Resource Materials Series. Mathematical Association of America, Washington, DC, 2006.

[10] Jose Gregorio Rodriguez and Margarita Toro. Virtual knot groups and combinatorial knots. *Sao Paulo J. Math. Sci.*, 3(2):299–316, 2009.

[11] Ayaka Shimizu. The warping degree of a link diagram. *Osaka J. Math.*, 48(1):209–231, 2011.

[12] Alexander Stoimenow, Vladimir Tchernov, and Alina Vdovina. The canonical genus of a classical and virtual knot. In *Geom. Dedicata*, volume 95, pages 215–225, 2002.

[13] M. V. Zenkina and V. O. Manturov. An invariant of links in a thickened torus. *Zap. Nauchn. Sem. S.-Peterburg. Otdel. Mat. Inst. Steklov. (POMI)*, 372(Geometriya i Topologiya. 11):5–18,203, 2009.

B.8 Chapter 8

[1] Colin C. Adams. *The knot book.* American Mathematical Society, Providence, RI, 2004.

[2] Peter R. Cromwell. *Knots and links.* Cambridge University Press, Cambridge, 2004.

[3] Naoko Kamada. On the Jones polynomials of checkerboard colorable virtual links. *Osaka J. Math.*, 39(2):325–333, 2002.

[4] Naoko Kamada. Span of the Jones polynomial of an alternating virtual link. *Algebraic & Geometric Topology*, 4(November):1083–1101, 2004.

[5] Louis H. Kauffman. State models and the Jones polynomial. *Topology*, 26(3):395–407, 1987.

[6] Kunio Murasugi. Jones polynomials of alternating links. *Trans. Amer. Math. Soc.*, 295(1):147–174, 1986.

[7] Kunio Murasugi. Jones polynomials and classical conjectures in knot theory. *Topology*, 26(1):187–194, 1987.

[8] Morwen B. Thistlethwaite. A spanning tree expansion of the Jones polynomial. *Topology*, 26(3):297–309, 1987.

B.9 Chapter 9

[1] J. Scott Carter. *How surfaces intersect in space*, volume 2 of *Series on Knots and Everything*. World Scientific, May 1995.

[2] Peter R. Cromwell. *Knots and links*. Cambridge University Press, Cambridge, 2004.

[3] Masao Hara, Seiichi Tani, and Makoto Yamamoto. Degrees of the Jones polynomials of certain pretzel links. *J. Knot Theory Ramifications*, 9(7):907–916, 2000.

[4] Naoko Kamada. On the Jones polynomials of checkerboard colorable virtual links. *Osaka J. Math.*, 39(2):325–333, 2002.

[5] Naoko Kamada. Span of the Jones polynomial of an alternating virtual link. *Algebraic & Geometric Topology*, 4(November):1083–1101, 2004.

[6] Louis H. Kauffman, Slavik Jablan, Ljiljana Radovic, and Radmila Sazdanovic. Reduced relative Tutte, Kauffman bracket and Jones polynomials of virtual link families. *J. Knot Theory Ramifications*, 22(4):1340003, April 2013.

[7] V. O. Manturov. Bifurcations, atoms and knots. *Vestnik Moskov. Univ. Ser. I Mat. Mekh.*, (1):3–8,71, 2000.

B.10 Chapter 10

[1] Kumud Bhandari, H. A. Dye, and Louis H. Kauffman. Lower bounds on virtual crossing number and minimal surface genus. In *The mathematics of knots*, volume 1 of *Contrib. Math. Comput. Sci.*, pages 31–43. Springer, Heidelberg, 2011.

[2] Joan S. Birman and Xiao-Song Lin. Knot polynomials and Vassiliev's invariants. *Invent. Math.*, 111(2):225–270, 1993.

[3] S. Chmutov, S. Duzhin, and J. Mostovoy. *Introduction to Vassiliev knot invariants*. Cambridge University Press, Cambridge, 2012.

[4] Micah W. Chrisman. Twist lattices and the Jones-Kauffman polynomial for long virtual knots. *J. Knot Theory Ramifications*, 19(5):655–675, 2010.

[5] H. A. Dye. Non-trivial realizations of virtual link diagrams. *J. Knot Theory Ramifications*, 15(8):21, 2006.

[6] H. A. Dye and Louis H. Kauffman. Minimal surface representations of virtual knots and links. *Algebr. Geom. Topol.*, 5(June):509–535, 2005.

[7] H. A. Dye and Louis H. Kauffman. Virtual crossing number and the arrow polynomial. *J. Knot Theory Ramifications*, 18(10):1335–1357, 2009.

[8] Heather A. Dye. Virtual knots undetected by 1- and 2-strand bracket polynomials. *Topology and its Applications*, 153:141–160, 2005.

[9] Young Ho Im, Sera Kim, and Kyeonghui Lee. Invariants of flat virtual links induced from Vassiliev invariants of degree one for virtual links. *J. Knot Theory Ramifications*, 20(12):1649–1667, December 2011.

[10] Young Ho Im, Kyeonghui Lee, and Sang Youl Lee. Index polynomial invariant of virtual links. *J. Knot Theory Ramifications*, 19(5):709–725, 2010.

[11] Young Ho Im, Kyeonghui Lee, and Heeok Son. An index polynomial invariant for flat virtual knots. *European J. Combin.*, 31(8):2130–2140, 2010.

[12] Young Ho Im and Sang Youl Lee. A four-variable index polynomial invariant of long virtual knots. *J. Knot Theory Ramifications*, 21(9):18,1250083, August 2012.

[13] Young Ho Im and Sang Youl Lee. On the index and arrow polynomials of periodic virtual links. *J. Knot Theory Ramifications*, 21(6):21,1250058, 2012.

[14] Atsushi Ishii. The pole diagram and the Miyazawa polynomial. *Internat. J. Math.*, 19(2):193–207, 2008.

[15] Atsushi Ishii, Naoko Kamada, and Seiichi Kamada. The virtual magnetic Kauffman bracket skein module and skein relations for the f-polynomial. *J. Knot Theory Ramifications*, 17(6):675–688, 2008.

[16] Atsushi Ishii, Naoko Kamada, and Seiichi Kamada. The Miyazawa polynomial for long virtual knots. *Topology Appl.*, 157(1):290–297, 2010.

[17] Teruhisa Kadokami. Detecting non-triviality of virtual links. *J. Knot Theory Ramifications*, 12(06):781–803, September 2003.

[18] Teruhisa Kadokami. Some numerical invariants of flat virtual links are always realized by reduced diagrams. *J. Knot Theory Ramifications*, 15(03):289–297, March 2006.

[19] Teruhisa Kadokami. Of 2-component flat virtual links with the supporting genus one. *J. Knot Theory Ramifications*, 17(5):633–647, 2008.

[20] Teruhisa Kadokami. The virtual crossing number of 2-component flat-virtual links with the supporting genus one. *J. Knot Theory Ramifications*, 17(5):633–647, 2008.

[21] Naoko Kamada. A relation of Kauffman's f-polynomials of virtual links. *Topology Appl.*, 146/147:123–132, 2005.

[22] Naoko Kamada. An index of an enhanced state of a virtual link diagram and Miyazawa polynomials. *Hiroshima Math. J.*, 37(3):409–429, 2007.

[23] Naoko Kamada. Miyazawa polynomials of virtual knots and virtual crossing numbers. In *Intelligence of low dimensional topology 2006*, volume 40 of *Ser. Knots Everything*, pages 93–100. World Sci. Publ., Hackensack, NJ, 2007.

[24] Naoko Kamada. Some relations on Miyazawa's virtual knot invariant. *Topology Appl.*, 154(7):1417–1429, 2007.

[25] Naoko Kamada and Yasuyuki Miyazawa. A 2-variable polynomial invariant for a virtual link derived from magnetic graphs. *Hiroshima Math. J.*, 35(2):309–326, 2005.

[26] Louis H. Kauffman. An extended bracket polynomial for virtual knots and links. *J. Knot Theory Ramifications*, 18(10):1369–1422, 2009.

[27] Kyeonghui Lee and Young Ho Im. Generalized index polynomials for virtual links. *J. Knot Theory Ramifications*, 21(14):16,1250128, December 2012.

[28] Yasuyuki Miyazawa. Magnetic graphs and an invariant for virtual links. *J. Knot Theory Ramifications*, 15(10):1319–1334, 2006.

[29] Yasuyuki Miyazawa. A multi-variable polynomial invariant for virtual knots and links. *J. Knot Theory Ramifications*, 17(11):1311–1326, 2008.

[30] Yasuyuki Miyazawa. A multi-variable polynomial invariant for unoriented virtual knots and links. *J. Knot Theory Ramifications*, 18(5):625–649, 2009.

[31] Yasuyuki Miyazawa. A virtual link polynomial and the virtual crossing number. *J. Knot Theory Ramifications*, 18(5):605–623, 2009.

[32] Yasuyuki Miyazawa. Link polynomials derived from magnetic graphs. *Topology Appl.*, 157(1):228–246, 2010.

B.11 Chapter 11

[1] Denis Afanasiev. On a generalization of the Alexander polynomial for long virtual knots. *J. Knot Theory Ramifications*, 18(10):1329–1333, 2009.

[2] Peter Andersson. The color invariant for knots and links. *Amer. Math. Monthly*, 102(5):442–448, 1995.

[3] J. Scott Carter. A survey of quandle ideas. In *Introductory lectures on knot theory*, volume 46 of *Ser. Knots Everything*, pages 22–53. World Sci. Publ., Hackensack, NJ, 2012.

[4] W. Edwin Clark, Mohamed Elhamdadi, Masahico Saito, and Timothy Yeatman. Quandle colorings of knots and applications. *J. Knot Theory Ramifications*, 23(6):1–29, 2014.

[5] Natasha Harrell and Sam Nelson. Quandles and linking number. *J. Knot Theory Ramifications*, 16(10):1283–1293, 2007.

[6] Allison Henrich and Sam Nelson. Semiquandles and flat virtual knots. *Pacific J. Math.*, 248(1):155–170, 2010.

[7] Mikami Hirasawa, Naoko Kamada, and Seiichi Kamada. Bridge presentations of virtual knots. *J. Knot Theory Ramifications*, 20(06):881–893, June 2011.

[8] David Joyce. A classifying invariant of knots, the knot quandle. *J. Pure Appl. Algebra*, 23(1):37–65, 1982.

[9] Seiichi Kamada. Knot invariants derived from quandles and racks. *Invariants of knots and 3-manifolds (Kyoto, 2001)*, 4(July 2002):103–117, 2002.

[10] Louis H. Kauffman. An affine index polynomial invariant of virtual knots. *J. Knot Theory Ramifications*, 22(4):30,1340007, 2013.

[11] Louis H. Kauffman and Pedro Lopes. On the minimum number of colors for knots. *Advances in Applied Mathematics*, 40(1):36–53, 2008.

[12] Ryszard L. Rubinsztein. Topological quandles and invariants of links. *J. Knot Theory Ramifications*, 16(06):789–808, August 2007.

[13] Masahico Saito. The minimum number of Fox colors and quandle cocycle invariants. *J. Knot Theory Ramifications*, 19(11):1449–1456, November 2010.

[14] Steven K. Winker. *Quandles, knot invariants and the n-fold branched cover.* PhD thesis.

B.12 Chapter 12

[1] Denis Afanasiev. On a generalization of the Alexander polynomial for long virtual knots. *J. Knot Theory Ramifications*, 18(10):1329–1333, 2009.

[2] Yongju Bae. Coloring link diagrams by Alexander quandles. *J. Knot Theory Ramifications*, 21(10):1250094, September 2012.

[3] J. S. Carter, D. S. Silver, and S. G. Williams. Three dimensions of knot coloring. *The American Mathematical Monthly*, 121(6):506–514, 2014.

[4] J. Scott Carter. A survey of quandle ideas. In *Introductory lectures on knot theory*, volume 46 of *Ser. Knots Everything*, pages 22–53. World Sci. Publ., Hackensack, NJ, 2012.

[5] Jessica Ceniceros and Sam Nelson. (T,S)-Racks and Their Link Invariants. page 15, November 2010.

[6] W. Edwin Clark, Mohamed Elhamdadi, Masahico Saito, and Timothy Yeatman. Quandle colorings of knots and applications. *J. Knot Theory Ramifications*, 23(6):1–29, 2014.

[7] R. H. Fox. A quick trip through knot theory. *Topology of 3-manifolds and related topics (Proc. The Univ. of Georgia Institute, 1961)*, pages 120–167, 1962.

[8] Seiichi Kamada. Knot invariants derived from quandles and racks. *Invariants of knots and 3-manifolds (Kyoto, 2001)*, 4(July 2002):103–117, 2002.

[9] Louis H. Kauffman and Pedro Lopes. On the minimum number of colors for knots. *Advances in Applied Mathematics*, 40(1):36–53, 2008.

[10] Akio Kawauchi. The first Alexander Z[Z]-modules of surface-links and of virtual links. In *The Zieschang Gedenkschrift*, volume 14 of *Geom. Topol. Monogr.*, pages 353–371. Geom. Topol. Publ., Coventry, 2008.

[11] Se-Goo Kim. Virtual knot groups and their peripheral structure. *J. Knot Theory Ramifications*, 9(6):797–812, 2000.

[12] Masahico Saito. The minimum number of Fox colors and quandle cocycle invariants. *Journal of Knot Theory and Its Ramifications*, 19(11):1449–1456, November 2010.

[13] Daniel S. Silver and Wilbur Whitten. Knot group epimorphisms. *J. Knot Theory Ramifications*, 15(02):153–166, February 2006.

B.13 Chapter 13

[1] J. Scott Carter, Daniel S. Silver, Susan G. Williams, Mohamed Elhamdadi, and Masahico Saito. Virtual knot invariants from group biquandles and their cocycles. *J. Knot Theory Ramifications*, 18(7):957–972, 2009.

[2] A. S. Crans, Allison Henrich, and Sam Nelson. Polynomial knot and link invariants from the virtual biquandle. *J. Knot Theory Ramifications*, 22(4):1–12, 2013.

[3] M. Elhamdadi and S. Nelson. *Quandles: an introduction to the algebra of knots*. American Mathematical Society, 2015.

[4] Naoko Kamada and Seiichi Kamada. Biquandles with structures related to virtual links and twisted links. *J. Knot Theory Ramifications*, 21(13):1–15, 2012.

[5] Louis H. Kauffman and David Radford. Bi-oriented quantum algebras , and a generalized Alexander polynomial for virtual links. *Diagrammatic morphisms and applications (San Francisco, CA, 2000)*, 318:113–140, 2003.

[6] Jorg Sawollek. On Alexander-Conway polynomials for virtual knots and links. *arXiv preprint math/9912173*, 1999.

B.14 Chapter 14

[1] S. Chmutov, S. Duzhin, and J. Mostovoy. *Introduction to Vassiliev knot invariants*. Cambridge University Press, Cambridge, 2012.

[2] Micah W. Chrisman. On the Goussarov-Polyak-Viro finite-type invariants and the virtualization move. *J. Knot Theory Ramifications*, 20(3):389–401, 2011.

[3] Micah W. Chrisman and Heather A. Dye. The three loop isotopy and framed isotopy invariants of virtual knots. *Topology and its Applications*, 173:107–134, August 2014.

[4] Micah Whitney Chrisman and Vassily Olegovich Manturov. Parity and exotic combinatorial formulae for finite-type invariants of virtual knots. *J. Knot Theory Ramifications*, 21(13):27,1240001, 2012.

[5] H. A. Dye. Vassiliev invariants from parity mappings. *J. Knot Theory Ramifications*, 22(4):21,1340008, 2013.

[6] Lena C. Folwaczny and Louis H. Kauffman. A linking number definition of the affine index polynomial and applications. *J. Knot Theory Ramifications*, 22(12):30,1341004, 2013.

[7] D. P. Ilyutko, V. O. Manturov, and I. M. Nikonov. Parity in knot theory and graph links. *Sovrem. Mat. Fundam. Napravl.*, 41:3–163, 2011.

[8] Denis Petrovich Ilyutko, Vassily Olegovich Manturov, and Igor Mikhailovich Nikonov. Virtual knot invariants arising from parities. In *Knots in Poland. III. Part 1*, volume 100 of *Banach Center Publ.*, pages 99–130. Polish Acad. Sci. Inst. Math., Warsaw, 2014.

[9] Aaron Kaestner and Louis H. Kauffman. Parity Biquandles. *Knots in Poland. III. Part 1*, 100:131–151, 2014.

[10] Louis H. Kauffman. Introduction to virtual knot theory. *J. Knot Theory Ramifications*, 21(13):37,1240007, 2012.

[11] V. O. Manturov. Parity in knot theory. *Mat. Sb.*, 201(5):65–110, 2010.

[12] V. O. Manturov. Parity, free knots, groups, and invariants of finite type. *Trans. Moscow Math. Soc.*, pages 157–169, 2011.

[13] V. O. Manturov. Free knots and parity. In *Introductory lectures on knot theory*, volume 46 of *Ser. Knots Everything*, pages 321–345. World Sci. Publ., Hackensack, NJ, 2012.

[14] Vassily Olegovich Manturov. Parity and projection from virtual knots to classical knots. *J. Knot Theory Ramifications*, 22(9):20,1350044, 2013.

[15] Olof-Petter Östlund. A diagrammatic approach to link invariants of finite degree. *Math. Scand.*, 94(2):295–319, 2004.

[16] Vladimir Turaev. Lectures on topology of words. *Jpn. J. Math.*, 2(1):1–39, 2007.

[17] Vladimir Turaev. Cobordism of knots on surfaces. *J. Topol.*, 1(2):285–305, 2008.

[18] Vladimir Turaev. Cobordisms of words. *Commun. Contemp. Math.*, 10(suppl. 1):927–972, 2008.

B.15 Chapter 15

[1] Dorothy Buck and Erica Flapan, editors. *Applications of knot theory*, volume 66 of *Proceedings of Symposia in Applied Mathematics*. American Mathematical Society, Providence, RI, 2009.

[2] J. Scott Carter, Daniel S. Silver, and Susan G. Williams. Invariants of links in thickened surfaces. *Algebr. Geom. Topol.*, 14(3):1377–1394, 2014.

[3] H. A. Dye and Louis H. Kauffman. Anyonic topological quantum computation and the virtual braid group. *J. Knot Theory Ramifications*, 20(1):91–102, 2011.

[4] Erica Flapan. Topological chirality and symmetries of non-rigid molecules. In *Applications of knot theory*, volume 66 of *Proc. Sympos. Appl. Math.*, pages 21–45. Amer. Math. Soc., Providence, RI, 2009.

[5] S. Grishanov, V. Meshkov, and A. Omelchenko. A topological study of textile structures. Part I: An introduction to topological methods. *Textile Research Journal*, 79(8):702–713, April 2009.

[6] S. Grishanov, V. Meshkov, and A. Omelchenko. A topological study of textile structures. Part II: Topological invariants in application to textile structures. *Textile Research Journal*, 79(9):822–836, May 2009.

[7] S. J. Lomonaco and L. H. Kauffman. Quantum knots and mosaics. *Quantum Information Processing*, pages 1–32, 2008.

[8] Anirban Pathak. *Elements of quantum computation and quantum communication.* CRC Press, 2013.

[9] De Witt Sumners. Lifting the curtain: using topology to probe the hidden action of enzymes. *Match*, (34):51–76, 1996.

Index